Introduction to Probability Theory

A First Course on
the Measure-Theoretic Approach

World Scientific Series on Probability Theory and Its Applications

Print ISSN: 2737-4467
Online ISSN: 2737-4475

Published:

Vol. 3 *Introduction to Probability Theory: A First Course on the Measure-Theoretic Approach*
by Nima Moshayedi (University of California, Berkeley, USA)

Vol. 2 *Introduction to Stochastic Processes*
by Mu-Fa Chen (Beijing Normal University, China) and
Yong-Hua Mao (Beijing Normal University, China)

Vol. 1 *Random Matrices and Random Partitions: Normal Convergence*
by Zhonggen Su (Zhejiang University, China)

World Scientific Series on
Probability Theory and Its Applications

Volume 3

Introduction to Probability Theory

A First Course on the Measure-Theoretic Approach

Nima Moshayedi
University of California, Berkeley, USA

World Scientific
NEW JERSEY • LONDON • SINGAPORE • BEIJING • SHANGHAI • HONG KONG • TAIPEI • CHENNAI • TOKYO

Published by

World Scientific Publishing Co. Pte. Ltd.
5 Toh Tuck Link, Singapore 596224
USA office: 27 Warren Street, Suite 401-402, Hackensack, NJ 07601
UK office: 57 Shelton Street, Covent Garden, London WC2H 9HE

Library of Congress Cataloging-in-Publication Data
Names: Moshayedi, Nima, author.
Title: Introduction to probability theory : a first course on the measure-theoretic approach / Nima Moshayedi, University of California, Berkeley, USA.
Description: New Jersey : World Scientific, [2022] | Series: World Scientific series on probability theory and its applications ; Volume 3 | Includes bibliographical references and index.
Identifiers: LCCN 2021045265 | ISBN 9789811246746 (hardcover) | ISBN 9789811243356 (ebook for institutions) | ISBN 9789811243363 (ebook for individuals)
Subjects: LCSH: Probabilities.
Classification: LCC QA273 .M785 2022 | DDC 519.2--dc23/eng/20211119
LC record available at https://lccn.loc.gov/2021045265

British Library Cataloguing-in-Publication Data
A catalogue record for this book is available from the British Library.

For any available supplementary material, please visit
https://www.worldscientific.com/worldscibooks/10.1142/12465#t=suppl

Desk Editors: Jayanthi Muthuswamy/Kwong Lai Fun

Typeset by Stallion Press
Email: enquiries@stallionpress.com

Printed in Singapore

Preface

The theory of probability plays an important role in many different scientific areas and concrete real life applications. Besides a lot of applications in the field of natural sciences, economics or finance, it can be also used to understand fundamental concepts and structures in mathematics itself. This book serves as a first introduction to the mathematical concepts of probability theory for undergraduate students with a solid background in real analysis and linear algebra. The way how we formulate probability theory in this book is based on the common language of *measure theory*. For readers who do not feel entirely comfortable with the measure-theoretic background used in this book, there is also an appendix which provides some of the most important notions and theorems needed to understand the constructions in the bulk. The book is primarily intended as a freestanding textbook for a corresponding course on probability theory or reference guide for any course in a similar direction. It covers all the basic notions of probability theory which are important for students to master the first challenges during their undergraduate and graduate studies in this direction. The book lays the foundations for future work or research using this subject by starting with basic notions and principles with applying these to examples and short exercises. By providing a list of important examples which illustrate the theory conveniently, we aim to crucially strengthen the understanding and the way of thinking about certain objects. The book is divided into two main parts. The first part covers the main concepts and mathematical objects of probability theory and the second part will dive into more sophisticated methods which develop the objects discussed before within more general circumstances.

Nima Moshayedi
University of California, Berkeley

About the Author

Nima Moshayedi is currently an SNSF postdoctoral fellow at the University of California, Berkeley. He was previously a postdoctoral researcher at the University of Zurich. He has completed his PhD studies in Mathematical Physics at the University of Zurich under the supervision of Prof. Alberto S. Cattaneo. His undergraduate studies in Mathematics and Physics were at the University of Zurich and ETH Zurich. His research is in algebraic and geometric aspects of Quantum Field Theory, especially with a focus on Symplectic Geometry, Poisson Geometry, Topological (Quantum) Field Theories, Perturbative Gauge Theories and the mathematical aspects of String Theory. His publications are featured in academic journals of the highest level, e.g., *Communications in Mathematical Physics*, *Letters in Mathematical Physics*, *Physics Letters B*, *Reviews in Mathematical Physics*, *Advances in Theoretical and Mathematical Physics*, *Annales Henri Poincaré* and *Pacific Journal of Mathematics*. He has done referee work for several academic journals, e.g., *Letters in Mathematical Physics*, *Journal of Mathematical Physics*, *Forum of Mathematics SIGMA*. He is a reviewer for the American Mathematical Society (AMS) and Zentralblatt Mathematik (zbMATH). Nima has also been an organizer of several research conferences, e.g., *Zurich Graduate Colloquium in Mathematics (ZGSM)*, *Higher Structures in QFT and String Theory—A Virtual Conference for Junior Researchers*. He has lectured on several courses/seminars for undergraduate and graduate students. He has also received several (competitive) grants from different institutions, e.g., Swiss National Science Foundation, SwissMAP, European Cooperation in Science and Technology (COST) and Candoc Research Grant (Forschungskredit) from the University of Zurich.

Acknowledgments

This book has started as lecture notes for a one-year course on probability theory given at the University of Zurich. I want to thank Ashkan Nikeghbali, Luca Spolaor and Brad Rodgers for comments on a very first draft of these notes. Moreover, I want to thank Rochelle Kronzek for giving me the opportunity to publish these notes as a book in the *World Scientific Series on Probability Theory and Its Applications* and Lai Fun Kwong for all the help during the publishing procedure. I also want to thank Jayanthi for the editing process of the book. Finally, I want to thank the University of Zurich and the University of California, Berkeley for providing an excellent working environment.

Contents

Chapter 2. The Modern Probability Language **27**

Part I
The Modern Probability Language

Introduction

We think of probability theory as a mathematical model for random events. Therefore, probability theory is widely used in many different topics, for example:

- Biology,
- Economics/Insurance/Stochastic finance theory,
- Physics (for example statistical mechanics/Quantum mechanics),
- Mathematics (Random matrix theory/Number theory/Group theory/etc.).

Random Experiments: *Random experiments* are experiments whose output can not be surely predicted in advance. The theory of probability aims towards a mathematical theory which describes such phenomena. There are four important objects for probability theory:

- **State space Ω:** The *state space* represents all possible outcomes of the experiment. There are many different state spaces:
  - (i) a toss of a coin: $\Omega = \{H, T\}$ (head or tail),
  - (ii) two successive tosses of a coin: $\Omega = \{HH, HT, TT, TH\}$,
  - (iii) a toss of two dice: $\Omega = \{(i, j) \mid 1 \leq i \leq 6, 1 \leq j \leq 6\}$,
  - (iv) the lifetime of a bulb: $\Omega = \mathbb{R}_{\geq 0}$.

- **Events:** An *event* is a property which can observed either to hold or not to hold after the experiment is done. In mathematical terms, an event is a subset of Ω. We shall denote by $\mathcal{A}$ the family of all events. If A and B are two events, then

(i) The contrary event is interpreted as the complement A^{C}.
(ii) The event A *or* B is interpreted as $A \cup B$.
(iii) The event A *and* B is interpreted as $A \cap B$.
(iv) The *sure* event is Ω.
(v) The *impossible* event is $\emptyset$.

- **Probability measure:** With each event A, one associates a number $\mathbb{P}(A)$ and call it the *probability of the event* A. This number measures the likelihood of the event A to be realized *a priori* before performing the experiment and $\mathbb{P}(A) \in [0,1]$. Imagine that these numbers are frequencies. Let us repeat the same experiment n times. Denote by $f_n(A)$ the frequency with which A is realized (i.e., the number of times the event A occurs divided by n) and we write $\mathbb{P}(A) = \lim_{n\to\infty} f_n(A)$. With this interpretation, we have to expect that

  (i) $\mathbb{P}(\Omega) = 1$,
  (ii) $\mathbb{P}(A \cup B) = \mathbb{P}(A) + \mathbb{P}(B)$ if $A \cap B = \emptyset$.

  In terms of mathematical modeling, i.e. considering a σ-algebra $\mathcal{A}$, we can think of P as a probability measure

$$\mathbb{P} : \mathcal{A} \to [0,1].$$

  In this book, a mathematical model for our random experiment is a triple (measure space)

$$(\Omega, \mathcal{A}, \mathbb{P})$$

  with $\Omega, \mathcal{A}, \mathbb{P}$ as we have seen.

- **Random variable:** A *random variable* is a quantity which depends on the outcome of the experiment. In mathematical terms, this is a map from the state space Ω into some space E (and E is often $\mathbb{R}$, $\mathbb{R}^d$, $\mathbb{N}$, $\mathbb{Z}$) and such a map $X : \Omega \to E$ should be measurable. We always think of E as a measure space, that is E is endowed with a σ-algebra $\mathcal{E}$, and from measure theory we know that one can transport $\mathbb{P}$ on E: for all $B \in \mathcal{E}, \mathbb{P}^X$ (or $\mathbb{P}_X$) is a probability measure on $\mathcal{E}$ such that for all $B \in \mathcal{E}$,

$$\mathbb{P}^X(B) = \mathbb{P}_X(B) = \mathbb{P}(X^{-1}(B)) = \mathbb{P}\left(\underbrace{\{\omega \in \Omega \mid X(\omega) \in B\}}_{\in \mathcal{A}}\right)$$

  and $\mathbb{P}_X$ is called the *law* or the *distribution* of X.

Example 1.0.0.1. Denote by $\mathcal{P}(A)$ (sometimes also 2^A) the *power set* of a set A. The data for the toss of two dice is given by

$$\begin{aligned} \Omega &= \{(i,j) \mid 1 \le i \le 6, 1 \le j \le 6\}, \\ \mathcal{A} &= \mathcal{P}(\Omega), \\ \mathbb{P}(\{\omega\}) &= \frac{1}{36}, \qquad \forall \omega \in \Omega, \\ \mathbb{P}(A) &= \frac{|A|}{36}, \qquad \forall A \subset \Omega. \end{aligned}$$

The map $X : \Omega \to (\mathbb{N}, \mathcal{P}(\mathbb{N}))$, $(i,j) \mapsto i+j$ is a random variable and for all $B \subset \mathbb{N}$,

$$\mathbb{P}_X(B) = \frac{|\{(i,j) \mid i+j \in B\}|}{36}.$$

For instance, if $B = \{2\} \subset \Omega$ then $\mathbb{P}_X(\{2\}) = \mathbb{P}(\{1,1\}) = \frac{1}{36}$.

To conclude the introduction, let us emphasize the fact that Ω can be more involved and the construction of $\mathbb{P}$ "completed".

Example 1.0.0.2. We throw a die until we get a 6. Here the choice of Ω is less obvious. The number of times, we have to throw the die, is not bounded. Let $\mathbb{N}^\times := \mathbb{N} \setminus \{0\}$. Then, a natural choice for Ω would be

$$\Omega = \{1,2,3,4,5,6\}^{\mathbb{N}^\times},$$

$\omega \in \Omega$ that is $\omega = (\omega_1, \ldots, \omega_n, \ldots)$ with $\omega_j = \{1,2,3,4,5,6\}$ for all $j \ge 1$. $\mathcal{A}$ will be the product σ-algebra, i.e., the smallest σ-algebra containing all sets of the form $\{\omega \mid \omega_1 = i_1, \ldots, \omega_n = i_n\}$, $n \ge 1$, $i_1, \ldots, i_n \in \{1,2,3,4,5,6\}$ and $\mathbb{P}$ is the unique probability measure on Ω.

Example 1.0.0.3. We are interested in the motion of a particle in space, which is subject to some random perturbations. If the time interval is $[0,1]$, a natural space of outcomes can be the space of continuous functions from $[0,1]$ into $\mathbb{R}^3$ and hence we set

$$\Omega = C^0([0,1], \mathbb{R}^3),$$

where $\omega \in \Omega$ is a possible trajectory of the particle: $\omega : [0,1] \to \mathbb{R}^3$. Take $\mathcal{A}$ to be the Borel σ-algebra when we consider the supremum-norm[1] on Ω. A famous example for $\mathbb{P}$ is the *Wiener measure* (denoted by $\mathbb{W}$). Under this measure, a typical trajectory is called a *Brownian motion.*

[1]This is the norm defined for $\omega \in \Omega$ by $\|\omega\|_\infty := \sup_{t\in[0,1]} |\omega(t)|$.

Chapter 1

Elements of Combinatorial Analysis and Simple Random Walks

In this chapter, we are going to give an introduction to combinatorial aspects of probability theory and describe several simple methods for the computation of different problems. We will also already give some of the most basic examples of discrete distributions and consider combinatorial aspects of random events as a first look at a simple random walks. We will look at the reflection principle and basic terminology of random walks as a combinatorial theory. Finally, we will consider probabilities of events for special random walks.

1.1. Summability Conditions

1.1.1. Basic definitions

Let I be an index set, and $(a_i)_{i\in I} \subset \overline{\mathbb{R}}^I_{\geq 0}$ (i.e., $a_i \in [0, \infty]$, for all $i \in I$). We will write

$$\sum_{i\in I} a_i := \sup_{\substack{J\subset I \\ J \text{ finite}}} \sum_{j\in J} a_j.$$

The family $(a_i)_{i\in I}$ is said to be summable if $\sum_{i\in I} a_i$ is finite. The following elementary properties are important for such a family $(a_i)_{i\in I}$.

(i) For $H \subseteq I$, we have

$$\sum_{h\in H} a_h \leq \sum_{i\in I} a_i$$

and $(a_h)_{h\in H}$ is summable if $(a_i)_{i\in I}$ is summable.

(ii) For a family $(a_i)_{i\in I}$ to be summable, it is necessary and sufficient that the Cauchy criterion is satisfied, i.e., for all $\varepsilon > 0$, there is a finite set $J \subset I$, such that with all finite subsets $K \subset I$ satisfying $K \cap J = \emptyset$, we get

$$\sum_{i\in K} a_i < \varepsilon.$$

(iii) If the family $(a_i)_{i\in I}$ is summable, the set $\{i \in I \mid a_i \neq 0\}$ is at most countable, as the union of the sets $I_n = \left\{i \in I \mid a_i \geq \frac{1}{n}\right\}$. Moreover, I_n has at most $\left[n\left(\sum_{i\in I} a_i\right)\right]$ elements, where $[x]$ denotes the integer part of x.

To avoid confusion, we want to emphasize that if $A_1, \ldots, A_n$ are disjoint sets, we will sometimes write $\sum_{n\geq 1} A_n$ instead of $\bigsqcup_{n\geq 1} A_n$.

Lemma 1.1.1.1. *Let $(a_i)_{i\in I} \subset \bar{\mathbb{R}}^I_{\geq 0}$ be a family of numbers in $\bar{\mathbb{R}}_{\geq 0}$. If $I = \sum_{\lambda\in L} H_\lambda$ and $q_\lambda = \sum_{i\in H_\lambda} a_i$, then*

$$\sum_{i\in I} a_i = \sum_{\lambda\in L} q_\lambda,$$

where L is an index set and $(H_\lambda)_{\lambda\in L}$ is a partition of I.

Exercise 1.1.1.1. Prove Lemma 1.1.1.1.

In the case $I = \mathbb{N}$, we say that $\sum_{n\in\mathbb{N}} u_n$ converges absolutely if $\sum_{n\in\mathbb{N}} |u_n|$ converges. In this case, we can modify the order in which the terms are taken without changing the convergence nor the sum of the series.

Theorem 1.1.1.2 (Stirling). *For all $n \in \mathbb{N}^\times := \mathbb{N}\setminus\{0\}$, we have*

$$n! = \kappa n^{n+\frac{1}{2}} \exp\left(-n + \frac{\theta(n)}{12n}\right),$$

where $1 - \frac{1}{12n+1} \leq \theta(n) \leq 1$ and $\kappa = \int_{-\infty}^{+\infty} e^{-\frac{1}{2}w^2} dw = \sqrt{2\pi}$.

1.2. Finite Probability Spaces

1.2.1. General probability spaces

Definition 1.2.1.1 (Probability measure). Let $(\Omega, \mathcal{A})$ be a measurable space. A *probability measure* defined on $\mathcal{A}$ is a map

$$\mathbb{P} : \mathcal{A} \to [0, 1],$$

such that

(i) $\mathbb{P}(\Omega) = 1$;
(ii) for all $(A_n)_{n \in \mathbb{N}}$, with $A_n \cap A_m = \emptyset$ whenever $n \neq m$, we have

$$\mathbb{P}\left(\bigcup_{n \in \mathbb{N}} A_n\right) = \sum_{n \in \mathbb{N}} \mathbb{P}(A_n).$$

Remark 1.2.1.2. We want to recall the following properties of a measure without a proof.

- For $A, B \in \mathcal{A}$ with $A \subset B$, we get $\mathbb{P}(A) \leq \mathbb{P}(B)$.
- For $A \in \mathcal{A}$, we get $\mathbb{P}(A^{\mathsf{C}}) = 1 - \mathbb{P}(A)$.
- If $(A_n)_{n \in \mathbb{N}} \subset \mathcal{A}$ and $A_n \uparrow A$ as $n \to \infty$, then $\mathbb{P}(A_n) \uparrow \mathbb{P}(A)$ as $n \to \infty$.
- If $(A_n)_{n \in \mathbb{N}} \subset \mathcal{A}$ and $A_n \downarrow A$ as $n \to \infty$, then $\mathbb{P}(A_n) \downarrow \mathbb{P}(A)$ as $n \to \infty$.

Theorem 1.2.1.1. *Let μ be a measure on a measurable space $(\Omega, \mathcal{A})$.*

(i) *If $(A_n)_{n \in \mathbb{N}} \subset \mathcal{A}$ is an increasing family of measurable sets, then*

$$\mu\left(\bigcup_{n \in \mathbb{N}} A_n\right) = \lim_{n \to \infty} \mu(A_n).$$

Conversely, if an additive function satisfies the property above, then it is a measure.

(ii) *Let μ be an additive, positive and bounded function on $\mathcal{A}$. Then σ-additivity is equivalent to say that for all decreasing sequences $(B_n)_{n \in \mathbb{N}} \subset \mathcal{A}$, we get the implication.*

$$\bigcap_{n \in \mathbb{N}} B_n = \emptyset \Rightarrow \lim_{n \to \infty} \mu(B_n) = 0.$$

1.2.2. Probability measures on finite spaces

There are two cases where $(\Omega, \mathcal{A})$ is a measurable space, either Ω is finite or $\mathcal{A}$ is finite. An important question is how to define probability measures on finite or countable spaces. If I is at most countable, then having a probability measure on its power set $\mathcal{P}(I)$ is equivalent to having a family $(a_i)_{i\in I}$ of positive real numbers such that $\sum_{i\in I} a_i = 1$. Moreover, for $A \subset I$, we can define a measure μ by

$$\mu(A) = \sum_{i\in A} a_i,$$

where $\mu(\{i\}) = a_i$ for all $i \in I$.

Example 1.2.2.1 (Poisson distribution). The Poisson distribution with parameter $\lambda > 0$, is a probability measure Π_λ on $(\mathbb{N}, \mathcal{P}(\mathbb{N}))$, defined for all $n \geq 0$ by

$$\Pi_\lambda(\{n\}) = e^{-\lambda}\frac{\lambda^n}{n!}.$$

For $A \subset \mathbb{N}$ we thus get

$$\Pi_\lambda(A) = \sum_{n\in A} e^{-\lambda}\frac{\lambda^n}{n!}.$$

In order to check that it is indeed a probability distribution, we need to check that

$$\sum_{i\geq 0} \Pi_\lambda(\{i\}) = 1.$$

Indeed, we have

$$\sum_{i\geq 0} e^{-\lambda}\frac{\lambda^i}{i!} = e^{-\lambda}\sum_{i\geq 0}\frac{\lambda^i}{i!} = 1.$$

Example 1.2.2.2 (Riemann zeta function). Another example can be obtained by considering the Riemann zeta function $\zeta(s) = \sum_{n\geq 1} n^{-s}$ for $s > 1$, which we can deform to a probability measure. Indeed, if we set $a_n := \frac{1}{\zeta(s)} \cdot \frac{1}{n^s}$ for all $n \geq 1$, we get

$$\sum_{n\geq 1} a_n = \frac{1}{\zeta(s)}\sum_{n\geq 1} n^{-s} = \frac{1}{\zeta(s)} \cdot \zeta(s) = 1.$$

Example 1.2.2.3 (Geometric distribution). Similarly one can define the geometric distribution with parameter $r \in (0, 1)$ on $(\mathbb{N}^{\times}, \mathcal{P}(\mathbb{N}^{\times}))$ by

$$\gamma_r(\{n\}) = (1-r)r^{n-1}.$$

Example 1.2.2.4 (Uniform distribution). For a finite state space Ω, we can define the uniform distribution for any measurable set $A \in \mathcal{A}$ by

$$\mathbb{P}(A) = \frac{|A|}{|\Omega|}.$$

For $\omega \in \Omega$ we get $\mathbb{P}(\{\omega\}) = \frac{1}{|\Omega|}$. Take for example $\Omega = \{1, 2, \dots, N\}$. Then $\mathbb{P}(\{k\}) = \frac{1}{N}$.

Remark 1.2.2.5. Note that there is no uniform measure on $\mathbb{N}$, since $\mathbb{N}$ is not finite. Thus from now on we shall only consider the case of a finite state space Ω or a finite σ-algebra $\mathcal{A}$.

Theorem 1.2.2.1. *Let $\mathcal{A}$ be a finite σ-algebra on a state space Ω. Then there exists a partition $A_1, \dots, A_n$ of Ω such that $A_k \in \mathcal{A}$ for all $k \in \{1, \dots, n\}$ and with the property that any $A \in \mathcal{A}$ can be written as a union of the events $A_1, \dots, A_n$. The sets are called the* atoms *of $\mathcal{A}$.*

Proof. Fix $\omega \in \Omega$. Define the set $C(\omega) := \bigcap_{\substack{B \in \mathcal{A},\\ \omega \in B}} B$. Then by finite intersection it follows that $C(\omega) \in \mathcal{A}$. In fact, $C(\omega)$ is the smallest element of $\mathcal{A}$ which contains ω. Moreover, it is easy to see that the relation $\sim$ given by $\omega \sim \omega'$ if and only if $C(\omega) = C(\omega')$ defines an equivalence relation on Ω and that $\omega' \in C(\omega)$ if and only if $C(\omega) = C(\omega')$. Hence if A is an equivalence class and $\omega \in A$, then $A = C(\omega)$. Therefore, one can take the events $A_1, \dots, A_n$ as the equivalence classes. □

1.3. Basics of Combinatorial Analysis

1.3.1. Sampling

Let us consider a population of size $n \in \mathbb{N}$, i.e., a set $S = \{S_1, \dots, S_n\}$ with n elements. We call any ordered sequence $(S_{i_1}, \dots, S_{i_r})$ of r

elements of S a sample of size r, drawn from this population. Then there are two possible procedures:

(i) Sampling with replacement, i.e., the same element can be drawn more than once.
(ii) Sampling without replacement, i.e., an element one chooses is removed from the population. In this case $r \leq n$.

In either case, our experiment is described by a sample space Ω in which each individual points represents a sample of size r. In the first case we get $|\Omega| = n^r$, where an element $\omega \in \Omega$ is of the form $\omega = (S_{i_1}, \ldots, S_{i_r})$. In the second case, we get

$$|\Omega| = n(n-1)\cdots(n-r+1) = \frac{n!}{(n-r)!}.$$

Example 1.3.1.1. If by a "ten letter word" is meant a (possibly meaningless) sequence of ten letters, then such a word represents a sample from the population of 26 letters. There are 26^{10} such words.

Example 1.3.1.2. Tossing a coin r times is one way of obtaining a sample of size r drawn from the population of the letters H and T. Usually we assign equal probabilities to all samples, namely n^{-r} is sampling with replacement and $\frac{1}{n(n-1)\cdots(n-r+1)}$ is a sampling without replacement.

Example 1.3.1.3. A random sample of size r with replacement is taken from a population of size n, and we assume that $r \leq n$. The probability p measuring the event where no element appears twice in the sample is given by

$$p = \frac{n(n-1)\cdots(n-r+1)}{n^r} \quad \left(\mathbb{P}(A) = \frac{|A|}{|\Omega|}\right).$$

Consequently, the probability of having four different numbers chosen from $\{0, 1, 2, 3, \ldots, 9\}$ is (here $n = 10$, $r = 4$)

$$\frac{10 \times 9 \times 8 \times 7}{10^4} \approx 0.5.$$

In sampling without replacement the probability for any fixed element of the population to be included in a random sample of size r is

$$1 - \underbrace{\frac{(n-1)(n-2)\cdots(n-r)}{n(n-1)\cdots(n-r+1)}}_{\text{probability of the complementary event}} = 1 - \frac{n-r}{n} = \frac{r}{n}.$$

We have used that for $A \in \mathcal{A}$, $\mathbb{P}(A) = 1 - \mathbb{P}(A^{\mathsf{C}})$. Similarly, in sampling with replacement the probability that a fixed element of the population to be included in a random sample of size r is

$$1 - \left(\frac{n-1}{n}\right)^r = 1 - \left(1 - \frac{1}{n}\right)^r.$$

1.3.2. Subpopulations

Let $S = \{S_1, \ldots, S_n\}$ be a population of size n. We call subpopulations of size r any set of r elements, distinct or not, chosen from S. Elements of a subpopulation are not ordered, i.e., two samples of size r from S correspond to the same subpopulation if they only differ by the order of the elements. Here we have two cases as well. We will first consider the case without replacement. We must have $r \leq n$. There are $\binom{n}{r} = \frac{n!}{r!(n-r)!}$ different configurations. Note that we have the symmetry $\binom{n}{r} = \binom{n}{n-r}$ and that we always use the convention $0! = 1$.

Example 1.3.2.1 (Poker). There are $\binom{52}{5} = 2'598'960$ hands at poker. Let us compute the probability p measuring the event where a hand at poker contains 5 different face values. The face values can be chosen in $\binom{13}{5}$ different ways, and corresponding to each case we can choose one of the four suits. It follows that

$$p = \frac{4^5 \cdot \binom{13}{5}}{\binom{52}{5}} \approx 0.5.$$

Example 1.3.2.2 (Senator problem). Each of the 50 states has two senators. We consider the event that in a committee of 50 senators chosen at random

- a given state is represented;
- all states are represented.

In the first case, it is better to calculate the probability q of the complementary event that the given state is not represented

$$q = \frac{\binom{98}{50}}{\binom{100}{50}} = \frac{50 \times 49}{100 \times 99}.$$

For the second case we note that a committee including a senator of all states can be chosen in 2^{50} different ways. The probability that all states are included is

$$\frac{2^{50}}{\binom{100}{50}} \approx 4.126 \cdot 10^{-14}.$$

Example 1.3.2.3 (An occupancy problem). Consider a random distribution of r balls in n cells. We want to find the probability p_k that a specified cell contains exactly k balls ($k = 0, 1, \dots, r$). We note that k balls can be chosen in $\binom{r}{k}$ different ways and the remaining $(r - k)$ balls can be placed into the remaining $(n - 1)$ cells in $(n - 1)^{r-k}$ so we get

$$p_k = \binom{r}{k}(n-1)^{r-k}\frac{1}{n^r} \quad \left(\mathbb{P}(A) = \frac{|A|}{|\Omega|}\right).$$

If we denote $\tilde{p} = \frac{1}{n}$, then

$$p_k = \binom{r}{k}\tilde{p}^k(1-\tilde{p})^{r-k}.$$

This is called the *binomial distribution* of parameters p and r. It is a probability distribution on $\{0, 1, \dots, r\}$ and

$$\sum_{k=0}^{r} p_k = \sum_{k=0}^{r}\binom{r}{k}\tilde{p}^k(1-\tilde{p})^{r-k} = (\tilde{p} + 1 - \tilde{p})^r = 1.$$

For example in a coin tossing game we have r tosses and we set $H \equiv 1$ and $T \equiv 0$. The state space is given by sequences

$$\Omega = \{(\omega_j)_{1 \le j \le r}\}, \ \omega_j \in \{0, 1\}.$$

Moreover, we have that

$$S_r = \sum_{j=1}^{r} \omega_j,$$

where S_r is the number of heads. If we assume that $\mathbb{P}(H) = p \in [0,1]$, then

$$\mathbb{P}(S_r = k) = \binom{r}{k} p^k (1-p)^{r-k}.$$

Theorem 1.3.2.1. *Let $r_1, \ldots, r_k$ be integers such that $r_1 + \cdots + r_k = n$ with $r_i \geq 0$, $r_i \in \mathbb{N}$, $1 \leq i \leq k$. The number of ways in which a population of n elements can be partitioned into k subpopulations of which the first contains r_1 elements, the second contains r_2 elements etc., is given by*

$$\frac{n!}{r_1! r_2! \cdots r_k!}$$

Proof. For the first part we note that

$$\frac{n!}{r_1! \cdots r_k!} = \binom{n}{r_1}\binom{n-r_1}{r_2}\binom{n-r_1-r_2}{r_3}\cdots\binom{n-r_1-r_2-\cdots-r_{k-2}}{r_k-1}.$$

For the second part we see that an induction on n shows that

$$(U_1 + \cdots + U_k)^n = \sum_{\substack{r_1 \geq 0, \ldots, r_k \geq 0 \\ r_1 + \cdots + r_k = n}} \frac{n!}{r_1! \cdots r_k!} U_1^{r_1} \cdots U_k^{r_k}. \qquad (1.3.2.1)$$

But the left-hand side is also

$$\sum_{1 \leq k_1, \ldots, k_n \leq k} U_{k_1} \cdots U_{k_n}.$$

Now ordering the terms and comparing with (1.3.2.1) yields the result. □

Example 1.3.2.4. A throw of twelve dice can result in 6^{12} different outcomes. The probability measuring the event that each face can appear twice is given by

$$\frac{12!}{2^6 \cdot 6^{12}}.$$

Now we want to turn to the case with replacement. Let us consider the construction for an occupancy problem similarly as in Example 1.3.2.3, i.e., our model is that of placing randomly r balls into n cells. Such an event is completely described by its occupancy numbers $r_1, \ldots, r_n$ where r_k stands for the number of balls in the kth cell. Every n-tupel of integers satisfying $r_1 + \cdots + r_k = r$ describes a possible configuration. Two distributions are distinguishable if the occupancy numbers are different. We denote by $A_{r,n}$ the number of distinguishable distributions. It is also the number of subpopulations of size r with replacement from a population $S = \{S_1, \ldots, S_n\}$ of size n and it is characterized by the number r_k of appearances of the individual S_k with the restriction $r_1 + \cdots + r_n = r$.

Theorem 1.3.2.2.

$$A_{r,n} = \binom{n+r-1}{r} = \binom{n+r-1}{n-r}.$$

Proof. There are two different proofs.

(i) We represent the balls by $\otimes$ and the cells by the n spaces between $n+1$ bars, e.g., the pictorial figure

$$| \otimes \otimes \otimes | \otimes | \; | \; | \; | \otimes \otimes \otimes \otimes |$$

is used as to illustrate a distribution of $r = 8$ balls in $n = 6$ cells. The occupancy numbers are then $3, 1, 0, 0, 0, 4$ and their sum is $3 + 1 + 0 + 0 + 0 + 4 = 8$. Such a symbol necessarily starts and ends with a bar, but the remaining $(n-1)$ bars and r balls can appear in an arbitrary order. We have that

$$\binom{n+r-1}{r}$$

is the number of ways of selecting r places (for the balls) out of $n+r-1$.

(ii) For $|t_i| < 1$ with $1 \le i \le n$, we have

$$\prod_{1\le i\le n} \frac{1}{1-t_i} = \prod_{1\le i\le n} \left(\sum_{m_i \ge 0} t_i^{m_i} \right) = \sum_{0 \le m_1,\ldots,m_n} t_1^{m_1} \cdot t_2^{m_2} \cdots t_n^{m_n}$$

$$= \sum_{r\ge 0} \sum_{\substack{m_1+\cdots+m_n=r \\ m_1\ge 0,\ldots,m_n\ge 0}} t_1^{m_1} \cdot t_2^{m_2} \cdots t_n^{m_n}. \tag{1.3.2.2}$$

Hence, if we have $t_1 = t_2 = \cdots = t_n = t$ and using (1.3.2.2), we get

$$\frac{1}{(1-t)^n} = \sum_{r\ge 0} \sum_{\substack{m_1+\cdots+m_n=r \\ m_1\ge 0,\ldots,m_n\ge 0}} t^r = \sum_{r\ge 0} A_{r,n} \cdot t^r. \tag{1.3.2.3}$$

On the other hand, using Taylor expansion, we get

$$\frac{1}{(1-t)^n} = \sum_{r\ge 0} \binom{n+r-1}{r} t^n, \tag{1.3.2.4}$$

Comparing (1.3.2.3) with (1.3.2.4), we can conclude that

$$A_{r,n} = \binom{n+r-1}{r}. \qquad \square$$

Example 1.3.2.5. The partial derivatives of order r of a smooth function $f(x_1, \ldots, x_n)$ of n variables do not depend on the order of differentiation but only on the number of times that each variable appears. Hence, there are $\binom{n+r-1}{r}$ different partial derivatives of order r.

Remark 1.3.2.6. $A_{r,n}$ is also the number of different integer solutions to the equation

$$r_1 + \cdots + r_n = r.$$

1.3.3. Combination of events

Let Σ_N be the symmetric group of order N. We will consider a permutation $\sigma \in \Sigma_N$ in the following way:

$$\sigma = \begin{pmatrix} 1 & 2 & \cdots & N \\ \sigma(1) & \sigma(2) & \cdots & \sigma(N) \end{pmatrix}.$$

We know that the symmetric group of order N has cardinality $|\Sigma_N| = N!$. Therefore, the probability assigned to a permutation is given by

$$\mathbb{P}(\sigma) = \frac{1}{N!}.$$

Let us now go back to a general probability space $(\Omega, \mathcal{A}, \mathbb{P})$. If $A_1, A_2 \in \mathcal{A}$ with $A = A_1 \cup A_2$, we get

$$\mathbb{P}(A_1 \cup A_2) = \mathbb{P}(A_1) + \mathbb{P}(A_2) - \mathbb{P}(A_1 \cap A_2).$$

It is natural to ask what we can do when we consider $\mathbb{P}(A)$ where $A = \bigcup_{i=1}^N A_i$, with $A_1, \ldots, A_N \subset \mathcal{A}$. Define $\mathbb{P}_i := \mathbb{P}(A_i)$ for $1 \leq i \leq N$ and $\mathbb{P}_{ij} := \mathbb{P}(A_i \cap A_j)$ for all $1 \leq i < j \leq N$. In general, for $1 \leq i_1 < \cdots < i_k \leq N$, we can define

$$\mathbb{P}_{i_1, \ldots, i_k} := \mathbb{P}\left(\bigcap_{j=1}^k A_{i_j} \right).$$

Moreover, let

$$\begin{aligned} S_1 &:= \sum_{1 \leq i \leq N} \mathbb{P}_i \\ S_2 &:= \sum_{1 \leq i < j \leq N} \mathbb{P}_{ij} \\ &\vdots \\ S_r &:= \sum_{1 \leq i_1 < \cdots < i_r \leq N} \mathbb{P}_{i_1, \ldots, i_r}. \end{aligned}$$

In particular, S_r has $\binom{N}{r}$ terms. Note that for $N = 2$ we get

$$\mathbb{P}(A) = S_1 - S_2.$$

Theorem 1.3.3.1 (Inclusion–Exclusion). *Let everything be as before. If we have*

$$\mathbb{P}\left(\bigcup_{i=1}^{N} A_i\right) = S_1 - S_2 + S_3 - S_4 + \cdots \pm S_N,$$

then we get

$$\mathbb{P}\left(\bigcup_{i=1}^{N} A_i\right) = \sum_{r=1}^{N}(-1)^{r-1}S_r.$$

Example 1.3.3.1 (Two decks of cards). Consider two equivalent decks consisting of N cards. Each are put into random order and matched against each other. If a card occupies the same place in both decks we speak of a match. We want to compute the probability of having at least one match. Let us number the cards with $1, \ldots, N$ and consider the permutation

$$\sigma = \begin{pmatrix} 1 & 2 & \cdots & N \\ \sigma(1) & \sigma(2) & \cdots & \sigma(N) \end{pmatrix},$$

where the first line denotes the first deck and the second line denotes the second deck. A match for k corresponds to $k = \sigma(k)$. In that case we call k a fix point of the permutation σ. We look for the number of permutations of $\{1, \ldots, N\}$ (out of the N) which have at least one fixed point. Let A_k be the event where $k = \sigma(k)$. Clearly, we have

$$\mathbb{P}_k := \mathbb{P}(A_k) = \frac{(N-1)!}{N!} = \frac{1}{N}.$$

Similarly, for $1 \leq i < j \leq N$, we have

$$\mathbb{P}_{ij} := \mathbb{P}(A_i \cap A_j) = \frac{(N-2)!}{N!} = \frac{1}{N(N-1)}.$$

More generally, for $1 \leq i_1 < \cdots < i_r \leq N$, we have

$$\mathbb{P}_{i_1,\ldots,i_r} = \frac{(N-r)!}{N!},$$

and hence

$$S_r = \sum_{1\le i_1<\cdots<i_r\le N} \mathbb{P}_{i_1,\ldots,i_r} = \sum_{1\le i_1<\cdots<i_r\le N} \frac{(N-r)!}{N!}$$
$$= \frac{(N-r)!}{N!} \sum_{1\le i_1<\cdots<i_r\le N} 1 = \frac{(N-r)!}{N!}\binom{N}{r} = \frac{1}{r!}. \quad (1.3.3.1)$$

If $\mathbb{P}_1$ denotes the probability of the least fixed point, then

$$\mathbb{P}_1 = \mathbb{P}\left(\bigcup_{i=1}^{N} A_i\right) = 1 - \frac{1}{2!} + \frac{1}{3!} - \cdots \pm \frac{1}{N!}.$$

1.4. Random Walks

1.4.1. The reflection principle

From a formal point of view, we shall be concerned with arrangements of finitely many $+1$ and -1. Consider $n = p + q$ symbols $\varepsilon_1, \ldots, \varepsilon_n$, where $\varepsilon_j \in \{-1, +1\}$ for all $1 \le j \le n$. Suppose that there are p-times $+1$ and q-times -1. Then $S_k = \varepsilon_1 + \cdots + \varepsilon_k$ represents the difference between p and q at the first k places. So we get

$$S_k - S_{k-1} = \varepsilon_k = \pm 1, \quad S_0 = 0, \quad S_n = p - q. \quad (1.4.1.1)$$

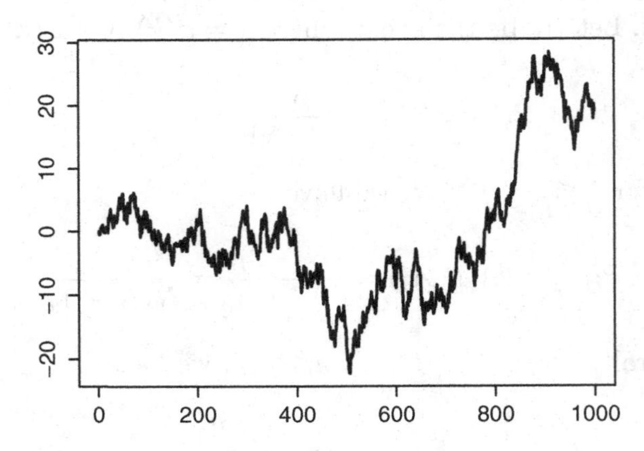

Fig. 1.1 Example for a random walk.

The arrangement $(\varepsilon_1, \ldots, \varepsilon_n)$ will be represented by a polygonal line whose kth side has slope ε_k, and whose kth vertex has ordinate S_k. Such lines will be called a *path.* We shall use (t, x) for coordinates on such a path.

Definition 1.4.1.1 (Path). Let $n > 0$ and x be integers. A path $(S_0, S_1, \ldots, S_t)$ from the origin to the point (t, x) is a polygonal line whose vertices have abscissas $0, 1, 2, \ldots, t$ and ordinates $S_0, S_1, \ldots, S_t$ satisfying (1.4.1.1) with $S_t = x$.

Remark 1.4.1.2. We shall refer to t as the length of the path. There are 2^t paths of length t. Moreover, we have

$$t = p + q,$$
$$x = p - q.$$

A path from the origin to an arbitrary point (t, x) exists only if t and x are of the form as in Definition 1.4.1.1. In this case, the number p takes the place of the amount of positive ε_k which can be chosen from the $t = p + q$ available places in

$$N_{t,x} = \binom{p+q}{p} = \binom{p+q}{q}$$

different ways. We use the convention that $N_{t,x} = 0$ whenever t and x are not of the form as in Definition 1.4.1.1. This will imply that $N_{t,x}$ represents the number of different paths from the origin to an arbitrary point (t, x).

Example 1.4.1.3 (Ballot theorem). Suppose that in a ballot,[1] candidate P scores p votes and candidate Q scores q votes, where $p > q$. The probability that throughout the counting there are always more votes for P than for Q equals $\frac{p-q}{p+q}$. The whole voting record may be represented by a path of length $p+q$ in which $\varepsilon_k = +1$, if the kth vote is for P and $\varepsilon_k = -1$ otherwise. Conversely, every path from the origin to the point $(p + q, p - q)$ can be interpreted as a voting with

[1] In the ballot theorem, it is implicitly assumed that all paths have equal probability.

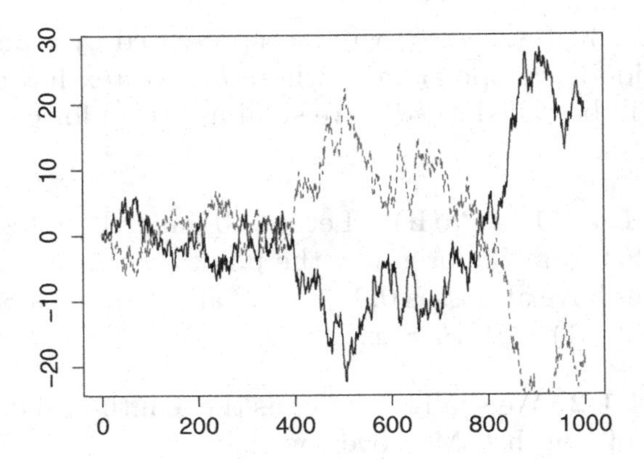

Fig. 1.2 Example for the reflection principle.

the given totals p and q. S_k will be the number of votes by which P leads just after the kth vote. The candidate P leads throughout the voting if $S_k > 0$ for all $1 \leq k \leq t$.

Let $A = (a, \alpha)$ and $B = (b, \beta)$ be two independent points with $b > a \geq 0$, $a, b \in \mathbb{N}$, $\alpha > 0$, $\beta > 0$, $\alpha, \beta \in \mathbb{N}$. By reflection of A on the t-axis we mean the point $A' = (a, -\alpha)$.

Lemma 1.4.1.1 (Reflection principle). The number of paths from A to B which touch or cross the t-axis equals the number of paths from A' to B.

Proof. Consider a path $(S_a = \alpha, S_{a+1}, \ldots, S_b = B)$ from A to B having one or more vertices on that axis. Let t be the abscissa of the first such vertex that is $S_a > 0, S_{a+1} > 0, \ldots, S_{k-1} > 0, S_k > 0$. Then $(-S_a, -S_{a+1}, \ldots, -S_{k-1}, S_k, S_{k+1}, \ldots, S_b)$ is a path leading from A' to B and having $T = (t, 0)$ as its vertex on the t-axis. This gives a one-to-one correspondence between all paths from A' to B and paths from A to B that have a vertex on the t-axis. □

Let us now prove the ballot theorem. Let t and x be positive integers. There are exactly $\frac{x}{t} N_{t,x}$ paths $(S_1, \ldots, S_t = x)$ such that

$$S_1 > 0,\ S_2 > 0, \ldots, S_t > 0.$$

Clearly, there exists exactly as many admissible paths as there are paths from $(1,1)$ to (t,x), which neither or cross the t-axis. From the previous lemma, the number of such paths equals

$$N_{t-1,x-1} - N_{t-1,x+1} = \binom{p+q-1}{p-1} - \binom{p+q-1}{p},$$

where we used that $N_{t,x} = \binom{p+q}{p}$. So we get

$$\begin{aligned}\binom{p+q-1}{p-1} - \binom{p+q-1}{p} &= \frac{(p+q-1)!}{(p-1)!q!} - \frac{(p+q-1)!}{p!(q-1)!} \\ &= \frac{p}{p+q} \cdot \frac{(p+q)!}{p!q!} - \frac{q}{p+q} \cdot \frac{(p+q)!}{p!q!} \\ &= \frac{p-q}{p+q} N_{t,x} = \frac{x}{t} N_{t,x}.\end{aligned}$$

1.4.2. Random walk terminology

We set $S_0 = 0$, $S_t = X_1 + \cdots + X_t$, $X_j \in \{-1,+1\}$ for all $1 \le j \le t$. Our state space would be Ω_t, containing all possible paths. Therefore, we get that $|\Omega| = 2^t$. We set our σ-algebra $\mathcal{A} = \mathcal{P}(\Omega_t)$. We consider $\mathbb{P}$ to be our uniform probability measure. Let us look at the event $\{S_t = r\}$ (at time t the particle is at the point r). We shall also speak of a *visit* to r at time t. The number $N_{t,x}$ of paths from the origin to $N_{t,r}$ is given by

$$\binom{t}{\frac{t+r}{2}},$$

with $t = p+q$, $r = p-q$ and hence $p = \frac{t+r}{2}$. Here we interpret $\binom{t}{\frac{t+r}{2}}$ as 0 if $t+r$ is not an even integer between 0 and t. Hence, we get that

$$\mathbb{P}_{t,r} := \mathbb{P}(S_t = r) = \binom{t}{\frac{t+r}{2}} \cdot 2^t.$$

A return to the origin occurs at time k if $S_k = 0$. Here k is necessarily an even integer, that we denote by $k = 2\nu$ with ν an integer.

The probability of a return to the origin is given by

$$\mathbb{P}_{2\nu,0}.$$

We denote it by $U_{2\nu}$, which is now given by

$$U_{2\nu} = \mathbb{P}(S_{2\nu} = 0) = \binom{2\nu}{\nu} \cdot 2^{-2\nu}.$$

Using Stirling's formula (Theorem 1.1.1.2), we get that $U_{2\nu} \sim \frac{1}{\sqrt{\pi r}}$. Among the returns to the origin, the first return receives special attention. A first return occurs at 2ν if

$$S_1 \neq 0,\ S_2 \neq 0, \ldots, S_{2\nu-1} \neq 0,\ S_{2\nu} = 0.$$

We denote the probability of this event by $f_{2\nu}$.

Lemma 1.4.2.1. *For all $n \geq 0$ we have*

$$\mathbb{P}(S_1 \neq 0, S_2 \neq 0, \ldots, S_{2n-1} \neq 0, S_{2n} \neq 0) = \mathbb{P}(S_{2n} = 0) = U_{2n}.$$

Remark 1.4.2.1. When the event on the left-hand side occurs, either all $S_j > 0$ or all $S_j < 0$. Since these events are equally probable, it follows that

$$\mathbb{P}(S_1 > 0, \ldots, S_{2n} > 0) = \frac{1}{2}U_{2n}.$$

Proof of Lemma 1.4.2.1. We have

$$\mathbb{P}(S_1 > 0, \ldots, S_{2n} > 0) = \sum_{r=1}^{\infty} \mathbb{P}(S_1 > 0, \ldots, S_{2n} = 2r),$$

where all the terms with $r > n$ are zero. By the ballot theorem, the number of paths $\mathbb{P}(S_1 > 0, \ldots, S_{2n-1} > 0, S_{2n} = 2r)$ is equal to $N_{2n-1,2r+1}$ and thus

$$\mathbb{P}(S_1 > 0, \dots, S_{2n-1} > 0, S_{2n} = 2r) = \frac{1}{2}(\mathbb{P}_{2n-1,2r-1} - \mathbb{P}_{2n-1,2r+1}),$$

$$\mathbb{P}(S_1 > 0, S_2 > 0, \dots, S_{2n} > 0) = \frac{1}{2}\sum_{r=1}^{n}(\mathbb{P}_{2n-1,2r-1} - \mathbb{P}_{2n-1,2r+1}),$$

$$\frac{1}{2}\left(\mathbb{P}_{2n-1,1} - \mathbb{P}_{2n-1,3} + \mathbb{P}_{2n-1,3} - \mathbb{P}_{2n-1,5} \pm \cdots + \mathbb{P}_{2n-1,2n-1} - \underbrace{\mathbb{P}_{2n-1,2n+1}}_{=0}\right)$$
$$= \frac{1}{2}\mathbb{P}_{2n-1,1} = \frac{1}{2}U_{2n}.$$

Therefore, we get

$$\mathbb{P}_{2n-1,1} = \mathbb{P}(S_{2n-1} = 1) = \binom{2n-1}{\frac{2n-1+1}{2}} 2^{-(2n-1)}$$
$$= 2^{-2n} \cdot 2 \cdot \frac{(2n-1)!}{n!(n-1)!} = 2^{2n}\binom{2n}{2} = U_{2n}. \qquad \square$$

Remark 1.4.2.2. Saying that the first return to the origin occurs at 2^n amounts $S_1 \neq 0, S_2 \neq 0, \dots, S_{2n-1} \neq 0, S_{2n} = 0$, we have

$$\mathbb{P}(S_1 \neq 0, S_2 \neq 0, \dots, S_{2n-1} \neq 0, S_{2n} = 0) = f_{2n},$$

and

$$\{S_1 \neq 0, S_2 \neq 0, \dots, S_{2n-1} \neq 0\}$$
$$= \{S_1 \neq 0, \dots, S_{2n-1} \neq 0, S_{2n} \neq 0\} \cup \{S_1 \neq 0, \dots, S_{2n-1} \neq 0, S_{2n} = 0\},$$

which implies that

$$U_{2n-2} = U_{2n} + f_{2n},$$

and hence for all $n \geq 1$ we have

$$f_{2n} = U_{2n-2} - U_{2n}.$$

Therefore, we get the relation

$$f_{2n} = \frac{1}{2n-1}U_{2n}.$$

Theorem 1.4.2.2. *The probability that up to time* $t = 2n$*, the last visit to the origin occurs at* $2k$ *is given by*

$$\alpha_{2k,2n} = U_{2k} N_{2n-2k}, \quad k = 0, 1, \ldots, n.$$

Proof. We are concerned with paths satisfying $S_{2k} = 0$, $S_{2k+1} \neq 0, \ldots, S_{2n} \neq 0$. The first $2k$ vertices of such paths can be chosen in $2^{2k} N_{2k}$ different ways. Taking the point $(2k, 0)$ as new origin and using Lemma 1.4.2.1, we see that the next $(2n - k)$ vertices can be chosen in $2^{2n-2k} N_{2n-2k}$ different ways. Therefore we get

$$\alpha_{2k,2n} = \frac{1}{2^{2n}} \left(2^{2k} N_{2k} 2^{2n-2k} N_{2n-2k} \right) = U_{2k} N_{2n-2k}. \qquad \square$$

Chapter 2

The Modern Probability Language

This is the main chapter of Part I, covering a basic introduction of the modern probability theory. We will discuss the concept of distributions and the notion of expectation at the beginning. Afterwards, the concept of moments, variance and covariance and several properties of those will be covered. We will continue with the concept of the characteristic function and independence, both for σ-algebras and for random variables. Furthermore, we look at the Borel–Cantelli lemma and move on to the weak and strong law of large numbers. The concept of different convergences will follow and finally we are going to spend time on the central limit theorem. After one has read this chapter, the more advanced structures and notions of probability theory of Part II will be accessible and lead to a fundamental understanding of modern probability theory.

2.1. General Definitions

2.1.1. Law of a random variable

Definition 2.1.1.1 (Random variable). Let $(\Omega, \mathcal{A}, \mathbb{P})$ be a probability space. Let $(E, \mathcal{E})$ be a measurable space. A measurable map $X : (\Omega, \mathcal{A}, \mathbb{P}) \to (E, \mathcal{E})$ is called a *random variable* with values in E.

Definition 2.1.1.2 (Law/Distribution). The *law* or *distribution* of a random variable is the image measure of $\mathbb{P}$ by X, and is usually denoted by $\mathbb{P}_X$. It is thus a probability measure on $(E, \mathcal{E})$.

In particular, for $B \in \mathcal{E}$, we have

$$\mathbb{P}_X(B) = \mathbb{P}(X^{-1}(B)) = \mathbb{P}(X \in B) = \mathbb{P}\left(\{\omega \in \Omega \mid X(\omega) \in B\}\right).$$

Let $\mathcal{B}(\mathbb{R}^d)$ be the Borel σ-algebra of $\mathbb{R}^d$. If μ is a probability measure on $(\mathbb{R}^d, \mathcal{B}(\mathbb{R}^d))$, (or even on a more general space $(E, \mathcal{E})$), there is a canonical way of constructing a random variable X such that $\mathbb{P}_X = \mu$ as a map

$$X : (\mathbb{R}^d, \mathcal{B}(\mathbb{R}^d), \mu) \to \mathbb{R}^d.$$

There are two special cases.

(i) *Discrete random variable*: Let E be a countable space and $\mathcal{E} := \mathcal{P}(E)$ its power set. The law of X is given by

$$\mathbb{P}_X := \sum_{x \in E} P(x)\delta_x,$$

where $P(x) := \mathbb{P}(X = x)$ and δ_x is the *Dirac measure* at x, meaning that for all $A \subset E$,

$$\delta_x(A) = \begin{cases} 1 & \text{if } x \in A, \\ 0 & \text{if } x \notin A. \end{cases}$$

Note that if $\mathbb{P}_X(E) = 1$, then

$$\sum_{x \in E} P(x)\delta_x(E) = \sum_{x \in E} P(x) = 1.$$

Indeed, for all $B \in \mathcal{E}$, we have that

$$\begin{aligned} \mathbb{P}_X(B) = \mathbb{P}(X \in B) = \mathbb{P}\left(\bigcup_{x \in B} \{X = x\}\right) &= \sum_{x \in B} \mathbb{P}(X = x) \\ &= \sum_{x \in E} P(x)\delta_x(B). \end{aligned}$$

(ii) *Continuous random variable*: A random variable X with values in $(\mathbb{R}^d, \mathcal{B}(\mathbb{R}^d))$ is said to have a density if $\mathbb{P}_X \ll \lambda$, where λ is the Lebesgue measure on $\mathbb{R}^d$. Here $\ll$ means *absolute continuity* of the left measure with respect to the right one. The *Radon–Nikodym theorem* (cf. Theorem 3.3.0.1) tells us then that there exists a measurable map $P : \mathbb{R}^d \to \mathbb{R}$ such that for all $B \in \mathcal{B}(\mathbb{R}^d)$ we have

$$\mathbb{P}_X(B) = \int_B P(x)\mathrm{d}x.$$

In particular, we get $\int_{\mathbb{R}^d} P(x)\mathrm{d}x = \mathbb{P}_X(\mathbb{R}^d) = 1$. Moreover, the map P is unique up to sets of Lebesgue measure zero. The map P is called the *density* of X. If $d = 1$, we get

$$\mathbb{P}(\alpha \leq X \leq \beta) = \mathbb{P}_X([\alpha, \beta]) = \int_\alpha^\beta P(x)\mathrm{d}x.$$

Definition 2.1.1.3 (Expected value/Expectation). Let $(\Omega, \mathcal{A}, \mathbb{P})$ be a probability space. Let X be a real-valued random variable (i.e., with values in $\mathbb{R}$). The *expectation* (sometimes also *expected value*) of such a random variable is defined as

$$\mathbb{E}[X] = \int_\Omega X(\omega)\mathrm{d}\mathbb{P}(\omega) = \int_{\mathbb{R}} x\mathrm{d}\mathbb{P}_X(x),$$

which is well-defined in the following two cases:

- if $x \geq 0$, which then gives $\mathbb{E}[X] \in [0, \infty)$;
- if $\mathbb{E}[X] = \int_\Omega |X(\omega)|\mathrm{d}\mathbb{P}(\omega) < \infty$.

We extend this definition to the case of a random variable $X = (X_1, \ldots, X_d)$ taking values in $\mathbb{R}^d$ by defining

$$\mathbb{E}[X] = (\mathbb{E}[X_1], \ldots, \mathbb{E}[X_d]),$$

provided that each $\mathbb{E}[X_i]$ is well-defined.

Remark 2.1.1.4. If $B \in \mathcal{A}$ and $X = \mathbb{1}_B$ (the *indicator function* of B), then

$$0 \leq \mathbb{E}[X] = \mathbb{E}[\mathbb{1}_B] = \mathbb{P}(B) \leq 1.$$

In general, $\mathbb{E}[X]$ is interpreted as the average or the mean of the random variable X. If X takes values in some discrete set

$\{x_1, \ldots, x_n, \ldots\}$ then

$$\mathbb{E}[X] = \sum_{n=1}^{\infty} x_n \mathbb{P}(X = x),$$

whenever it is well-defined.

The expectation is a special case of an integral with respect to a positive measure. In particular, the following points hold for the expectation:

- For all integrable random variables X, Y and $a, b \in \mathbb{R}$ we have

$$\mathbb{E}[aX + bY] = a\mathbb{E}[X] + b\mathbb{E}[Y].$$

- If C is a constant and $\mathbb{E}[X] = C$, then

$$\int_{\Omega} C \mathrm{d}\mathbb{P}(\omega) = C\mathbb{P}(\Omega) = C.$$

- If $X \geq 0$ and $\mathbb{E}[X] \geq 0$ and if $X \leq Y$ are both integrable then

$$\mathbb{E}[X] \leq \mathbb{E}[Y].$$

- *Monotone convergence* (cf. Theorem B.1.0.3). If $(X_n)_{n \geq 1}$ is a sequence of real-valued random variables, and if $X_n \geq 0$ for all $n \geq 1$ and $X_n \uparrow X$ as $n \to \infty$, then

$$\mathbb{E}[X_n] \uparrow \mathbb{E}[X] \quad \text{as } n \to \infty.$$

- *Fatou's lemma* (cf. Lemma B.1.0.7). If $(X_n)_{n \geq 1}$ is a sequence of real-valued random variables with $X_n \geq 0$ for all $n \geq 1$, then

$$\mathbb{E}\left[\liminf_{n \to \infty} X_n\right] \leq \liminf_{n \to \infty} \mathbb{E}[X_n].$$

- *Dominated convergence* (cf. Theorem B.3.0.1). If $(X_n)_{n \geq 1}$ is a sequence of real-valued random variables with $|X_n| \leq Z$ for all $n \geq 1$, such that $\mathbb{E}[Z] < \infty$, for another real-valued random variable Z, and $X_n \xrightarrow{n \to \infty} X$ almost everywhere (usually abbreviated a.e.), then

$$\mathbb{E}[X_n] \xrightarrow{n \to \infty} \mathbb{E}[X].$$

Remark 2.1.1.5. In probability theory we speak of *almost sure convergence* and write a.s., rather than almost everywhere. However, the meaning is the same, it just indicates the fact that one is working with probability measures. If $X_n \xrightarrow{n\to\infty} X$ a.s., then we mean

$$\mathbb{P}\left(\{\omega \in \Omega \mid X_n(\omega) \xrightarrow{n\to\infty} X(\omega)\}\right) = 1.$$

Proposition 2.1.1.1. *Let X be a random variable with values in a measurable space $(E, \mathcal{E})$. If $f : E \to [0, \infty]$ is measurable, then*

$$\mathbb{E}[f(X)] = \int_E f(x) \mathrm{d}\mathbb{P}_X(x).$$

Similarly, if $f : E \to \mathbb{R}$ is such that $\mathbb{E}[f(X)] < \infty$, then

$$\mathbb{E}[f(X)] = \int_E f(x) \mathrm{d}\mathbb{P}_X(x).$$

Remark 2.1.1.6. In such a case, $f(X)$ is also a random variable.

Proof of Proposition 2.1.1.1. In the case where $f = \mathbb{1}_B$ with $B \in \mathcal{E}$, we get that

$$\mathbb{E}[f(X)] = \mathbb{P}(X \in B) = \mathbb{P}_X(B)$$

by definition of the distribution of a random variable. Then, by linearity, the result is true for positive simple functions. Further, we can use the fact that for any measurable $f \geq 0$, there is a sequence $(f_n)_{n\in\mathbb{N}}$, where the f_n's are simple and positive such that $f_n \uparrow f$ as $n \to \infty$. Finally, applying the *monotone convergence theorem* yields the result. □

Remark 2.1.1.7. One often uses Proposition 2.1.1.1 to compute the law of a random variable X. If one is able to write $\mathbb{E}[X] = \int f \mathrm{d}\nu$ for a sufficiently large class of functions f, then one can deduce that $\mathbb{P}_X = \nu$. The idea is to be able to take $f = \mathbb{1}_B$, since then $\mathbb{E}[f(X)] = \mathbb{P}_X(B) = \nu(B)$.

Example 2.1.1.8. Assume that $\mathbb{P}_X$ is absolutely continuous with density $h(x) = \frac{1}{\sqrt{2\pi}} \mathrm{e}^{-\frac{x^2}{2}}$ for $x \in \mathbb{R}$ and $Y = X^2$. Then one can ask about the distribution of Y. Let $f : \mathbb{R} \to \mathbb{R}_{\geq 0}$ be a measurable map. Then

$$\mathbb{E}[f(Y)] = \mathbb{E}[f(X^2)] = \int_{-\infty}^{\infty} f(x^2)\frac{1}{\sqrt{2\pi}}\mathrm{e}^{-\frac{x^2}{2}}\,\mathrm{d}x.$$

We can write

$$\int_{-\infty}^{\infty} f(x^2)\frac{1}{\sqrt{2\pi}}\mathrm{e}^{-\frac{x^2}{2}}\,\mathrm{d}x = 2\int_{0}^{\infty} f(x^2)\frac{1}{\sqrt{2\pi}}\mathrm{e}^{-\frac{x^2}{2}}\,\mathrm{d}x.$$

Now we can set $y = x^2$. Then $\mathrm{d}y = 2x\mathrm{d}x$ and hence $\mathrm{d}x = \frac{\mathrm{d}y}{2\sqrt{y}}$. Now we can write

$$2\int_{0}^{\infty} f(y)\frac{\mathrm{e}^{-\frac{y}{2}}}{2\sqrt{2\pi y}}\mathrm{d}y = \int_{0}^{\infty} f(y)\frac{\mathrm{e}^{-\frac{y}{2}}}{\sqrt{2\pi y}}\mathrm{d}y,$$

which implies that

$$\mathrm{d}\nu(y) = \frac{\mathrm{e}^{-\frac{y}{2}}}{\sqrt{2\pi y}}\mathbb{1}_{\{y>0\}}\mathrm{d}y.$$

So we see that the distribution of Y is given by $\frac{\mathrm{e}^{-\frac{y}{2}}}{\sqrt{2\pi y}}\mathbb{1}_{\{y>0\}}$.

Proposition 2.1.1.2. *Let $X = (X_1, \ldots, X_d) \in \mathbb{R}^d$ be an $\mathbb{R}^d$-valued random variable. Assume that X has density $P(x_1, \ldots, x_d)$. Then for all $j \in \{1, \ldots, n\}$, X_j has density*

$$P_j(x) = \int_{\mathbb{R}^{d-1}} P(x_1, \ldots, x_{j-1}, x_j, x_{j+1}, \ldots, x_d)\mathrm{d}x^1 \cdots \mathrm{d}x^{j-1}\mathrm{d}x^{j+1} \cdots \mathrm{d}x^d.$$

Example 2.1.1.9. Let $d = 2$ and $X = (X_1 = X, X_2 = Y)$. Then $P_1(x) = \int_{\mathbb{R}} P(x, y)\mathrm{d}y$ and $P_2(x) = \int_{\mathbb{R}} P(x, y)\mathrm{d}x$.

Proof of Proposition 2.1.1.2. Let $\pi_j : (x_1, \dots, x_d) \mapsto x_j$ be the projection onto the jth factor. Using *Fubini's theorem*, we get for all Borel measurable $f : \mathbb{R} \to \mathbb{R}_{\geq 0}$ that

$$\mathbb{E}\,[f(X_j)] = \mathbb{E}[f(\pi_j(X))] = \int_{\mathbb{R}^d} f(x_j) P(x_1, \dots, x_d) \mathrm{d}x^1 \cdots \mathrm{d}x^d$$

$$= \int_{\mathbb{R}} f(x_j) \underbrace{\left(\int_{\mathbb{R}^{d-1}} P(x_1, \dots, x_{j-1}, x_j, x_{j+1}, \dots, x_d) \mathrm{d}x^1 \cdots \mathrm{d}x^{j-1} \mathrm{d}x^{j+1} \cdots \mathrm{d}x^d \right) \mathrm{d}x^j}_{\mathrm{d}\nu(x_j) = P(x_j)\mathrm{d}x^j}.$$

By renaming $x_j = y$, we get

$$\mathbb{E}[f(X_j)] = \int_{\mathbb{R}} f(y) P_j(y) \mathrm{d}y.$$

Hence, the distribution of X_j has density $P_j(y)$ on $\mathbb{R}$. □

Remark 2.1.1.10. If $X = (X_1, \dots, X_d) \in \mathbb{R}^d$ is an $\mathbb{R}^d$-valued random variable, the distributions $\mathbb{P}_{X_j}$ are called the *margins* of X. Proposition 2.1.1.2 shows us that the margins are determined by

$$\mathbb{P}_{X=(X_1,\dots,X_d)},$$

but the converse is wrong. For example, take Q to be a density on $\mathbb{R}$ and observe that $P(x_1, x_2) = Q(x_1)Q(x_2)$ is also a density on $\mathbb{R}^2$. We have already seen that we can construct (in a canonical way) a random variable $X = (X_1, X_2) \in \mathbb{R}^2$ such that $\mathbb{P}_X$ has $P(x_1, x_2)$ as density. Now the margins of X, namely $\mathbb{P}_{X_1}$ and $\mathbb{P}_{X_2}$, have density $q(x)$. We now observe that the *random variable*'s $X = (X_1, X_2)$ and $X' = (X_1, X_1)$ have the same margin but they are different. In particular, $\mathbb{P}_X$ has support in $\mathbb{R}^2$, while $\mathbb{P}_{X'}$ has support in the diagonal of $\mathbb{R}^2$, which is of Lebesgue measure zero in $\mathbb{R}^2$. In general, we have $\mathbb{P}_X \neq \mathbb{P}_{X'}$.

2.2. Classical Probability Distributions

Let $(\Omega, \mathcal{A}, \mathbb{P})$ be a probability space and let $X : (\Omega, \mathcal{A}, \mathbb{P}) \to (E, \mathcal{E})$ be a random variable taking values in some measurable space $(E, \mathcal{E})$.

2.2.1. Discrete distributions

2.2.1.1. *Uniform distribution*

Assume that $|E| < \infty$. A random variable X with values in E is said to be *uniform* on E, if for all $x \in E$ we have

$$\mathbb{P}(X = x) = \frac{1}{|E|}.$$

2.2.1.2. *Bernoulli distribution*

A random variable X with values in $\{0, 1\}$ is said to be *Bernoulli distributed* with parameter $p \in [0, 1]$ if we have

$$\mathbb{P}(X = 1) = p, \qquad \mathbb{P}(X = 0) = 1 - p.$$

The random variable X can be interpreted as the outcome of a coin toss. The expectation of X is then given by

$$\mathbb{E}[X] = 0 \cdot \mathbb{P}(X = 0) + 1 \cdot \mathbb{P}(X = 1).$$

2.2.1.3. *Binomial distribution*

A random variable X taking values in $\{0, 1, \ldots, n\}$ is said to be *binomially distributed* with parameters $n \in \mathbb{N} \setminus \{0\}$ and $p \in [0, 1]$ if we have

$$\mathbb{P}(X = k) = \binom{n}{k} p^k (1 - p)^{n-k}.$$

We will abbreviate this distribution by $\mathcal{B}(n, p)$.

The random variable X is interpreted as the number of heads of the n tosses of the previous case. One has to check that it is indeed a probability distribution, i.e., that the sum $\sum_{k=0}^{n} \mathbb{P}(X = k) = 1$. Indeed, we have

$$\sum_{k=0}^{n} \mathbb{P}(X = k) = \sum_{k=0}^{n} \binom{n}{k} p^k (1 - p)^{n-k} = (p + (1 - p))^n = 1.$$

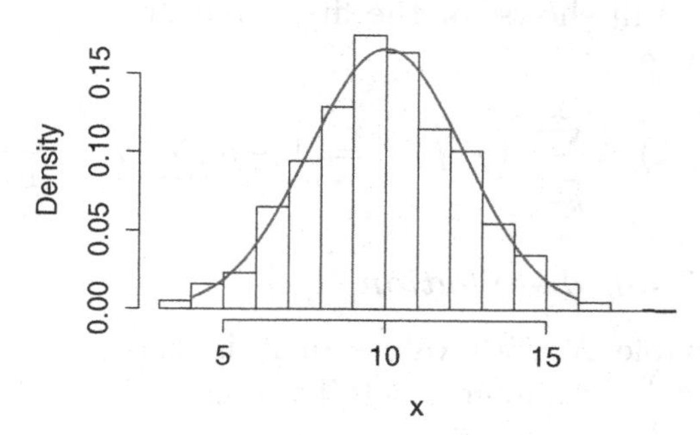

Fig. 2.1 Histogram of a binomially distributed random variable.

The expected value for the binomial distribution is given by

$$\begin{aligned}
\mathbb{E}[X] = \sum_{k=0}^{n} k\mathbb{P}(X=k) &= \sum_{k=0}^{n} k\binom{n}{k} p^k (1-p)^{n-k} \\
&= np\sum_{k=0}^{n} k\frac{(n-1)!}{(n-k)!k!} p^{k-1}(1-p)^{(n-1)-(k-1)} \\
&= np\sum_{k=1}^{n} \frac{(n-1)!}{(n-k)!(k-1)!} p^{k-1}(1-p)^{(n-1)-(k-1)} \\
&= np\sum_{k=1}^{n} \binom{n-1}{k-1} p^{k-1}(1-p)^{(n-k)-(k-1)} \\
&= np\sum_{l=0}^{n-1} \binom{n-1}{l} p^{l}(1-p)^{(n-1)-l} \\
&= np\sum_{l=0}^{m} \binom{m}{l} p^{l}(1-p)^{m-l} = np(p+(1-p))^m = np,
\end{aligned}$$

where we have defined $l := k-1$ and $m := n-1$.

2.2.1.4. *Geometric distribution*

A random variable X with values in $\mathbb{N}$ is said to be *geometrically distributed* with parameter $p \in [0,1]$ if we have

$$\mathbb{P}(X = k) = (1-p)p^k.$$

The random variable X can be interpreted as the number of heads obtained before tail shows for the first time. It is also a probability distribution, since

$$\sum_{k=0}^{\infty} \mathbb{P}(X = k) = \sum_{k=0}^{\infty} (1-p)p^k = (1-p) \sum_{k=0}^{\infty} p^k = \frac{1-p}{1-p} = 1.$$

2.2.1.5. *Poisson distribution*

A random variable X with values in $\mathbb{N}$ is said to be *Poisson distributed* with real parameter $\lambda > 0$ if we have

$$\mathbb{P}(X = k) = \mathrm{e}^{-\lambda} \frac{\lambda^k}{k!}, \qquad k \in \mathbb{N}.$$

The Poisson distribution is very important, both from the point of view of applications and from the theoretical point of view. Intuitively, it describes the number of rare events that have occurred during a long period. If $X_n \sim \mathcal{B}(n, p_n)$ and if $np_n \xrightarrow{n\to\infty} \lambda > 0$, i.e., $p_n \sim \frac{\lambda}{n}$ for $n \geq 1$, then for every $k \in \mathbb{N}$ we get

$$\mathbb{P}(X_n = k) \xrightarrow{n\to\infty} \mathrm{e}^{-\lambda} \frac{\lambda^k}{k!}.$$

The expected value is then given by

$$\mathbb{E}[X] = \sum_{k=0}^{\infty} k \frac{\lambda^k}{k!} \mathrm{e}^{-\lambda} = \lambda \mathrm{e}^{-\lambda} \sum_{k=1}^{\infty} \frac{\lambda^{k-1}}{(k-1)!} = \lambda \mathrm{e}^{-\lambda} \sum_{j=0} \frac{\lambda^j}{j!} = \lambda.$$

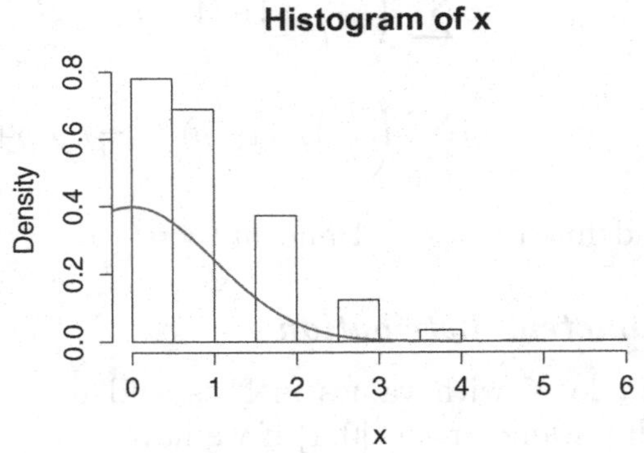

Fig. 2.2 Histogram of a Poisson distributed random variable.

Remark 2.2.1.1. Note that we have used the notation $X \sim \mathcal{D}$ to indicate that the random variable X follows the distribution $\mathcal{D}$. We will use this notation frequently from now on.

2.2.2. Absolutely continuous distributions

Let now $E \subset \mathbb{R}$. We want to analyze the densities $P(x)$ of a certain distributed random variable in the continuous case.

2.2.2.1. *Uniform distribution on an interval*

The density of a continuous, uniformly distributed random variable X on an interval $[a, b] \subset \mathbb{R}$ is given by

$$P(x) = \frac{1}{b-a}\mathbb{1}_{[a,b]}(x).$$

We need to check that it is a probability density, which now means to check that $\int_{\mathbb{R}} P(x)\mathrm{d}x = 1$. Indeed, we have

$$\begin{aligned}\int_{-\infty}^{\infty} P(x)\mathrm{d}x &= \int_{-\infty}^{\infty} \frac{1}{b-a}\mathbb{1}_{[a,b]}(x)\mathrm{d}x \\ &= \frac{1}{b-a}\int_{-\infty}^{\infty} \mathbb{1}_{[a,b]}(x)\mathrm{d}x \\ &= \frac{1}{b-a}(b-a) = 1.\end{aligned}$$

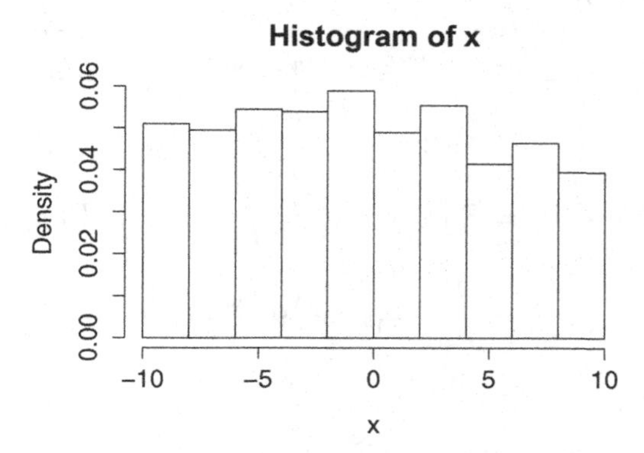

Fig. 2.3 Histogram of a uniformly distributed random variable on a given interval.

In fact, if X is uniform on $[a,b]$, then $|X| \leq |a|+|b| < \infty$ a.s. and thus $\mathbb{E}[|a|+|b|] = |a|+|b| < \infty$ which implies that $\mathbb{E}[X] < \infty$. The expectation is thus well-defined and is given by

$$\mathbb{E}[X] = \int_{-\infty}^{\infty} xP(x)\mathrm{d}x = \int_{-\infty}^{\infty} \frac{1}{b-a}\mathbb{1}_{[a,b]}(x)\mathrm{d}x = \frac{1}{b-a}\int_a^b x\mathrm{d}x$$
$$= \frac{1}{b-a}\frac{1}{2}(b^2-a^2) = \frac{a+b}{2}.$$

2.2.2.2. *Exponential distribution*

The density of a random variable X which is *exponentially distributed* with parameter $\lambda > 0$ is given by

$$P(x) = \lambda \mathrm{e}^{-\lambda x}\mathbb{1}_{\mathbb{R}_{\geq 0}}(x),$$

with $X \geq 0$ a.s. The expectation is then given by

$$\mathbb{E}[X] = \int_{-\infty}^{\infty} xP(x)\mathrm{d}x = \int_0^{\infty} x\lambda \mathrm{e}^{-\lambda x}\mathrm{d}x = \lambda\int_0^{\infty} x\mathrm{e}^{-\lambda x}\mathrm{d}x.$$

With $u = \lambda x$ we get $\mathrm{d}x = \frac{\mathrm{d}u}{\lambda}$ and hence

$$\lambda\int_0^{\infty} \frac{u}{\lambda}\mathrm{e}^{-u}\frac{\mathrm{d}u}{\lambda} = \frac{1}{\lambda}\int_0^{\infty} u\mathrm{e}^{-u}\mathrm{d}u = \frac{1}{\lambda}.$$

If $a,b > 0$, then

$$\mathbb{P}(X > a+b) = \int_{a+b}^{\infty} \lambda\mathrm{e}^{-\lambda x}\mathrm{d}x = \lambda\left[-\frac{1}{\lambda}\mathrm{e}^{-\lambda x}\right]_{a+b}^{\infty}$$
$$= \mathrm{e}^{-\lambda(a+b)} = \mathrm{e}^{-\lambda a}\mathrm{e}^{-\lambda b} = \mathbb{P}(X > a)\mathbb{P}(X > b).$$

Note that

$$\mathbb{P}(X < 0) = \mathbb{E}\left[\mathbb{1}_{\{X<0\}}\right] = \int_{-\infty}^{\infty} \mathbb{1}_{\{x<0\}}P(x)\mathrm{d}x$$
$$= \int_{-\infty}^{\infty} \mathbb{1}_{\{x<0\}}\lambda\mathrm{e}^{-\lambda x}\mathbb{1}_{\{x\geq 0\}}\mathrm{d}x = 0$$

and also that

$$\mathbb{P}(X = x) = \int_{-\infty}^{\infty} \mathbb{1}_{\{y=x\}}P(y)\mathrm{d}y = 0.$$

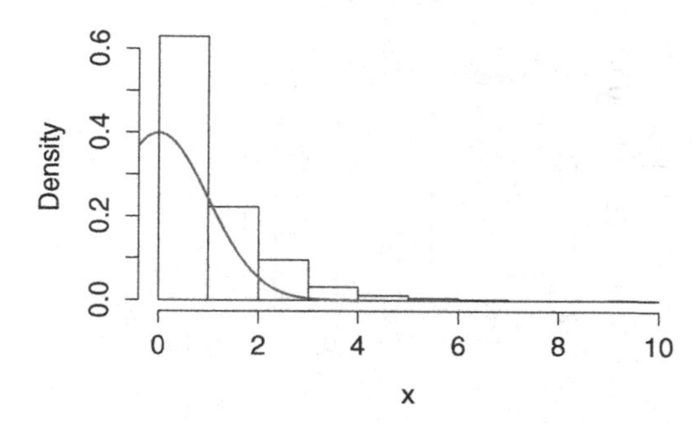

Fig. 2.4 Histogram of an exponentially distributed random variable.

2.2.2.3. *Gaussian (normal) distribution*

The density of a random variable X which is *Gaussian distributed* (or also *normally distributed*) with parameters $m \in \mathbb{R}$ and $\sigma \in \mathbb{R}_{\geq 0}$ is given by

$$P(x) = \frac{1}{\sigma\sqrt{2\pi}} \exp\left(-\frac{(x-m)^2}{2\sigma^2}\right).$$

This is the most important distribution in probability theory and we will abbreviate it by $\mathcal{N}(m, \sigma^2)$. We need to check that $P(x)$ is a probability density, i.e.,

$$\int_{-\infty}^{\infty} \frac{1}{\sigma\sqrt{2\pi}} \exp\left(-\frac{(x-m)^2}{2\sigma^2}\right) \mathrm{d}x = 1.$$

We redefine $u := x - m$ and hence $\mathrm{d}u = \mathrm{d}x$. So we get

$$\int_{-\infty}^{\infty} \frac{1}{\sigma\sqrt{2\pi}} \mathrm{e}^{-\frac{u^2}{2\sigma^2}} \mathrm{d}u.$$

Now we define $t := \frac{u}{\sigma}$ and hence $\mathrm{d}u = \sigma \mathrm{d}t$. Now we get

$$\int_{-\infty}^{\infty} \frac{1}{\sigma\sqrt{2\pi}} \mathrm{e}^{-\frac{t^2}{2}} \sigma \mathrm{d}t = \frac{1}{\sqrt{2\pi}} \int_{-\infty}^{\infty} \mathrm{e}^{-\frac{t^2}{2}} \mathrm{d}t = \frac{\sqrt{2\pi}}{\sqrt{2\pi}} = 1.$$

We have used the fact that $\int_{-\infty}^{\infty} \mathrm{e}^{-\frac{x^2}{2}} \mathrm{d}x = \sqrt{2\pi}$, by change of coordinates from Cartesian coordinates to polar coordinates. Consider

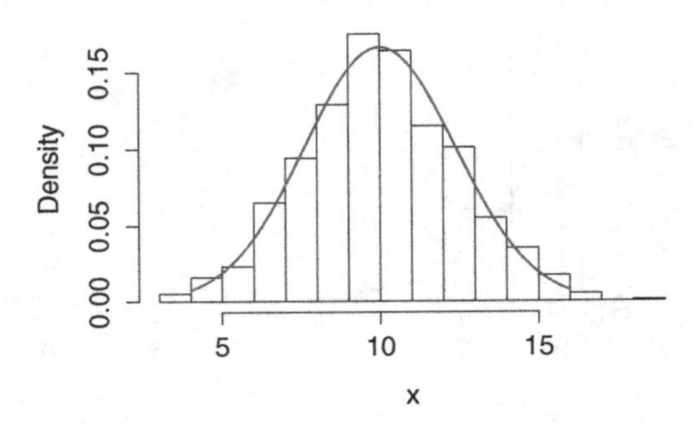

Fig. 2.5 Histogram of a Gaussian distributed random variable.

$\mathcal{N}(0,1)$ with density $P(x) = \frac{1}{\sqrt{2\pi}} e^{-\frac{x^2}{2}}$. It is called the *standard* Gaussian distribution (here we have $m = 0$, $\sigma = 1$). We note that if X is distributed according to $\mathcal{N}(m, \sigma^2)$, then

$$\mathbb{E}[X] = m, \qquad \mathbb{E}[(X - m)^2] = \sigma^2.$$

Indeed, we have

$$\mathbb{E}[|X|] = \int_{-\infty}^{\infty} |x| \frac{1}{\sigma\sqrt{2\pi}} \exp\left(-\frac{(x-m)^2}{2\sigma^2}\right) \mathrm{d}x < \infty$$

and therefore

$$\mathbb{E}[X] = \int_{-\infty}^{\infty} x \frac{1}{\sigma\sqrt{2\pi}} \exp\left(-\frac{(x-m)^2}{2\sigma^2}\right) \mathrm{d}x.$$

We redefine $u := x - m$ and hence $\mathrm{d}u = \mathrm{d}x$. So we get

$$\int_{-\infty}^{\infty} \frac{(u+m)}{\sigma\sqrt{2\pi}} \mathrm{e}^{-\frac{u^2}{2\sigma^2}} \mathrm{d}u = \underbrace{\int_{-\infty}^{\infty} \frac{u}{\sigma\sqrt{2\pi}} \mathrm{e}^{-\frac{u^2}{2\sigma^2}} \mathrm{d}u}_{=0} + \underbrace{m \int_{-\infty}^{\infty} \frac{1}{\sigma\sqrt{2\pi}} \mathrm{e}^{-\frac{u^2}{2\sigma^2}} \mathrm{d}u}_{=m}.$$

Thus, we get $\mathbb{E}[X] = m$.

Exercise 2.2.2.1. Show similarly as above that $\mathbb{E}[(X - m^2)] = \sigma^2$.

2.2.3. The distribution function

Let $X : \Omega \to \mathbb{R}$ be a real-valued random variable.

Definition 2.2.3.1 (Distribution function). The *distribution function* of X is defined by

$$\begin{aligned} F_X &: \mathbb{R} \to [0, 1], \\ t &\mapsto F_X(t) := \mathbb{P}(X \le t) = \mathbb{P}_X((-\infty, t)). \end{aligned}$$

In fact, F_X is *increasing* and *right continuous*, meaning that

$$\lim_{t \to -\infty} F_X(t) = 0, \qquad \lim_{t \to \infty} F_X(t) = 1.$$

Thus, we can write

$$\mathbb{P}(a \le X \le b) = F_X(b) - \underbrace{F_X(a^-)}_{\lim_{\substack{t \to a \\ t < a}} F_X(t)} \qquad \text{and therefore}$$

$$\mathbb{P}(a < X < b) = F_X(b^-) - F_X(a).$$

In particular, for a single value $a \in \mathbb{R}$, we get

$$\mathbb{P}(X = a) = F_X(a) - F_X(a^-),$$

which is called the *jump* of the function F_X.

Exercise 2.2.3.2. Using the monotone convergence theorem, show that if X and Y are two random variables, such that $F_X(t) = F_Y(t)$, then $\mathbb{P}_X = \mathbb{P}_Y$.

Note that if F is an increasing and right continuous function, then the set

$$A := \{a \in \mathbb{R} \mid F(a) \ne F(a^-)\}$$

is at most countable. Moreover, if $\mathbb{P}_X$ is absolutely continuous, then

$$\mathbb{P}_X(\{a\}) = \mathbb{P}(X = a) = 0,$$

which implies that for all $a \in \mathbb{R}$ we have $F_X(a) = F_X(a^-)$ and hence F_X is continuous. Another point of view is to say that, if $P(x)$ is the density of $\mathbb{P}_X$, then

$$F_X(t) = \mathbb{P}_X((-\infty, t]) = \int_{\mathbb{R}} \mathbb{1}_{(-\infty,t]}(x) P(x) \mathrm{d}x = \int_{-\infty}^{t} P(x) \mathrm{d}x,$$

is a continuous function of t.

2.2.4. σ-algebras generated by a random variable

Let $(\Omega, \mathcal{A}, \mathbb{P})$ be a probability space. Let X be a random variable taking values in some measurable space $(E, \mathcal{E})$, i.e., $X : (\Omega, \mathcal{A}, \mathbb{P}) \to (E, \mathcal{E})$. The σ-algebra generated by X, which we denote by $\sigma(X)$, is by definition the smallest σ-algebra, which makes X measurable, so we have

$$\sigma(X) := \{A = X^{-1}(B) \mid B \in \mathcal{E}\}.$$

Remark 2.2.4.1. One can of course extend this definition to the case of a family of random variables $(X_i)_{i \in I}$ for some index set I, taking values in measurable spaces $(E_i, \mathcal{E}_i)_{i \in I}$. In this case we have

$$\sigma((X_i)_{i \in I}) := \sigma(\{X_i^{-1}(B) \mid B_i \in \mathcal{E}_i, i \in I\}).$$

Proposition 2.2.4.1. *Let $(\Omega, \mathcal{A}, \mathbb{P})$ be a probability space. Let X be a random variable with values in a measurable space $(E, \mathcal{E})$ and let Y be a real-valued random variable. Then the following conditions are equivalent.*

(i) *Y is $\sigma(X)$-measurable.*

(ii) *There exists a measurable map $f : (E, \mathcal{E}) \to (\mathbb{R}, \mathcal{B}(\mathbb{R}))$, such that*

$$Y = f(X).$$

Proof. Note that we have the following cases:

$$X : (\Omega, \mathcal{A}, \mathbb{P}) \to (E, \mathcal{E}), \quad X : (\Omega, \sigma(X), \mathbb{P}) \to (E, \mathcal{E}),$$

$$Y : (\Omega, \mathcal{A}, \mathbb{P}) \to (\mathbb{R}, \mathcal{B}(\mathbb{R})).$$

- (ii) $\Rightarrow$ (i): This follows from the fact that the composition of two measurable maps is measurable.
- (i) $\Rightarrow$ (ii): Assume that Y is $\sigma(X)$-measurable. For simplicity, assume moreover that Y is simple, i.e.,

$$Y = \sum_{i=1}^{n} \lambda_i \mathbb{1}_{A_i}(x), \quad \lambda_i \in \mathbb{R},\ A_i \in \sigma(X)\ 1 \leq i \leq n.$$

Now by definition of $\sigma(X)$, there exists a $B_i \in \mathcal{E}$ such that $A_i \in X_i^{-1}(B_i)$ for all $1 \leq i \leq n$. Therefore, we have

$$Y = \sum_{i=1}^{n} \lambda_i \mathbb{1}_{A_i} = \sum_{i=1}^{n} \lambda_i \mathbb{1}_{B_i} \circ X = f \circ X,$$

where $f = \sum_{i=1}^{n} \lambda_i \mathbb{1}_{B_i}$ is $\mathcal{E}$-measurable. More generally, if Y is $\mathcal{E}$-measurable, there exists a sequence (Y_n) of simple functions such that Y_n is $\sigma(X)$-measurable and $Y_n \xrightarrow{n\to\infty} Y$. The above implies $Y_n = f_n(X)$ when $f_n : E \to \mathbb{R}$ is a measurable map. For $x \in E$, we can set

$$f(x) := \begin{cases} \lim_{n\to\infty} f_n(x) & \text{if the limit exists,} \\ 0 & \text{otherwise.} \end{cases}$$

Hence, f is measurable. Moreover, for all $\omega \in \Omega$ we get

$$X(\omega) \in \left\{ x \in E \,\middle|\, \lim_{n\to\infty} f_n(x) \text{ exists} \right\},$$

since $\lim_{n\to\infty} f_n(X(\omega)) = \lim_{n\to\infty} Y_n(\omega) = Y(\omega)$ and $f(X(\omega)) = \lim_{n\to\infty} f_n(X(\omega))$. Therefore, we get that $Y = f(X)$. □

2.3. Moments of Random Variables

2.3.1. Moments and variance

Let $(\Omega, \mathcal{A}, \mathbb{P})$ be a probability space. Let X be a random variable and let $p \geq 1$ be an integer (or even a real number).

Definition 2.3.1.1 (pth moment). The *pth moment* of X is defined as $\mathbb{E}[X^p]$, which is well-defined when $X \geq 0$ or

$$\mathbb{E}[|X|^p] = \int_{\Omega} |X(\omega)|^p \mathrm{d}\mathbb{P}(\omega) < \infty.$$

Remark 2.3.1.2. Note that when $p = 1$, we get the usual expectation of X.

Definition 2.3.1.3 (Centered). We say that X is *centered*, if $\mathbb{E}[X] = 0$.

Recall the spaces $L^p(\Omega, \mathcal{A}, \mathbb{P}) := \mathcal{L}^p(\Omega, \mathcal{A}, \mathbb{P})/\sim = \{f : \Omega \to \mathbb{R} \mid \int_\Omega |f|^p \mathrm{d}\mathbb{P} < \infty\}/\sim$ for $p \in [1, \infty)$ where $\sim$ denotes the equivalence relation where two functions agree almost surely. From *Hölder's inequality* we can observe that

$$\mathbb{E}[|XY|] \leq \mathbb{E}[|X|^p]^{\frac{1}{p}} \mathbb{E}[|Y|^q]^{\frac{1}{q}}, \tag{2.3.1.1}$$

whenever $p, q \geq 1$ and $\frac{1}{p} + \frac{1}{q} = 1$. If we take $Y = 1$ in (2.3.1.1), we obtain

$$\mathbb{E}[|X|] \leq \mathbb{E}[|X|^p]^{\frac{1}{p}},$$

which means $\|X\|_1 \leq \|X\|_p$. This can be extended as $\|X\|_r \leq \|X\|_p$ for any $r \leq p$. So it follows that $L^p(\Omega, \mathcal{A}, \mathbb{P}) \subset L^r(\Omega, \mathcal{A}, \mathbb{P})$. For $p = q = 2$, we get the *Cauchy–Schwarz inequality*

$$\mathbb{E}[|XY|] \leq \mathbb{E}[|X|^2]^{\frac{1}{2}} \mathbb{E}[|Y|^2]^{\frac{1}{2}}. \tag{2.3.1.2}$$

Again, setting $Y = 1$ in (2.3.1.2), we get $\mathbb{E}[|X|]^2 \leq \mathbb{E}[|X|^2]$.

Definition 2.3.1.4 (Variance/Standard deviation). Let $(\Omega, \mathcal{A}, \mathbb{P})$ be a probability space. Consider a random variable $X \in L^2(\Omega, \mathcal{A}, \mathbb{P})$. The *variance* of X is defined as

$$\mathbb{V}(X) := \mathbb{E}[(X - \mathbb{E}[X])^2],$$

and the *standard deviation* of X is defined by

$$\sigma_X := \sqrt{\mathbb{V}(X)}.$$

Remark 2.3.1.5. Informally, the variance represents the deviation of X around its mean $\mathbb{E}[X]$. Note that $\mathbb{V}(X) = 0$ if and only if X is a.s. constant.

Proposition 2.3.1.1. *For any random variable X we have*

$$\mathbb{V}(X) = \mathbb{E}[X^2] - \mathbb{E}[X]^2, \tag{2.3.1.3}$$

$$\mathbb{E}[(X - a)^2] = \mathbb{V}(X) + (\mathbb{E}[X] - a)^2, \qquad \forall a \in \mathbb{R}. \tag{2.3.1.4}$$

Remark 2.3.1.6. Note that a combination of the statement of Proposition 2.3.1.1 gives

$$\mathbb{V}(X) = \inf_{a \in R} \mathbb{E}[(X-a)^2].$$

Proof of Proposition 2.3.1.1. Note that we have

$$\begin{aligned}\mathbb{V}(X) &= \mathbb{E}[(X-\mathbb{E}[X])^2] = \mathbb{E}[X^2 - \{\mathbb{E}[X]X + \mathbb{E}[X]^2\}] \\ &= \mathbb{E}[X^2] - 2\mathbb{E}[X]\mathbb{E}[X] + \mathbb{E}[X]^2 = \mathbb{E}[X^2] - \mathbb{E}[X]^2. \quad (2.3.1.5)\end{aligned}$$

Moreover, we have

$$\mathbb{E}[(X-a)^2] = \mathbb{E}[(X-\mathbb{E}[X]) + (\mathbb{E}[X]-a)^2] = \mathbb{V}(X) + (\mathbb{E}[X]-a)^2,$$

which implies that for all $a \in \mathbb{R}$ we have

$$\mathbb{E}[(X-a)^2] \geq \mathbb{V}(X),$$

with equality whenever $a = \mathbb{E}[X]$. $\square$

Remark 2.3.1.7. It follows that if X is centered (Definition 2.3.1.3), we get that $\mathbb{V}(X) = \mathbb{E}[X^2]$.

Proposition 2.3.1.2 (Markov inequality). *For any positive random variable $X \geq 0$ and $a \geq 0$ we have*

$$\mathbb{P}(X > a) \leq \frac{1}{a}\mathbb{E}[X].$$

Proof. Indeed, we have

$$\mathbb{P}(X \geq a) = \mathbb{E}[\mathbb{1}_{\{X \geq a\}}] \leq \mathbb{E}\left[\frac{X}{a}\underbrace{\mathbb{1}_{\{X \geq a\}}}_{\leq 1}\right] \leq \mathbb{E}\left[\frac{X}{a}\right].$$

$\square$

Proposition 2.3.1.3 (Tchebisheff inequality). *For any random variable X and all $a > 0$ we have*

$$\mathbb{P}(|X - \mathbb{E}[X]| > a) \leq \frac{1}{a^2}\mathbb{V}(X).$$

Proof. This follows from Proposition 2.3.1.2 because $|X - \mathbb{E}[X]|$ is a positive random variable and hence

$$\mathbb{P}(|X - \mathbb{E}[X]| \geq a) = \mathbb{P}(|X - \mathbb{E}[X]|^2 \geq a^2) \leq \frac{1}{a^2} \underbrace{\mathbb{E}[|X - \mathbb{E}[X]|^2]}_{=:\mathbb{V}(X)}.$$

□

Definition 2.3.1.8 (Covariance). Let $(\Omega, \mathcal{A}, \mathbb{P})$ be a probability space. Consider two random variables $X, Y \in L^2(\Omega, \mathcal{A}, \mathbb{P})$. The *covariance* of X and Y is defined as

$$\mathbb{CV}(X, Y) := \mathbb{E}[(X - \mathbb{E}[X])(Y - \mathbb{E}[Y])] = \mathbb{E}[XY] - \mathbb{E}[X]\mathbb{E}[Y].$$

Remark 2.3.1.9. If $X = (X_1, \ldots, X_d) \in \mathbb{R}^d$ is an $\mathbb{R}^d$-valued random variable such that for all $1 \leq i \leq d$ we have $X_i \in L^2(\Omega, \mathcal{A}, \mathbb{P})$, then the *covariance matrix* of X is defined as

$$\mathbf{\Sigma}_X := (\mathbb{CV}(X_i, X_j))_{1 \leq i,j \leq d}.$$

Remark 2.3.1.10. Informally speaking, the covariance between X and Y measures the correlation between X and Y.

Note that $\mathbb{CV}(X, X) = \mathbb{V}(X)$ and using the Cauchy–Schwarz inequality, we get

$$|\mathbb{CV}(X, Y)| \leq \sqrt{\mathbb{V}(X)} \cdot \sqrt{\mathbb{V}(Y)}.$$

Clearly, the map $(X, Y) \mapsto \mathbb{CV}(X, Y)$ defines a bilinear form on $L^2(\Omega, \mathcal{A}, \mathbb{P})$. We can also see that $\mathbf{\Sigma}_X$ is symmetric and positive, i.e., if $\lambda_1, \ldots, \lambda_d \in \mathbb{R}$ and $\lambda := (\lambda_1, \ldots, \lambda_d)^T$, then

$$\langle \mathbf{\Sigma}_X \lambda, \lambda \rangle = \sum_{i,j=1}^{d} \lambda_i \lambda_j \mathbb{CV}(X_i, X_j) \geq 0.$$

Here we have denoted by $(\dots)^T$ the transpose. Therefore, we get

$$\begin{aligned}\sum_{i,j=1}^{d} \lambda_i\lambda_j \mathbb{CV}(X_i, X_j) &= \mathbb{V}\left(\sum_{j=1}^{d} \lambda_j X_j\right) = \mathbb{E}\left[\left(\sum_{j=1}^{d} \lambda_j X_j - \mathbb{E}\left[\sum_{j=1}^{d} \lambda_j X_j\right]\right)^2\right] \\ &= \mathbb{E}\left[\left(\sum_{j=1}^{d} \lambda_j X_j - \sum_{j=1}^{d} \lambda_j \mathbb{E}[X_j]\right)^2\right] = \mathbb{E}\left[\left(\sum_{j=1}^{d} (X_j - \mathbb{E}[X_j])\right)^2\right] \\ &= \mathbb{E}\left[\sum_{j=1} \lambda_j (X_j - \mathbb{E}[X_j]) \sum_{i=1}^{d} \lambda_i (X_i - \mathbb{E}[X_i])\right] \\ &= \sum_{i,j=1}^{d} \lambda_i \lambda_j \mathbb{E}[(X_i - \mathbb{E}[X_i])(X_j - \mathbb{E}[X_j])] \geq 0.\end{aligned}$$

Exercise 2.3.1.11. Let $X = (X_1, \dots, X_d) \in \mathbb{R}^d$ be an $\mathbb{R}^d$-valued random variable. Show that if A is a matrix of size $n \times d$, and $Y := AX$, we have

$$\mathbf{\Sigma}_Y = A\mathbf{\Sigma}_X A^T.$$

Remark 2.3.1.12. Let $X = (X_1, \dots, X_d)^T$ and $XX^T = (X_i X_j)_{1\leq i,j\leq n}$. Then, informally, we get

$$\mathbf{\Sigma}_X = \mathbb{E}[XX^T] = (\mathbb{E}[X_i X_j])_{1\leq i,j\leq n},$$

and for $Y := AX$ we get

$$\mathbf{\Sigma}_Y = \mathbb{E}[AXX^T A^T] = A\mathbb{E}[XX^T]A^T = A\mathbf{\Sigma}_X A^T.$$

2.3.2. Linear regression

Let $(\Omega, \mathcal{A}, \mathbb{P})$ be a probability space. Let $X, Y_1, \dots, Y_n$ be random variables in $L^2(\Omega, \mathcal{A}, \mathbb{P})$. We want the best approximation of X as an affine function of $Y_1, \dots, Y_n$. More precisely, we want to minimize the expression

$$\mathbb{E}[(X - \mathbb{E}[\beta_0 + \beta_1 Y_1 + \cdots + \beta_n Y_n])^2]$$

over all possible choices of $(\beta_0, \beta_1, \dots, \beta_n) \in \mathbb{R}^{n+1}$.

Proposition 2.3.2.1. *Let $(\Omega, \mathcal{A}, \mathbb{P})$ be a probability space. Let $X, Y \in L^1(\Omega, \mathcal{A}, \mathbb{P})$ be two random variables. Then*

$$\inf_{(\beta_0,\beta_1,\ldots,\beta_n)\in\mathbb{R}^{n+1}} \mathbb{E}[(X - (\beta_0 + \beta_1 Y_1 + \cdots + \beta_n Y_n))^2] = \mathbb{E}[(X - Z)^2],$$

where $Z := \mathbb{E}[X] + \sum_{j=1}^n \alpha_j (Y_j - \mathbb{E}[Y_j])$ and the α_j's are solutions to the system

$$\sum_{j=1}^n \alpha_j \mathbb{CV}(Y_j, Y_k) = \mathbb{CV}(X, Y_k)_{1\leq k\leq n}.$$

Moreover, if $\mathbf{\Sigma}_Y$ is invertible, we have $\alpha = \mathbb{CV}(X, Y)\mathbf{\Sigma}_Y^{-1}$, where

$$\mathbb{CV}(X, Y) = \begin{pmatrix} \mathbb{CV}(X, Y_1) \\ \vdots \\ \mathbb{CV}(X, Y_n) \end{pmatrix}.$$

Proof. Let H be the linear subspace of $L^2(\Omega, \mathcal{A}, \mathbb{P})$ spanned by $\{1, Y_1, \ldots, Y_n\}$. Then we know that the random variable Z, which minimizes

$$\|X - U\|_2 = \mathbb{E}[(X - U)^2]$$

for $U \in H$, is the orthogonal projection of X on H. We can thus write

$$Z = \alpha_0 + \sum_{j=1}^n \alpha_j (Y_j - \mathbb{E}[Y_j]).$$

The orthogonality of $X - Z$ to H can be written as $\mathbb{E}[(X - Z) \cdot 1] = 0$. Therefore $\mathbb{E}[X] = \mathbb{E}[Z]$ and thus $\alpha_0 = \mathbb{E}[X]$. Moreover, we get $\mathbb{E}[(X - Z)(Y_k - \mathbb{E}[Y_k])] = 0$, which implies that for all $1 \leq k \leq n$ we get $\mathbb{E}[(X - \mathbb{E}[X] + \mathbb{E}[Z] - Z)] \cdot (Y_k - \mathbb{E}[Y_k]) = 0$. □

Remark 2.3.2.1. Note that when $n = 1$, we have

$$Z = \mathbb{E}[X] + \frac{\mathbb{CV}(X, Y)}{\mathbb{V}(Y)}(Y - \mathbb{E}[Y]).$$

2.4. The Characteristic Function

Definition 2.4.0.1 (Characteristic function). Let $(\Omega, \mathcal{A}, \mathbb{P})$ be a probability space. Let X be a random variable with values in $\mathbb{R}^d$, i.e., $X : (\Omega, \mathcal{A}, \mathbb{P}) \to \mathbb{R}^d$. Then we can look at the *characteristic function* of X, which is given by the Fourier transform

$$\begin{aligned} &\Phi_X : \mathbb{R}^d \to \mathbb{C}, \\ &\xi \mapsto \Phi_X(\xi) = \mathbb{E}\left[e^{i\langle \xi, X\rangle}\right] = \int_{\mathbb{R}^d} e^{i\langle \xi, x\rangle} d\mathbb{P}_X(x) \end{aligned} \tag{2.4.0.1}$$

where $\langle \xi, X\rangle = \sum_{k=1}^d \xi_k X_k$.

Remark 2.4.0.2. For $d = 1$ and $\xi \in \mathbb{R}$, we get

$$\Phi_X(\xi) = \mathbb{E}\left[e^{i\xi X}\right] = \int_{\mathbb{R}} e^{i\xi x} d\mathbb{P}_X(x).$$

Remark 2.4.0.3. The function Φ_X is clearly bounded since

$$|\Phi_X(\xi)| = \left|\mathbb{E}\left[e^{i\langle \xi, X\rangle}\right]\right| \leq \mathbb{E}\left[\underbrace{\left|e^{i\langle \xi, X\rangle}\right|}_{=1}\right] \leq 1.$$

Moreover, Φ_X is a continuous function since we know that $e^{i\langle \xi, x\rangle}$ is a continuous function of $\xi \in \mathbb{R}^d$ for every $x \in \mathbb{R}^d$, so we have

$$\left|e^{i\langle \xi, X\rangle}\right| \leq 1, \qquad \mathbb{E}[1] = 1 < \infty.$$

The claim follows.

Theorem 2.4.0.1. *The characteristic function uniquely characterizes probability distributions, meaning that for two random variables X and Y satisfying*

$$\Phi_X(\xi) = \Phi_Y(\xi), \qquad \forall \xi \in \mathbb{R}^d,$$

we get that

$$\mathbb{P}_X = \mathbb{P}_Y.$$

Lemma 2.4.0.2. *Let X be a random variable which is $\mathcal{N}(0,\sigma^2)$ distributed. Then*

$$\Phi_X(\xi) = \exp\left(-\frac{\sigma^2\xi^2}{2}\right), \quad \xi \in \mathbb{R}.$$

Proof. By the definition of the characteristic function and the density of a $\mathcal{N}(0,\sigma^2)$-distributed random variable, we get

$$\Phi_X(\xi) = \int_{-\infty}^{\infty} \mathrm{e}^{i\xi x}\mathrm{e}^{-\frac{x^2}{2\sigma^2}}\frac{\mathrm{d}x}{\sigma\sqrt{2\pi}}.$$

Assume for simplicity $\sigma = 1$ (change of variables: $\xi = \frac{x}{\sigma}$). Therefore, we have

$$\begin{aligned}\Phi_X(\xi) &= \int_{-\infty}^{\infty} \frac{1}{\sqrt{2\pi}}\mathrm{e}^{-\frac{x^2}{2}}\cos(\xi x)\mathrm{d}x + \underbrace{i\int_{-\infty}^{\infty} \frac{1}{\sqrt{2\pi}}\mathrm{e}^{-\frac{x^2}{2}}\sin(\xi x)\mathrm{d}x}_{0,\quad\text{by parity}} \\ &= \int_{-\infty}^{\infty} \frac{1}{\sqrt{2\pi}}\mathrm{e}^{-\frac{x^2}{2}}\cos(\xi x)\mathrm{d}x.\end{aligned}$$

Hence, we have

$$\frac{\mathrm{d}\Phi_X}{\mathrm{d}\xi}(\xi) = -\int_{-\infty}^{\infty} \frac{1}{\sqrt{2\pi}}x\mathrm{e}^{-\frac{x^2}{2}}\sin(\xi x)\mathrm{d}x,$$

where we have used the fact that $\mathrm{e}^{-\frac{x^2}{2}}\cos(\xi x)$ is smooth in both variables and that $\left|x\sin(\xi x)\mathrm{e}^{-\frac{x^2}{2}}\right| \leq |x|\mathrm{e}^{-\frac{x^2}{2}}$, which is integrable on $\mathbb{R}$. Using integration by parts we get

$$\frac{\mathrm{d}\Phi_X}{\mathrm{d}\xi}(\xi) = \underbrace{\left[\frac{1}{\sqrt{2\pi}}\mathrm{e}^{-\frac{x^2}{2}}\sin(\xi x)\right]_{-\infty}^{\infty}}_{=0} - \xi\int_{-\infty}^{\infty} \frac{1}{\sqrt{2\pi}}\mathrm{e}^{-\frac{x^2}{2}}\cos(\xi x)\mathrm{d}x.$$

Thus, we have the following *Cauchy problem*:

$$\begin{cases}\frac{\mathrm{d}\Phi_X}{\mathrm{d}\xi}(\xi) = -\xi\Phi_X(\xi), \\ \Phi_X(0) = 1.\end{cases}$$

Solving this ODE, we get

$$\Phi_X(\xi) = \mathrm{e}^{-\frac{-\xi^2}{2}}.$$

□

Proposition 2.4.0.3. *Let $X = (X_1, \dots, X_d) \in \mathbb{R}^d$ be an $\mathbb{R}^d$-valued random variable such that $\mathbb{E}[\|X\|^2] < \infty$, where $\|\cdot\|$ denotes the Euclidean norm. Then*

$$\lim_{\|\xi\|\to 0} \Phi_X(\xi) = 1 + i\sum_{j=1}^{d} \mathbb{E}[X_j]\xi_j - \frac{1}{2}\sum_{j=1}^{d}\sum_{k=1}^{d} \xi_j\xi_k\mathbb{E}[X_jX_k] + O(\|\xi\|^2), \tag{2.4.0.2}$$

Proof. Note that we can write

$$\frac{\partial \Phi_X(\xi)}{\partial \xi_j} = i\mathbb{E}\left[X_j e^{i\langle \xi, X\rangle}\right].$$

This follows from the differentiation under the integral sign with $\left|X_j e^{i\langle \xi, X\rangle}\right| \leq |X_j|$, which is integrable. Since

$$\mathbb{E}[|X_iX_k|] \leq \mathbb{E}[|X_j|^2]^{\frac{1}{2}}\mathbb{E}[|X_k|^2]^{\frac{1}{2}} < \infty,$$

we have

$$\frac{\partial^2 \Phi_X(\xi)}{\partial \xi_j \partial \xi_k} = -\mathbb{E}\left[\underbrace{X_jX_k e^{i\langle \xi, X\rangle}}_{\leq |X_jX_k| \in L^1(\mathbb{R})}\right].$$

Taking $\xi = 0$, we get that $\frac{\partial \Phi_X}{\partial \xi_j}(0) = i\mathbb{E}[X_j]$ and $\frac{\partial^2 \Phi_X}{\partial \xi_j \partial \xi_k}(0) = -\mathbb{E}[X_jX_k]$. Further, we can see that Equation (2.4.0.2) is the Taylor expansion at order two around zero of the C^2-function $\Phi_X(\xi)$. □

Remark 2.4.0.4. Using the proof of Proposition 2.4.0.3, we can obtain that when $d = 1$, we have

$$\mathbb{E}[|X|] < \infty \Rightarrow \mathbb{E}[X] = i\Phi'_X(0) \quad \text{and} \quad \mathbb{E}[X^2] < \infty \Rightarrow \mathbb{E}[X^2] = -\Phi''_X(0).$$

2.5. Independence

2.5.1. Independent events

Let $(\Omega, \mathcal{A}, \mathbb{P})$ be a probability space.

Definition 2.5.1.1 (Independent events). Two events $A, B \in \mathcal{A}$ are said to be *independent* if

$$\mathbb{P}(A \cap B) = \mathbb{P}(A)\mathbb{P}(B).$$

Example 2.5.1.2 (Throw of a die). Consider the state space $\Omega = \{1, 2, 3, 4, 5, 6\}$ for the toss of a die. For $\omega \in \Omega$ we then get $\mathbb{P}(\{\omega\}) = \frac{1}{6}$. Moreover, consider the events $A = \{1, 2\}$ and $B = \{1, 3, 5\}$. Then we have

$$\mathbb{P}(A \cap B]) = \mathbb{P}(\{1\}) = \frac{1}{6} \quad \text{and} \quad \mathbb{P}(A) = \frac{1}{3}, \ \mathbb{P}(B) = \frac{1}{2}.$$

Therefore, we get

$$\mathbb{P}(A \cap B) = \mathbb{P}(A)\mathbb{P}(B).$$

Hence, we get that A and B are independent.

Definition 2.5.1.3 (Independence of events). We say that the n events $A_1, \ldots, A_n \in \mathcal{A}$ are *independent* if for all $\{j_1, \ldots, j_l\} \subset \{1, \ldots, n\}$ we have

$$\mathbb{P}(A_{j_1} \cap A_{j_2} \cap \cdots \cap A_{j_l}) = \mathbb{P}(A_{j_1}) \cdots \mathbb{P}(A_{j_l}).$$

Remark 2.5.1.4. Note that it is not enough to have $\mathbb{P}(A_1 \cap \cdots \cap A_n) = \mathbb{P}(A_1) \cdots \mathbb{P}(A_n)$ in order to have independence. It is also not enough to check that for all $\{i, j\} \subset \{1, \ldots, n\}$ we have $\mathbb{P}(A_i \cap A_j) = \mathbb{P}(A_i)\mathbb{P}(A_j)$. For instance, consider two tosses of a coin (H = heads, T = tails) and consider three events A, B, C given by

$$A := \{H \text{ at the first throw}\}, \quad B := \{T \text{ at the first throw}\},$$
$$C := \{\text{same outcome for both tosses}\}.$$

The events A, B, C are two by two independent but A, B, C are not independent events in the sense of Definition 2.5.1.3.

Proposition 2.5.1.1. *The n events $A_1, \ldots, A_n \in \mathcal{A}$ are independent if and only if*

$$\mathbb{P}(B_1 \cap \cdots \cap B_n) = \mathbb{P}(B_1) \cdots \mathbb{P}(B_n), \tag{2.5.1.1}$$

for all $B_i \in \sigma(A_i) = \{\emptyset, A_i, A_i^{\mathsf{C}}, \Omega\}$ with $i \in \{1, \ldots, n\}$.

Proof. If the above condition is satisfied and if $\{j_1, \ldots, j_l\} \subset \{1, \ldots, n\}$, then for $i \in \{j_1, \ldots, j_l\}$ take $B_i = A_i$ and for $i \notin \{j_1, \ldots, j_l\}$ take $B_i = \Omega$. So it follows that

$$\mathbb{P}(A_{j_1} \cap \cdots \cap A_{j_l}) = \mathbb{P}(A_{j_1}) \cdots \mathbb{P}(A_{j_l}).$$

Conversely, assume that $A_1, \ldots, A_n \in \mathcal{A}$ are independent in order to deduce (2.5.1.1). We can assume that for all $i \in \{1, \ldots, n\}$ we have

$B_i \neq \emptyset$ (for otherwise (2.5.1.1) is trivially satisfied). If $\{j_1, \ldots, j_l\} = \{i \mid B_i \neq \Omega\}$, we have to check that

$$\mathbb{P}(B_{j_1} \cap \cdots \cap B_{j_l}) = \mathbb{P}(B_{j_1}) \cdots \mathbb{P}(B_{j_l}),$$

as soon as $B_{j_k} = A_{j_k}$ or $B_{j_k} = A_{j_k}^{\complement}$. Finally, it is enough to show that if $C_1, \ldots, C_p$ are independent events, then

$$C_1^{\complement}, C_2, \ldots, C_p$$

are also independent, but if $1 \notin \{i_1, \ldots, i_q\}$, for all $\{i_1, \ldots, i_q\} \subset \{1, \ldots, p\}$, then, from Definition 2.5.1.3, we have

$$\mathbb{P}(C_{i_1} \cap \cdots \cap C_{i_q}) = \mathbb{P}(C_{i_1}) \cdots \mathbb{P}(C_{i_q}).$$

If $1 \in \{i_1, \ldots, i_q\}$, say, e.g., $1 = i_1$, then

$$\begin{aligned}
\mathbb{P}(C_{i_1}^{\complement} \cap C_{i_2} \cap \cdots \cap C_{i_q}) &= \mathbb{P}(C_{i_1} \cap \cdots \cap C_{i_q}) - \mathbb{P}(C_1 \cap C_{i_2} \cap \cdots \cap C_{i_q}) \\
&= \mathbb{P}(C_{i_2}) \cdots \mathbb{P}(C_{i_q}) - \mathbb{P}(C_1)\mathbb{P}(C_{i_2}) \cdots \mathbb{P}(C_{i_q}) \\
&= (1 - \mathbb{P}(C_1))\mathbb{P}(C_{i_2}) \cdots \mathbb{P}(C_{i_q}) \\
&= \mathbb{P}(C_1^{\complement})\mathbb{P}(C_{i_2}) \cdots \mathbb{P}(C_{i_q}).
\end{aligned}$$

□

2.5.2. Conditional probability

Definition 2.5.2.1 (Conditional probability). Let $(\Omega, \mathcal{A}, \mathbb{P})$ be a probability space. Let $A, B \in \mathcal{A}$ be two events such that $\mathbb{P}(B) > 0$. The *conditional probability* of A given B is then defined as

$$\mathbb{P}(A \mid B) = \frac{\mathbb{P}(A \cap B)}{\mathbb{P}(B)}.$$

Theorem 2.5.2.1. *Let $(\Omega, \mathcal{A}, \mathbb{P})$ be a probability space. Let $A, B \in \mathcal{A}$ be two events such that $\mathbb{P}(B) > 0$. Then the following conditions hold:*

(i) *A and B are independent if and only if*

$$\mathbb{P}(A \mid B) := \mathbb{P}(A).$$

(ii) *The map*

$$\begin{aligned} \mathbb{Q}_B : \mathcal{A} &\to [0,1], \\ A &\mapsto \mathbb{Q}_B(A) := \mathbb{P}(A \mid B) \end{aligned} \tag{2.5.2.1}$$

defines a new probability measure on $\mathcal{A}$ called the conditional probability given B.

Proof. (i) If A and B are independent events, then by definition

$$\mathbb{P}(A \mid B) = \frac{\mathbb{P}(A \cap B)}{\mathbb{P}(B)} = \frac{\mathbb{P}(A)\mathbb{P}(B)}{\mathbb{P}(B)} = \mathbb{P}(A).$$

Conversely, if $\mathbb{P}(A \mid B) = \mathbb{P}(A)$, we get that

$$\mathbb{P}(A \cap B) = \mathbb{P}(A)\mathbb{P}(B),$$

and thus A and B are independent.

(ii) Note that we have

$$\mathbb{Q}_B(\Omega) = \mathbb{P}(\Omega \mid B) = \frac{\mathbb{P}(\Omega \cap B)}{\mathbb{P}(B)} = \frac{\mathbb{P}(B)}{\mathbb{P}(B)} = 1.$$

Now take a disjoint family $(A_n)_{n\geq 1} \subset \mathcal{A}$ of events. Then we have

$$\begin{aligned} \mathbb{Q}_B\left(\bigcup_{n\geq 1} A_n\right) &= \mathbb{P}\left(\bigcup_{n\geq 1} A_n \,\middle|\, B\right) = \frac{\mathbb{P}\left(\left(\bigcup_{n\geq 1} A_n\right) \cap B\right)}{\mathbb{P}(B)} = \mathbb{P}\left(\bigcup_{n\geq 1} (A_n \cap B)\right) \\ &= \sum_{n\geq 1} \frac{\mathbb{P}(A_n \cap B)}{\mathbb{P}(B)} = \sum_{n\geq 1} \mathbb{Q}_B(A_n). \end{aligned}$$

□

Theorem 2.5.2.2. *Let $(\Omega, \mathcal{A}, \mathbb{P})$ be a probability space. Let $A_1, \dots, A_n \in \mathcal{A}$ with*

$$\mathbb{P}(A_1 \cap \cdots \cap A_n) > 0.$$

Then we have

$$\begin{aligned} \mathbb{P}(A_1 \cap \cdots \cap A_n) = \mathbb{P}(A_1)\mathbb{P}(A_2 \mid A_1)\mathbb{P}(A_3 \mid A_1 \cap A_2) \\ \cdots \mathbb{P}(A_n \mid A_1 \cap \cdots \cap A_{n-1}). \end{aligned}$$

Proof. We prove this by induction. For $n = 2$ it is just the definition of conditional probability (Definition 2.5.2.1). Next we want to go from $n-1$ to n. Therefore, set $B := A_1 \cap \cdots \cap A_{n-1}$. Then we get

$$\begin{aligned}\mathbb{P}(B \cap A_n) &= \mathbb{P}(A_n \mid B)\mathbb{P}(B)\\ &= \mathbb{P}(A_n \mid B)\mathbb{P}(A_1)\mathbb{P}(A_| A_1) \cdots \mathbb{P}(A_{n-1} \mid A_1 \cap \cdots \cap A_{n-2}). \square\end{aligned}$$

Theorem 2.5.2.3. *Let $(\Omega, \mathcal{A}, \mathbb{P})$ be a probability space and let $(E_n)_{n\geq 1}$ be a finite or countable measurable partition of Ω such that $\mathbb{P}(E_n) > 0$ for all $n \geq 1$. Then we get*

$$\mathbb{P}(A) = \sum_{n\geq 1} \mathbb{P}(A \mid E_n)\mathbb{P}(E_n), \qquad \forall A \in \mathcal{A}.$$

Proof. First, note that

$$A = A \cap \Omega = A \cap \left(\bigcup_{n\geq 1} E_n \right) = \bigcup_{n\geq 1} (A_n \cap E_n).$$

Now since the $(A \cap E_n)_{n\geq 1}$ are all disjoint, we can write

$$\mathbb{P}(A) = \sum_{n\geq 1} \mathbb{P}(A \cap E_n) = \sum_{n\geq 1} \mathbb{P}(A \mid E_n)\mathbb{P}(E_n). \qquad \square$$

Theorem 2.5.2.4 (Bayes). *Let $(\Omega, \mathcal{A}, \mathbb{P})$ be a probability space and let $(E_n)_{n\geq 1}$ be a finite or countable partition of Ω and assume that $\mathbb{P}(E_n) > 0$ for all $n \geq 1$ and $\mathbb{P}(A)$ for all $A \in \mathcal{A}$. Then*

$$\mathbb{P}(E_n \mid A) = \frac{\mathbb{P}(A \mid E_n)\mathbb{P}(E_n)}{\sum_{n\geq 1} \mathbb{P}(A \mid E_n)\mathbb{P}(E_n)}.$$

Proof. By Theorem 2.5.2.3, we know that

$$\mathbb{P}(A) = \sum_{n\geq 1} \mathbb{P}(A \mid E_n)\mathbb{P}(E_n), \quad \mathbb{P}(E_n \mid A) = \frac{\mathbb{P}(E_n \cap A)}{\mathbb{P}(A)},$$

$$\mathbb{P}(A \mid E_n) = \frac{\mathbb{P}(A \cap E_n)}{\mathbb{P}(E_n)}.$$

Therefore, combining everything, we get

$$\mathbb{P}(E_n \mid A) = \frac{\mathbb{P}(E_n \cap A)}{\mathbb{P}(A)} = \frac{\mathbb{P}(A \mid E_n)\mathbb{P}(E_n)}{\sum_{n\geq 1} \mathbb{P}(A \mid E_n)\mathbb{P}(E_n)}.$$

$\square$

2.5.3. Independent random variables and independent σ-algebras

Let $(\Omega, \mathcal{A}, \mathbb{P})$ be a probability space.

Definition 2.5.3.1 (Independence of σ-algebras). We say that the sub-σ-algebras $\mathcal{B}_1, \ldots, \mathcal{B}_n$ of $\mathcal{A}$ are independent if for all $A_1 \in \mathcal{B}_1, \ldots, A_n \in \mathcal{B}_n$ we have

$$\mathbb{P}(A_1 \cap \cdots \cap A_n) = \mathbb{P}(A_1) \cdots \mathbb{P}(A_n).$$

Let $X_1, \ldots, X_n$ be n random variables with values in measurable spaces $(E_1, \mathcal{E}_1), \ldots, (E_n, \mathcal{E}_n)$, respectively.

Definition 2.5.3.2 (Independent random variables). We say that n random variables $X_1, \ldots, X_n$ are independent if the σ-algebras $\sigma(X_1), \ldots, \sigma(X_n)$ are independent.

Remark 2.5.3.3. This is equivalent to the fact that for all $F_1 \in \mathcal{E}_1, \ldots, F_n \in \mathcal{E}_n$ we have

$$\mathbb{P}(\{X_1 \in F_1\} \cap \cdots \cap \{X_n \in F_n\}) = \mathbb{P}(X_1 \in F_1) \cdots \mathbb{P}(X_n \in F_n)$$

which comes from the fact that for all $i \in \{1, \ldots, n\}$ we have that $\sigma(X_i) = \{X_i^{-1}(F) \mid F \in \mathcal{E}_i\}$.

Remark 2.5.3.4. If $\mathcal{B}_1, \ldots, \mathcal{B}_n$ are n independent sub σ-algebras and if $X_1, \ldots, X_n$ are independent random variables such that X_i is $\mathcal{B}_i$-measurable for all $i \in \{1, \ldots, n\}$, then $X_1, \ldots, X_n$ are independent random variables. This comes from the fact that for all $i \in \{1, \ldots, n\}$ we have that $\sigma(X_i) \subset \mathcal{B}_i$.

Remark 2.5.3.5. The n events $A_1, \ldots, A_n \in \mathcal{A}$ are independent if and only if $\sigma(A_1), \ldots, \sigma(A_n)$ are independent.

Theorem 2.5.3.1 (Independence of random variables). *Let $X_1, \ldots, X_n$ be n random variables. Then $X_1, \ldots, X_n$ are independent if and only if the law of the vector $(X_1, \ldots, X_n)$ is the product of the* *laws of* $X_1, \ldots, X_n$*, i.e.,*

$$\mathbb{P}_{(X_1,\ldots,X_n)} = \mathbb{P}_{X_1} \otimes \cdots \otimes \mathbb{P}_{X_n}.$$

Moreover, for every measurable map $f_i : (E_i, \mathcal{E}_i) \to \mathbb{R}_{\geq 0}$ *defined on a measurable space* $(E_i, \mathcal{E}_i)$ *for all* $i \in \{1, \ldots, n\}$*, we have*

$$\mathbb{E}\left[\prod_{i=1}^{n} f_i(X_i)\right] = \prod_{i=1}^{n} \mathbb{E}[f_i(X_i)].$$

Proof. Let $F_i \in \mathcal{E}_i$ for all $i \in \{1, \ldots, n\}$. Note that we have

$$\mathbb{P}_{(X_1,\ldots,X_n)}(F_1 \times \cdots \times F_n) = \mathbb{P}(\{X_1 \in F_1\} \cap \cdots \cap \{X_n \in F_n\})$$

and on the other hand

$$\begin{aligned}(\mathbb{P}_{X_1} \otimes \cdots \otimes \mathbb{P}_{X_n})(F_1 \times \cdots \times F_n) &= \mathbb{P}_{X_1}(F_1) \cdots \mathbb{P}_{X_n}(F_n) \\ &= \prod_{i=1}^{n} \mathbb{P}_{X_i}(F_i) = \prod_{i=1}^{n} \mathbb{P}(X_i \in F_i).\end{aligned}$$

If $X_1, \ldots, X_n$ are independent, then

$$\begin{aligned}\mathbb{P}_{(X_1,\ldots,X_n)}(F_1 \times \cdots \times F_n) = \prod_{i=1}^{n} \mathbb{P}[X_i \in F_i] = (\mathbb{P}_{X_1} \otimes \cdots \otimes \mathbb{P}_{X_n}) \\ \times (F_1 \times \cdots \times F_n),\end{aligned}$$

which implies that $\mathbb{P}_{(X_1,\ldots,X_n)}$ and $\mathbb{P}_{X_1} \otimes \cdots \otimes \mathbb{P}_{X_n}$ are equal on rectangles. Hence, the *monotone class theorem* (Theorem A.9.0.19) implies that

$$\mathbb{P}_{(X_1,\ldots,X_n)} = \mathbb{P}_{X_1} \otimes \cdots \otimes \mathbb{P}_{X_n}.$$

Conversely, if $\mathbb{P}_{(X_1,\ldots,X_n)} = \mathbb{P}_{X_1} \otimes \cdots \otimes \mathbb{P}_{X_n}$, then for all $F_i \in \mathcal{E}_i$, with $i \in \{1, \ldots, n\}$, we get that

$$\mathbb{P}_{(X_1,\ldots,X_n)}(F_1 \times \cdots \times F_n) = (\mathbb{P}_{X_1} \otimes \cdots \otimes \mathbb{P}_{X_n})(F_1 \times \cdots \times F_n)$$

and therefore

$$\mathbb{P}(\{X_1 \in F_1\} \cap \cdots \cap \{X_n \in F_n\}) = \mathbb{P}(X_1 \in F_1) \cdots \mathbb{P}(X_n \in F_n).$$

This implies that $X_1, \dots, X_n$ are independent. For the second assumption we get

$$\mathbb{E}\left[\prod_{i=1}^n f_i(X_i)\right] = \int_{E_1\times\cdots\times E_n} \prod_{i=1}^n f_i(X_i) \underbrace{P_{X_1}\mathrm{d}x_1 \cdots P_{X_n}\mathrm{d}x_n}_{\mathbb{P}_{X_1,\dots,X_n}(\mathrm{d}x_1\cdots\mathrm{d}x_n)}$$

$$= \prod_{i=1}^n \int_{E_i} f_i(x_i) P_{X_i}\mathrm{d}x_i = \prod_{i=1}^n \mathbb{E}[f_i(X_i)],$$

where we have used the first part of the theorem and Fubini's theorem. □

Remark 2.5.3.6. We can see from the proof of Theorem 2.5.3.1 that as soon as for all $i \in \{1, \dots, n\}$ we have $\mathbb{E}[|f_i(X_i)|] < \infty$, it follows that

$$\mathbb{E}\left[\prod_{i=1}^n f_i(X_i)\right] = \prod_{i=1}^n \mathbb{E}[f_i(X_i)].$$

Indeed, the previous result shows that

$$\mathbb{E}\left[\prod_{i=1}^n |f_i(X_i)|\right] = \prod_{i=1}^n \mathbb{E}[|f_i(X_i)|] < \infty$$

and thus we can apply Fubini's theorem. In particular if $X_1, \dots, X_n \in L^1(\Omega, \mathcal{A}, \mathbb{P})$ and independent, we get that

$$\mathbb{E}\left[\prod_{i=1}^n X_i\right] = \prod_{i=1}^n \mathbb{E}[X_i].$$

Corollary 2.5.3.2. *Let X_1 and X_2 be two independent random variables in $L^2(\Omega, \mathcal{A}, \mathbb{P})$. Then we get*

$$\mathbb{CV}(X_1, X_2) = 0.$$

Proof. Recall that if $X \in L^2(\Omega, \mathcal{A}, \mathbb{P})$, we also have that $X \in L^1(\Omega, \mathcal{A}, \mathbb{P})$. Thus

$$\mathbb{CV}(X_1, X_2) = \mathbb{E}[X_1X_2] - \mathbb{E}[X_1]\mathbb{E}[X_2] = \mathbb{E}[X_1]\mathbb{E}[X_2] - \mathbb{E}[X_1]\mathbb{E}[X_2] = 0.$$

□

Remark 2.5.3.7. Note that the converse is not true! Indeed, let $X_1 \sim \mathcal{N}(0,1)$ (in fact, we can also take any symmetric random variable in $L^2(\Omega, \mathcal{A}, \mathbb{P})$ with density $P(x)$, such that $P(-x) = P(x)$). Recall that being in $L^2(\Omega, \mathcal{A}, \mathbb{P})$ simply means

$$\mathbb{E}[X^2] = \int_{\mathbb{R}} x^2 P(x)\mathrm{d}x < \infty,$$

which implies that $P(x)=P(-x)$ and thus $\mathbb{E}[X^2]=\int_{\mathbb{R}} x^2P(x)\mathrm{d}x=0$. Now consider a random variable Y with values in $\{-1,+1\}$. Then we get $\mathbb{P}(Y = 1) = \mathbb{P}(Y = -1) = \frac{1}{2}$ and thus Y is independent of X_1. Define moreover $X_2 := YX_1$ and observe then

$$\mathbb{CV}(X_1, X_2) = \mathbb{E}[X_1X_2] - \mathbb{E}[X_1]\mathbb{E}[X_2] = \mathbb{E}[YX_1^2] - \mathbb{E}[YX_1]\mathbb{E}[X_1]$$

and hence

$$\underbrace{\mathbb{E}[Y]\mathbb{E}[X_1^2]}_{=0} - \underbrace{\mathbb{E}[Y]\mathbb{E}^2[X_1]}_{=0} = 0.$$

If X_1 and X_2 are independent, we note that $|X_1|$ and $|X_2|$ would also be independent. But $|X_2| = |Y||X_1| = |X_1|$. This would mean that $|X_1|$ is independent of itself. So it follows that $|X_1|$ is equal to a constant a.s. If $c = \mathbb{E}[|X_1|]$, and we want to look at $\mathbb{E}[(|X_1| - c)^2]$, we now know that $|X_1| - c$ is independent of itself. Therefore, we get

$$\mathbb{E}[(|X_1| - c)^2] = \mathbb{E}[|X_1| - c]\mathbb{E}[|X_1| - c] = 0 \Rightarrow |X_1| = c \quad \text{a.s.}$$

This is a contradiction since $|X_1|$ is the absolute value of a standard Gaussian distribution, which has a density given by

$$P(x) = \frac{1}{\sqrt{2\pi}}\mathrm{e}^{-\frac{x^2}{2}}.$$

Corollary 2.5.3.3. *Let $X_1, \ldots, X_n$ be n real random variables, i.e., with values in $\mathbb{R}$.*

(i) *Assume that for $i \in \{1, \ldots, n\}$, $\mathbb{P}_{X_i}$ has density P_i and that the random variables $X_1, \ldots, X_n$ are independent. Then the law of $(X_1, \ldots, X_n)$ also has density given by*

$$P(x_1, \ldots, x_n) = \prod_{i=1}^{n} P_i(x_i).$$

(ii) *Conversely, assume that the law of* $(X_1, \ldots, X_n)$ *has density* $P(x_1, \ldots, x_n) = \prod_{i=1}^n Q_i(x_i)$, *where* Q_i *is Borel measurable and positive. Then the random variables* $X_1, \ldots, X_n$ *are independent and the law of* X_i *has density* $P_i = c_i Q_i$, *with* $c_i > 0$ *for* $i \in \{1, \ldots, n\}$.

Proof. We only need to show (ii). From Fubini's theorem we get

$$\int_{\mathbb{R}} \prod_{i=1}^{n} Q_i(x_i) \mathrm{d}x_i = \prod_{i=1}^{n} \int_{\mathbb{R}} Q_i(x_i) \mathrm{d}x_i = \int_{\mathbb{R}^n} P(x_1, \ldots, x_n) \mathrm{d}x_1 \cdots \mathrm{d}x_n = 1,$$

which implies that $K_i := \int_{\mathbb{R}} Q_i(x_i) \mathrm{d}x_i \in (0, \infty)$, for all $i \in \{1, \ldots, n\}$. Now we know that the law of X_i has density P_i given by

$$\begin{aligned} P_i(x_i) &= \int_{\mathbb{R}^{n-1}} P(x_1, \ldots, x_{i-1}, x_i, x_{i+1}, \ldots, x_n) \\ &\qquad \mathrm{d}x_1 \cdots \mathrm{d}x_{i-1} \mathrm{d}x_{i+1} \cdots \mathrm{d}x_n \\ &= \left(\prod_{j \neq i} K_j \right) Q_i(x_i) = \frac{1}{K_i} Q_i(x_i). \end{aligned} \tag{2.5.3.1}$$

We can rewrite

$$P(x_1, \ldots, x_n) = \prod_{i=1}^{n} Q_i(x_i) = \prod_{i=1}^{n} P_i(x_i).$$

Hence, we get $P(x_1, \ldots, x_n) = \mathbb{P}_{X_1} \otimes \cdots \otimes \mathbb{P}_{X_n}$ and therefore X_1, $\ldots, X_n$ are independent. □

Example 2.5.3.8. Let U be a random variable with exponential distribution. Let V be a uniform random variable on $[0, 1]$. We assume that U and V are independent. Define the random variables $X := \sqrt{U} \cos(2\pi V)$ and $Y := \sqrt{U} \sin(2\pi V)$. We can show that X and Y are independent. Indeed, for a measurable function $\varphi : \mathbb{R}^2 \to \mathbb{R}_{\geq 0}$ we get

$$\mathbb{E}[\varphi(X,Y)] = \int_0^\infty \int_0^1 \varphi(\sqrt{u}\cos(2\pi v), \sqrt{u}\sin(2\pi v))\mathrm{e}^u \mathrm{d}u\mathrm{d}v$$
$$= \frac{1}{\sqrt{\pi}} \int_0^\infty \int_0^{2\pi} \varphi(r\cos(\theta), r\sin(\theta)) r\mathrm{e}^{-r^2} \mathrm{d}r\mathrm{d}\theta, \tag{2.5.3.2}$$

which implies that (X,Y) has density $\frac{\mathrm{e}^{-x^2}\mathrm{e}^{-y^2}}{\pi}$ on $\mathbb{R}\times\mathbb{R}$. Together with Corollary 2.5.3.3 we get that X and Y are independent and X and Y have the same density $P(x) = \frac{1}{\sqrt{\pi}}\mathrm{e}^{-x^2}$. This means that X and Y are independent.

Remark 2.5.3.9. We write $X \overset{\text{law}}{=} Y$ to say that $\mathbb{P}_X = \mathbb{P}_Y$. Thus in the example above we would have

$$X \overset{\text{law}}{=} Y \sim \mathcal{N}(0, 1/2).$$

Proposition 2.5.3.4. *Let $X_1, \ldots, X_n$ be n real-valued random variables. Then the following conditions are equivalent:*

(i) *$X_1, \ldots, X_n$ are independent.*
(ii) *For $X = (X_1, \ldots, X_n) \in \mathbb{R}^n$, we have*

$$\Phi_X(\xi_1, \ldots, \xi_n) = \prod_{i=1}^n \Phi_{X_i}(\xi_i).$$

(iii) *For all $a_1, \ldots, a_n \in \mathbb{R}$, we have*

$$\mathbb{P}(X_1 \leq a_1, \ldots, X_n \leq a_n) = \prod_{i=1}^n \mathbb{P}(X_i \leq a_i).$$

(iv) *If $f_1, \ldots, f_n : \mathbb{R} \to \mathbb{R}_{\geq 0}$ are continuous, measurable maps with compact support, then*

$$\mathbb{E}\left[\prod_{i=1}^n f_i(X_i)\right] = \prod_{i=1}^n \mathbb{E}[f_i(X_i)].$$

Proof. First we show (i) $\Rightarrow$ (ii). By definition of the characteristic function and the i.i.d. property, we get

$$\Phi_X(\xi_1, \dots, \xi_n) = \mathbb{E}\left[\mathrm{e}^{i(\xi_1 X_1 + \dots + \xi_n X_n)}\right] = \mathbb{E}\left[\mathrm{e}^{i\xi_1 X_1} \dots \mathrm{e}^{i\xi_n X_n}\right]$$
$$= \prod_{i=1}^{n} \mathbb{E}[\mathrm{e}^{i\xi X_1}] = \prod_{i=1}^{n} \Phi_{X_i}(\xi_i),$$

where the map $t \mapsto \mathrm{e}^{it}$ is measurable and bounded. Next, we show (ii) $\Rightarrow$ (i). Note that by Theorem 2.5.3.1 we have $\mathbb{P}_X = \mathbb{P}_Y$ if

$$\Phi_X(\xi_1, \dots, \xi_n) = \Phi_Y(\xi_1, \dots, \xi_n).$$

Now if $\Phi_X(\xi_1, \dots, \xi_n) = \prod_{i=1}^{n} \Phi_{X_i}(\xi_i)$, we note that $\prod_{i=1}^{n} \Phi_{X_i}(\xi_i)$ is the characteristic function of the probability distribution if the probability distribution is $\mathbb{P}_{X_1} \otimes \dots \otimes \mathbb{P}_{X_n}$. Now from injectivity it follows that $\mathbb{P}_{(X_1, \dots, X_n)} = \mathbb{P}_{X_1} \otimes \dots \otimes \mathbb{P}_{X_n}$, which implies that $X_1, \dots, X_n$ are independent. □

Proposition 2.5.3.5. *Let $\mathcal{B}_1, \dots, \mathcal{B}_n \subset \mathcal{A}$ be sub-σ-algebras of $\mathcal{A}$. For every $i \in \{1, \dots, n\}$, let $\mathcal{C}_i \subset \mathcal{B}_i$ be a family of subsets of Ω such that $\mathcal{C}_i$ is stable under finite intersection and $\sigma(\mathcal{C}_i) = \mathcal{B}_i$. Assume that for all $C_i \in \mathcal{C}_i$ with $i \in \{1, \dots, n\}$ we have*

$$\mathbb{P}\left(\prod_{i=1}^{n} C_i\right) = \prod_{i=1}^{n} \mathbb{P}(C_i).$$

Then $\mathcal{B}_1, \dots, \mathcal{B}_n$ are independent σ-algebras.

Proof. Let us fix $C_2 \in \mathcal{C}_2, \dots, C_n \in \mathcal{C}_n$ and define

$$M_1 := \{B_1 \in \mathcal{B}_1 \mid \mathbb{P}(B_1 \cap C_2 \cap \dots \cap C_n) = \mathbb{P}(B_1)\mathbb{P}(C_2) \cdots \mathbb{P}(C_n)\}.$$

Now since $\mathcal{C}_1 \subset M_1$ and M_1 is a monotone class, we get $\sigma(\mathcal{C}_1) = \mathcal{B}_1 \subset M_1$ and thus $\mathcal{B}_1 = M_1$. Let now $B_1 \in \mathcal{B}_1$, $C_3 \in \mathcal{C}_3, \dots, C_n \in \mathcal{C}_n$ and define

$$M_2 := \{B_2 \in \mathcal{B}_2 \mid \mathbb{P}(B_2 \cap B_1 \cap C_3 \cap \dots \cap C_n) = \mathbb{P}(B_2)\mathbb{P}(B_1)\mathbb{P}(C_3) \cdots \mathbb{P}(C_n)\}.$$

Again, since $\mathcal{C}_2 \subset M_2$, we get $\sigma(\mathcal{C}_2) = \mathcal{B}_2 \subsetneq M_2$ and thus $B_2 = M_2$. By induction we complete the proof. □

Remark 2.5.3.10. A consequence of Proposition 2.5.3.5 is given as follows. Let $\mathcal{B}_1, \ldots, \mathcal{B}_n$ be n independent σ-algebras and let $0 = n_0 < n_1 < \cdots < n_{p-1} < n_p = n$. Then the σ-algebras

$$\mathcal{D}_1 = \mathcal{B}_1 \vee \cdots \vee \mathcal{B}_n = \sigma(\mathcal{B}_1, \ldots, \mathcal{B}_n) = \sigma\left(\bigcup_{k=1}^{n} \mathcal{B}_k\right),$$

$$\mathcal{D}_2 = \mathcal{B}_{n_1+1} \vee \cdots \vee \mathcal{B}_{n_2},$$

$$\vdots$$

$$\mathcal{D}_p = \mathcal{B}_{n_{p-1}+1} \vee \cdots \vee \mathcal{B}_{n_p},$$

are also independent. Indeed, we can apply Proposition 2.5.3.5 to the class of sets

$$C_j = \{B_{n_{j-1}+1} \cap \cdots \cap B_{n_j} \mid B_i \in \mathcal{C}_i, i \in \{n_{j-1}+1, \ldots, n_j\}\}.$$

In particular, if $X_1, \ldots, X_n$ are independent random variables, then

$$Y_1 = (X_1, \ldots, X_n), \ldots, \quad Y_p = (X_{n_1}, \ldots, X_{n_p}),$$

are also independent.

Example 2.5.3.11. Let X_1, X_2, X_3, X_4 be real-valued independent random variables. Then $Z_1 := X_1 X_3$ and $Z_2 := X_2^3 + X_4$ are independent and $Z_3 := \sigma(X_1, X_3)$ and $Z_4 := \sigma(X_2, X_4)$ are measurable. From before we can observe that $\sigma(X_1, X_3)$ and $\sigma(X_2, X_4)$ are independent if for $X : \Omega \to \mathbb{R}$ we have that Y is $\sigma(X)$-measurable if and only if $Y = f(X)$ with f being a measurable map, i.e., if Y is $\sigma(X_1, \ldots, X_n)$-measurable, then $Y = f(X_1, \ldots, X_n)$.

Proposition 2.5.3.6 (Independence for an infinite family). *Let $(\mathcal{B}_i)_{i \in I}$ be an infinite family of sub- σ-algebras of $\mathcal{A}$. We say that the family $(\mathcal{B}_i)_{i \in I}$ is independent if for all $\{i_1, \ldots, i_p\} \in I$, $\mathcal{B}_{i_1}, \ldots, \mathcal{B}_{i_p}$ are independent. If $(X_i)_{i \in I}$ is a family of random variables we say that they are independent if $(\sigma(X_i))_{i \in I}$ is independent.*

Proposition 2.5.3.7. *Let $(X_n)_{n \geq 1}$ be a sequence of independent random variables. Then for all $p \in \mathbb{N}$ we get that $\sigma(X_1, \ldots, X_p)$ and $\sigma(X_{p+1}, \ldots, X_n)$ are independent.*

Proof. Apply Proposition 2.5.3.5 to the σ-algebras $\mathcal{C}_1 = \sigma(X_1, \ldots, X_p) \in \mathcal{B}_1$ and $\mathcal{C}_2 = \bigcup_{k=p+1}^{\infty} \sigma(X_{p+1}, \ldots, X_n) \in \mathcal{B}_2$. □

2.5.4. The Borel–Cantelli lemma

Let $(\Omega, \mathcal{A}, \mathbb{P})$ be a probability space. Let $(A_n)_{n\in\mathbb{N}}$ be a sequence of events in $\mathcal{A}$. Recall that we can write

$$\limsup_{n\to\infty} A_n = \bigcap_{n=0}^{\infty}\left(\bigcup_{k=n}^{\infty} A_k\right) \quad \text{and} \quad \liminf_{n\to\infty} A_n = \bigcup_{n=0}^{\infty}\left(\bigcap_{k=n}^{\infty} A_k\right).$$

Moreover, both are again measurable sets. For $\omega \in \limsup_n A_n$ we get that $\omega \in \bigcup_{k=n}^{\infty} A_k$, for all $n \geq 0$. Moreover, for all $n \geq 0$, there exists a $k \geq n$ such that, $\omega \in A_n$ and ω is in infinitely many A_k's. For $\omega \in \liminf_n A_n$, we get that for all $n \geq 0$ such that $\omega \in \bigcap_{k=n}^{\infty} A_k$, there exists $n \geq 0$, such that for all $k \geq n$ we have $\omega \in A_k$, which shows that $\liminf_n A_n \subset \limsup_n A_n$.

Lemma 2.5.4.1 (Borel–Cantelli). Let $(A_n)_{n\in\mathbb{N}} \in \mathcal{A}$ be a family of measurable sets.

(i) If $\sum_{n\geq 1} \mathbb{P}(A_n) < \infty$, then

$$\mathbb{P}\left(\limsup_{n\to\infty} A_n\right) = 0,$$

which means that the set $\{n \in \mathbb{N} \mid \omega \in A_n\}$ is a.s. finite.

(ii) If $\sum_{n\geq 1} \mathbb{P}(A_n) = \infty$, and if the events $(A_n)_{n\in\mathbb{N}}$ are independent, then

$$\mathbb{P}\left(\limsup_{n\to\infty} A_n\right) = 1,$$

which means that the set $\{n \in \mathbb{N} \mid \omega \in A_n\}$ is a.s. finite.

Proof. (i) If $\sum_{n\geq 1} \mathbb{P}(A_n) < \infty$, then, by Fubini's theorem, we get

$$\mathbb{E}\left[\sum_{n\geq 1} \mathbb{1}_{A_n}\right] = \sum_{n\geq 1} \mathbb{P}(A_n),$$

which implies that $\sum_{n\geq 1} \mathbb{1}_{A_n} < \infty$ and $\mathbb{1}_{A_n} \neq 0$ a.s. for finite numbers of n.

(ii) Fix $n_0 \in \mathbb{N}$ and note that for all $n \geq n_0$ we have

$$\mathbb{P}\left(\bigcap_{k=n_0}^{n} A_k^{\complement}\right) = \prod_{k=n_0}^{n} \mathbb{P}(A_k^{\complement}) = \prod_{k=n_0}^{n} \mathbb{P}(1 - A_n).$$

Now we see that

$$\sum_{n\geq 1} \mathbb{P}(A_n) = \infty$$

and thus

$$\mathbb{P}\left(\bigcap_{k=n_0}^{n} A_k^{\complement}\right) = 0.$$

Since this is true for every n_0 we have that

$$\mathbb{P}\left(\bigcup_{n=0}^{\infty}\bigcap_{k=n_0}^{\infty} A_k^{\complement}\right) \leq \sum_{n\geq 1} \mathbb{P}(A_k^{\complement}) = 0.$$

Hence we get

$$\mathbb{P}\left(\bigcup_{n=0}^{\infty}\bigcap_{k=n_0}^{\infty} A_k^{\complement}\right) = \mathbb{P}\left(\bigcap_{n=0}^{\infty}\bigcup_{k=n}^{\infty} A_k\right) = \mathbb{P}\left(\limsup_{n\to\infty} A_n\right) = 1.$$

□

We can use Lemma 2.5.4.1 in order to prove different mathematical statements.

Example 2.5.4.1 (Nonexistence of probability measures). We can show that there does not exist a probability measure on $\mathbb{N}$ such that the probability of the set of multiples of an integer n is $\frac{1}{n}$ for $n \geq 1$. Let us assume that such a probability measure exists. Let $\mathcal{P}$ denote the set of prime numbers. For $p \in \mathcal{P}$, we write $A_p := p\mathbb{N}$ for the set of all multiples of p. We first show that the sets $(A_p)_{p\in\mathcal{P}}$ are independent. Indeed, let $p_1, \ldots, p_n \in \mathcal{P}$ be distinct. Then we have

$$\mathbb{P}(p_1\mathbb{N} \cap \cdots \cap p_n\mathbb{N}) = \mathbb{P}(p_1 \cdots p_n\mathbb{N}) = \frac{1}{p_1 \cdots p_n} = \mathbb{P}(p_1\mathbb{N}) \cdots \mathbb{P}(p_n\mathbb{N}).$$

Moreover, it is known that

$$\sum_{p\in\mathcal{P}} \mathbb{P}(p\mathbb{N}) = \sum_{p\in\mathcal{P}} \frac{1}{p} = \infty.$$

The second part of Lemma 2.5.4.1 implies that all integers n belong to infinitely many A_p's. So it follows that n is divisible by infinitely many distinct prime numbers which is clearly impossible.

Example 2.5.4.2 (Computing limits). Let X be an exponentially distributed random variable with parameter $\lambda = 1$. Thus, we know that X has density $\mathrm{e}^{-x}\mathbb{1}_{\mathbb{R}_{\geq 0}}(x)$. Now consider a sequence $(X_n)_{n\geq 1}$ of independent random variables with the same distribution as X, i.e., for all $n \geq 1$, we have $X_n \sim X$. Then $\limsup_n \frac{X_n}{\log(n)} = 1$ a.s., i.e., there exists an $N \in \mathcal{A}$ such that $\mathbb{P}(N) = 0$ and we get

$$\limsup_{n\to\infty} \frac{X_n(\omega)}{\log(n)} = 1, \qquad \forall \omega \notin N.$$

Therefore, we can compute the probability

$$\mathbb{P}(X > t) = \int_t^\infty \mathrm{e}^{-x}\mathrm{d}x = \mathrm{e}^{-t}.$$

Now let $\varepsilon > 0$ and consider the sets $A_n := \{X_n > (1+\varepsilon)\log(n)\}$ and $B_n := \{X_n > \log(n)\}$. Then

$$\begin{aligned}\mathbb{P}(A_n) = \mathbb{P}(X_n > (1+\varepsilon)\log(n)) &= \mathbb{P}(X > (1+\varepsilon)\log(n)) \\ &= \mathrm{e}^{-(1+\varepsilon)\log(n)} = \frac{1}{n^{1+\varepsilon}}.\end{aligned}$$

This implies that

$$\sum_{n\geq 1} \mathbb{P}(A_n) < \infty.$$

Using Lemma 2.5.4.1, we get that $\mathbb{P}(\limsup_{n\to\infty} A_n) = 0$. Let us define

$$N_\varepsilon := \limsup_{n\to\infty} A_n.$$

Then we have $\mathbb{P}(N_\varepsilon) = 0$ for all $\omega \notin N_\varepsilon$, which implies that there exists an $n_0(\omega)$ such that for all $n \geq n_0(\omega)$ we have

$$X_n(\omega) \leq (1+\varepsilon)\log(n)$$

and thus for all $\omega \notin N_\varepsilon$, we get $\limsup_{n\to\infty} \frac{X_n(\omega)}{\log(n)} \leq 1+\varepsilon$. Moreover, let

$$N' := \bigcup_{\varepsilon\in\mathbb{Q}_{\geq 0}} N_\varepsilon.$$

Then we get $\mathbb{P}(N') \leq \sum_{\varepsilon\in\mathbb{Q}_{\geq 0}} \mathbb{P}(N_\varepsilon) = 0$ for all $\omega \notin N'$. Hence, we get

$$\limsup_{n\to\infty} \frac{X_n(\omega)}{\log(n)} \leq 1.$$

Now we note that the B_n's are independent by the fact that $B_n \in \sigma(X_n)$ since the X_n's are independent. Moreover, we have

$$\mathbb{P}(B_n) = \mathbb{P}(X_n > \log(n)) = \mathbb{P}(X > \log(n)) = \frac{1}{n},$$

which gives that

$$\sum_{n\geq 1} \mathbb{P}(B_n) = \infty.$$

Now we can use Lemma 2.5.4.1 again to obtain

$$\mathbb{P}\left(\limsup_{n\to\infty} B_n\right) = 1.$$

If we denote $N'' := (\limsup_{n\to\infty} B_n)^{\mathsf{C}}$, then for all $\omega \notin N''$ we get that $X_n(\omega) > \log(n)$ for infinitely many n. So it follows that for all $\omega \notin N''$ we have

$$\limsup_{n\to\infty} \frac{X_n(\omega)}{\log(n)} \geq 1.$$

Finally, take $N := N' \cup N''$ to obtain $\mathbb{P}(N) = 0$. Thus, for all $\omega \notin N$ we get

$$\limsup_{n\to\infty} \frac{X_n(\omega)}{\log(n)} = 1.$$

2.5.5. Sums of independent random variables

Let us first define the convolution of two probability measures. If μ and ν are two probability measures on $\mathbb{R}^d$, we denote by $\mu * \nu$ the image of the measure $\mu \otimes \nu$ by the map

$$\mathbb{R}^d \times \mathbb{R}^d \to \mathbb{R}^d,$$
$$(x, y) \mapsto x + y.$$

Moreover, for all measurable maps $\varphi : \mathbb{R}^d \to \mathbb{R}_{\geq 0}$, we have

$$\int_{\mathbb{R}^d} \varphi(z)(\mu * \nu)(\mathrm{d}z) = \iint_{\mathbb{R}^d \times \mathbb{R}^d} \varphi(x + y)\mu(\mathrm{d}x)\nu(\mathrm{d}y).$$

Proposition 2.5.5.1. *Let $(\Omega, \mathcal{A}, \mathbb{P})$ be a probability space. Let X and Y be two independent random variables with values in $\mathbb{R}^d$. Then the following conditions hold:*

(i) *The law of $X + Y$ is given by $\mathbb{P}_X * \mathbb{P}_Y$. In particular, if X has density f and Y has density g, then $X + Y$ has density $f * g$, where $*$ now denotes the* convolution product, *which is given by*

$$f * g(\xi) := \int_{\mathbb{R}^d} f(x)g(\xi - x)\mathrm{d}x.$$

(ii) $\Phi_{X+Y}(\xi) = \Phi_X(\xi)\Phi_Y(\xi)$.

(iii) *If X and Y are in $L^2(\Omega, \mathcal{A}, \mathbb{P})$, we get*

$$K_{X+Y} = \mathbf{\Sigma}_X + \mathbf{\Sigma}_Y.$$

In particular, when $d = 1$, we obtain

$$\mathbb{V}(X + Y) = \mathbb{V}(X) + \mathbb{V}(Y).$$

Proof. (i) If X and Y are independent random variables, then $\mathbb{P}_{(X,Y)} = \mathbb{P}_X \otimes \mathbb{P}_Y$. Consequently, for all measurable maps $\varphi : \mathbb{R}^d \to$

$\mathbb{R}_{\geq 0}$, we have

$$\begin{aligned}\mathbb{E}[\varphi(X+Y)] &= \iint_{\mathbb{R}^d\times\mathbb{R}^d} \varphi(X+Y)\mathbb{P}_{(X,Y)}(\mathrm{d}x\mathrm{d}y)\\ &= \iint_{\mathbb{R}^d\times\mathbb{R}^d} \varphi(X+Y)\mathbb{P}_X(\mathrm{d}x)\mathbb{P}_Y(\mathrm{d}y)\\ &= \int_{\mathbb{R}^d} \varphi(\xi)(\mathbb{P}_X * \mathbb{P}_Y)(\mathrm{d}\xi).\end{aligned}$$

Now since X and Y have densities f and g, respectively, we get

$$\begin{aligned}\mathbb{E}[\varphi(Z=X+Y)] &= \iint_{\mathbb{R}^d\times\mathbb{R}^d} \varphi(X+Y)f(x) * g(y)\mathrm{d}x\mathrm{d}y\\ &= \int_{\mathbb{R}^d} \varphi(\xi)\left(\int_{\mathbb{R}^d} f(x)g(\xi-x)\mathrm{d}x\right)\mathrm{d}\xi. \qquad (2.5.5.1)\end{aligned}$$

Since identity (2.5.5.1) is true for all measurable maps $\varphi : \mathbb{R}^d \to \mathbb{R}_{\geq 0}$, the random variable $Z := X + Y$ has density

$$h(\xi) = (f * g)(\xi) = \int_{\mathbb{R}^d} f(x)g(\xi - x)\mathrm{d}x.$$

(ii) By definition of the characteristic function and the independence property, we get

$$\begin{aligned}\Phi_{X+Y}(\xi) &= \mathbb{E}\left[\mathrm{e}^{i\xi(X+Y)}\right] = \mathbb{E}\left[\mathrm{e}^{i\xi X}\mathrm{e}^{i\xi Y}\right] = \mathbb{E}\left[\mathrm{e}^{i\xi X}\right]\mathbb{E}\left[\mathrm{e}^{i\xi Y}\right]\\ &= \Phi_X(\xi)\Phi_Y(\xi).\end{aligned}$$

(iii) If $X = (X_1, \ldots, X_d)$ and $Y = (Y_1, \ldots, Y_d)$ are independent random variables on $\mathbb{R}^d$, we get that $\mathbb{CV}(X_i, Y_j) = 0$, for all $0 \leq i, j \leq d$. By using the multi-linearity of the covariance, we get that

$$\mathbb{CV}(X_i + Y_i, X_j + Y_j) = \mathbb{CV}(X_i, X_j) + \mathbb{CV}(Y_j + Y_j),$$

and hence $K_{X+Y} = \mathbf{\Sigma}_X + \mathbf{\Sigma}_Y$. For $d = 1$ we get

$$\begin{aligned}\mathbb{V}(X+Y) &= \mathbb{E}[((X+Y)-\mathbb{E}[X+Y])^2] = \mathbb{E}[((X-\mathbb{E}[X])+(Y-\mathbb{E}[Y]))^2]\\ &= \underbrace{\mathbb{E}[(X-\mathbb{E}[X])^2]}_{\mathbb{V}(X)} + \underbrace{\mathbb{E}[(Y-\mathbb{E}[Y])^2]}_{\mathbb{V}(Y)} +\\ &\quad \underbrace{2\mathbb{E}[(X-\mathbb{E}[X])(Y-\mathbb{E}[Y])]}_{2\mathbb{CV}(X,Y)}.\end{aligned}$$

Now since $\mathbb{CV}(X, Y) = 0$, we get the result. □

Theorem 2.5.5.2 (Weak law of large numbers). *Let $(\Omega, \mathcal{A}, \mathbb{P})$ be a probability space. Let $(X_n)_{n\geq 1}$ be a sequence of independent random variables. Moreover, let $\mu := \mathbb{E}[X_n]$ for all $n \geq 1$ and assume $\mathbb{E}[(X_n - \mu)^2] \leq C$ for all $n \geq 1$ and for some constant $C < \infty$. We also denote $S_n := \sum_{j=1}^n X_j$ and $\tilde{X}_n := \frac{S_n}{n}$ for all $n \geq 1$. Then for all $\varepsilon > 0$ we get*

$$\mathbb{P}(|\tilde{X}_n - \mu| > \varepsilon) \xrightarrow{n\to\infty} 0.$$

Thus, we also have

$$\mathbb{E}[S_n] = \frac{1}{n}\mathbb{E}\left[\sum_{j=1}^n X_j\right] = \frac{1}{n} n\mathbb{E}[X_j] = \mathbb{E}[X_j].$$

Proof. We note that

$$\mathbb{E}[(S_n - n\mu)^2] = \sum_{j=1}^n \mathbb{E}[(X_j - \mu)^2] \leq nC.$$

Hence, for $\varepsilon > 0$ we get by Markov's inequality (Proposition 2.3.1.2)

$$\mathbb{P}(|\tilde{X} - \mu| > \varepsilon) = \mathbb{P}((S_n - n\mu)^2 > (n\varepsilon)^2) \leq \frac{\mathbb{E}[(S_n - n\mu)^2]}{n^2\varepsilon^2} \leq \frac{C}{n\varepsilon^2} \xrightarrow{n\to\infty} 0.$$

□

Corollary 2.5.5.3. *Let $(\Omega, \mathcal{A}, \mathbb{P})$ be a probability space. Let $(A_n)_{n\geq 1} \in \mathcal{A}$ be a sequence of independent events with the same probabilities, i.e., $\mathbb{P}(A_i) = \mathbb{P}(A_j)$, for all $i, j \geq 1$. Then*

$$\lim_{n\to\infty} \frac{1}{n}\sum_{i=1}^n \mathbb{1}_{A_i} = \mathbb{P}(A_1) \quad a.s.$$

Proof. Note that by Theorem 2.5.5.2, we get for a sequence of independent random variables $(X_n)_{n\geq 1}$ with the same expectation for all $n \geq 1$

$$\lim_{n\to\infty} \mathbb{E}\left[\frac{1}{n}\sum_{j=1}^n X_j\right] = \mathbb{E}[X_1],$$

and thus we can take $X_j = \mathbb{1}_{A_j}$, since we know that $\mathbb{E}[\mathbb{1}_A] = \mathbb{P}(A)$.

□

2.6. Finding the Distribution of Some Random Variables

2.6.1. The case of sums of independent random variables

Let X be a Poisson distributed random variable with parameter $\lambda > 0$ (see Example 1.2.2.1). We already know that for all $\xi \in \mathbb{R}$, we have $\mathbb{E}\left[e^{i\xi X}\right] = \exp(-\lambda(1 - e^{i\xi}))$. Let $X_1, \ldots, X_n$ be n independent random variables for $n \in \mathbb{N}$ with X_j being a Poisson distributed random variable with parameter $\lambda_j > 0$ for all $1 \leq j \leq n$. Define further the sum $S_n := \sum_{j=1}^{n} X_j$. We want to figure out what the law of S_n is. Note that we have

$$\begin{aligned}
\mathbb{E}\left[e^{i\xi S_n}\right] &= \mathbb{E}\left[e^{i\xi(X_1+\cdots+X_n)}\right] = \mathbb{E}\left[e^{i\xi X_1} \cdots e^{i\xi X_n}\right] \\
&= \mathbb{E}\left[e^{i\xi X_1}\right] \cdots \mathbb{E}\left[e^{i\xi X_n}\right] \\
&= \exp(-\lambda_1(1 - e^{i\xi})) \cdots \exp(-\lambda_n(1 - e^{i\xi})) \\
&= \exp(-(\lambda_1 + \cdots + \lambda_n)(1 - e^{i\xi})) \\
&= \exp(-\Lambda(1 - e^{i\xi})),
\end{aligned}$$

with $\Lambda := \lambda_1 + \cdots + \lambda_n$. Since the characteristic function uniquely characterizes the probability distributions, we can conclude that

$$S_n \sim \Pi(\Lambda).$$

Let now X be a random variable with $X \sim \mathcal{N}(m, \sigma^2)$. Then we have seen in Section 2.4 that

$$\mathbb{E}\left[e^{i\xi X}\right] = e^{i\xi m} e^{-i\sigma^2 \frac{\xi^2}{2}}.$$

Now let $X_1, \ldots, X_n$ be n independent random variables for $n \in \mathbb{N}$, such that $X_j \sim \mathcal{N}(m_j, \sigma_j^2)$ for all $0 \leq j \leq n$. Set again $S_n = \sum_{j=1}^{n} X_j$. Then, we get

$$\mathbb{E}\left[e^{i\xi S_n}\right] = e^{im_1\xi} e^{-\sigma_1^2 \frac{\xi^2}{2}} \cdots e^{im_n\xi} e^{-\sigma_n^2 \frac{\xi^2}{2}} = e^{i(m_1+\cdots+m_n)\xi} e^{-(\sigma_1^2+\cdots+\sigma_n^2)\frac{\xi^2}{2}},$$

which implies, because of the same argument as above, that

$$S_n \sim \mathcal{N}(m_1 + \cdots + m_n, \sigma_1^2 + \cdots + \sigma_n^2).$$

2.6.2. Using change of variables

Let $g : \mathbb{R}^n \to \mathbb{R}^n$ be a measurable function given as $g(x) = (g_1(x), \dots, g_n(x))$ for $x \in \mathbb{R}^n$. Then the Jacobian of g is given by

$$J_g(x) = \left(\frac{\partial g_i(x)}{\partial x_j}\right)_{1 \leq i,j \leq n}.$$

Recall that for $g : G \subset \mathbb{R}^n \to \mathbb{R}^n$, where G is a open subset of $\mathbb{R}^n$, with J_g injective such that $\det(J_g(x)) \neq 0$ for all $x \in G$, we have for every measurable and positive map $f : \mathbb{R}^n \to \mathbb{R}_{\geq 0}$, or for every integrable $f = \mathbb{1}_G$, that

$$\int_{g(G)} f(y)\mathrm{d}y = \int_G f(g(x))|\det(J_g(x))|\mathrm{d}x,$$

where $g(G) = \{y \in \mathbb{R}^n \mid \exists x \in G : g(x) = y\}$.

Theorem 2.6.2.1. *Let $(\Omega, \mathcal{A}, \mathbb{P})$ be a probability space. Let $X = (X_1, \dots, X_n)$ be a random variable on $\mathbb{R}^n$ for $n \in \mathbb{N}$, having a joint density f. Let $g : \mathbb{R}^n \to \mathbb{R}^n \in C^1(\mathbb{R}^n)$ be an injective measurable map such that* $\det(J_g(x)) \neq 0$ *for all $x \in \mathbb{R}^n$. Then $Y := g(X)$ has the density*

$$f_Y(y) = \begin{cases} f_X(g^{-1}(y))|\det(J_{g^{-1}}(y))|, & y \in g(x) \\ 0, & \text{otherwise} \end{cases}$$

Proof. Let $B \in \mathcal{B}(\mathbb{R}^n)$ be a Borel set and $A := g^{-1}(B)$. Then we have

$$\begin{aligned}\mathbb{P}(X \in A) = \int_A f_X(x)\mathrm{d}x &= \int_{g^{-1}(B)} f_X(x)\mathrm{d}x \\ &= \int_B f_X(g^{-1}(x))|\det(J_{g^{-1}}(x))|\mathrm{d}x.\end{aligned}$$

But we know that $\mathbb{P}(Y \in B) = \mathbb{P}(X \in A)$, for all $B \in \mathcal{B}(\mathbb{R}^n)$ with

$$\mathbb{P}(Y \in B) = \int_B f_X(g^{-1}(x))|\det(J_{g^{-1}}(x))|\mathrm{d}x.$$

It follows that Y has density given by

$$f_Y(x) = f_X(g^{-1}(x))|\det(J_{g^{-1}}(x))|.$$

□

Example 2.6.2.1. Let X and Y be two independent $\mathcal{N}(0,1)$ distributed random variables. We want to know what is the joint union distribution of the random variable $(U,V)=(X+Y,X-Y)$. Therefore, let $g:\mathbb{R}^2\to\mathbb{R}^2$ be given by $(x,y)\mapsto(x+y,x-y)$. The inverse $g^{-1}:\mathbb{R}^2\to\mathbb{R}^2$ is then given by $(u,v)\mapsto\left(\frac{u+v}{2},\frac{u-v}{2}\right)$. We have the following Jacobian equations:

$$J_{g^{-1}}=\begin{pmatrix}\frac{1}{2} & \frac{1}{2}\\ \frac{1}{2} & -\frac{1}{2}\end{pmatrix},\quad \det(J_{g^{-1}})=-\frac{1}{2}.$$

Moreover, we get

$$\begin{aligned}f_{(U,V)}(u,v)&=f_{(X,Y)}\left(\frac{u+v}{2},\frac{u-v}{2}\right)\det(J_{g^{-1}})\\&=\left(\frac{1}{\sqrt{2\pi}}\mathrm{e}^{-\frac{1}{2}\left(\frac{u+v}{2}\right)^2}\frac{1}{\sqrt{2\pi}}e^{-\frac{1}{2}\left(\frac{u-v}{2}\right)^2}\right)\left(\frac{1}{2}\right)\\&=\frac{1}{\sqrt{4\pi}}\mathrm{e}^{-\frac{u^2}{4}}\frac{1}{\sqrt{4\pi}}\mathrm{e}^{-\frac{v^2}{4}}.\end{aligned}$$

Thus U and V are independent and $U\overset{\text{law}}{=}V\sim\mathcal{N}(0,2)$.

Example 2.6.2.2. Let (X,Y) be a random variable on $\mathbb{R}^2$ with joint density f. We want to find the density of the product $Z:=XY$. In this case, consider $h:\mathbb{R}^2\to\mathbb{R}^2$, given by $(x,y)\mapsto xy$. We then define $g:\mathbb{R}^2\to\mathbb{R}^2$, given by $(x,y)\mapsto(xy,x)$. We write $S_0=\{(x,y)\mid x=0,y\in\mathbb{R}\}$ and $S_1=\mathbb{R}^2\setminus S_0$. Now g is injective from S_1 to $\mathbb{R}\setminus\{0\}$ and $g^{-1}(u,v)=\left(v,\frac{u}{v}\right)$. The Jacobian is thus given by

$$J_{g^{-1}}(u,v)=\begin{pmatrix}0 & \frac{1}{v}\\ 1 & -\frac{u}{v^2}\end{pmatrix},\quad \det(J_{g^{-1}}(u,v))=-\frac{1}{v}.$$

Moreover, we have

$$f_{(U,V)}(u,v)=f_{(X,Y)}\left(u,\frac{u}{v}\right)\frac{1}{|v|}\mathbb{1}_{V\neq0}.$$

Therefore, we get

$$f_U(u)=\int_{\mathbb{R}}f_{(X,Y)}(u,v)\frac{1}{|v|}\mathrm{d}v.$$

2.7. Convergence of Random Variables

2.7.1. Types of convergences

We have already seen the notion of almost sure convergence. There are different types of convergences for random variables in probability theory. Let us recall the notion of almost sure (a.s.) convergence.

Definition 2.7.1.1 (Almost sure convergence). Let $(\Omega, \mathcal{A}, \mathbb{P})$ be a probability space. Let $(X_n)_{n\geq 1}$ be a sequence of random variables and let X be a random variable with values in $\mathbb{R}^d$. Then

$$\lim_{\substack{n\to\infty \\ a.s.}} X_n = X \Longleftrightarrow \mathbb{P}\left(\left\{\omega \in \Omega \,\middle|\, \lim_{n\to\infty} X_n(\omega) = X(\omega)\right\}\right) = 1.$$

Remark 2.7.1.2. Another very important convergence type is the L^p-convergence which is similarly defined as in measure theory. Recall that convergence in L^p for $p \in [1, \infty)$ in the probability language means

$$\lim_{\substack{n\to\infty \\ L^p}} X_n = X \Longleftrightarrow \lim_{n\to\infty} \mathbb{E}\left[|X_n - X|^p\right] = 0.$$

Definition 2.7.1.3 (Convergence in probability). Let $(\Omega, \mathcal{A}, \mathbb{P})$ be a probability space. We say that the sequence $(X_n)_{n\geq 1}$ converges in probability to X if for all $\varepsilon > 0$

$$\lim_{\substack{n\to\infty \\ \mathbb{P}}} X_n = X \Longleftrightarrow \lim_{n\to\infty} \mathbb{P}(|X_n - X| > \varepsilon) = 0.$$

Proposition 2.7.1.1. *Let $(\Omega, \mathcal{A}, \mathbb{P})$ be a probability space. Let $\mathcal{L}^0_{\mathbb{R}^d}(\Omega, \mathcal{A}, \mathbb{P})$ be the space of random variables with values in $\mathbb{R}^d$ and let $L^0_{\mathbb{R}^d}(\Omega, \mathcal{A}, \mathbb{P})$ be the quotient of $\mathcal{L}^0_{\mathbb{R}^d}(\Omega, \mathcal{A}, \mathbb{P})$ by the equivalence relation $X \sim Y :\Longleftrightarrow X = Y$ a.s. Then the map*

$$\begin{aligned} \mathsf{d} : L^0_{\mathbb{R}^d}(\Omega, \mathcal{A}, \mathbb{P}) \times L^0_{\mathbb{R}^d}(\Omega, \mathcal{A}, \mathbb{P}) &\to \mathbb{R}_{\geq 0} \\ (X, Y) &\mapsto \mathsf{d}(X, Y) := \mathbb{E}[|X - Y| \wedge 1] \end{aligned}$$

defines a distance (metric) in $L^0_{\mathbb{R}^d}(\Omega, \mathcal{A}, \mathbb{P})$, which is compatible with convergence in probability, i.e.,

$$\lim_{\substack{n\to\infty \\ \mathbb{P}}} X_n = X \Longleftrightarrow \lim_{n\to\infty} \mathsf{d}(X_n, X) = 0.$$

Moreover, $L^0_{\mathbb{R}^d}(\Omega, \mathcal{A}, \mathbb{P})$ is complete for the metric d.

Proof. It is easy to see that d defines a metric. If $\lim_{\substack{n\to\infty \\ \mathbb{P}}} X_n = X$, then for all $\varepsilon > 0$ we get

$$\lim_{n\to\infty} \mathbb{P}(|X_n - X| > \varepsilon) = 0.$$

Fix $\varepsilon > 0$. Then

$$\mathbb{E}[|X_n - X| \wedge 1] = \mathbb{E}[(|X_n - X| \wedge 1)\mathbb{1}_{|X_n - X|\leq\varepsilon}]$$
$$+ \mathbb{E}[(|X_n - X| \wedge 1)\mathbb{1}_{|X_n - X|>\varepsilon}] \leq \varepsilon + \mathbb{E}[\mathbb{1}_{|X_n-X|>\varepsilon}] \xrightarrow{n\to\infty} 0,$$

for ε arbitrary small. Conversely, assume that $\lim_{n\to\infty} \mathsf{d}(X_n, X) = 0$. Then for all $\varepsilon \in (0, 1)$ we have

$$\mathbb{P}(|X_n - X| > \varepsilon) \leq \frac{1}{\varepsilon}\mathbb{E}[|X_n - X| \wedge 1] = \frac{1}{\varepsilon}\mathsf{d}(X_n, X).$$

Next we show completeness: Let $(X_k)_{k\geq 0}$ be a Cauchy sequence for d. Then, there exists a subsequence $Y_k := X_{n_k}$ such that $\mathsf{d}(Y_k, Y_{k+1}) \leq \frac{1}{2^k}$. It follows that

$$\mathbb{E}\left[\sum_{k=1}^{\infty}(|Y_k - Y_{k+1}| \wedge 1)\right] = \sum_{k=1}^{\infty}\mathsf{d}(Y_k, Y_{k+1}) < \infty,$$

which implies that $\sum_{k=1}^{\infty}(|Y_k - Y_{k+1}| \wedge 1) < \infty$ and hence $\sum_{k=1}^{\infty} |Y_k - Y_{k+1}| < \infty$. Hence, the random variable $X := Y_1 + \sum_{k=1}^{\infty} Y_{k+1} - Y_k = X_{n_1} + \sum_{k=1}^{\infty} X_{n_k+1} - X_{n_k}$ is well-defined and $(X_n)_{n\geq 1}$ converges to X. □

Proposition 2.7.1.2. *Let $(\Omega, \mathcal{A}, \mathbb{P})$ be a probability space. If $(X_n)_{n\geq 1}$ converges a.s. or in L^p to X, it also converges in probability to X. Conversely, if $(X_n)_{n\geq 1}$ converges to X in probability, then there exists a subsequence $(X_{n_k})_{k\geq 1}$ of $(X_n)_{n\geq 1}$ such that*

$$\lim_{\substack{n\to\infty \\ a.s.}} X_n = X.$$

Proof. Consider $\mathsf{d}(X_n, X)$. We need to prove that $\lim_{\substack{n\to\infty \\ a.s.}} X_n = X$ or $\lim_{\substack{n\to\infty \\ L^p}} X_n = X$, which implies that $\lim_{n\to\infty} \mathsf{d}(X_n, X) = 0$. If $\lim_{\substack{n\to\infty \\ a.s.}} X_n = X$, we can apply *Lebesgue's dominated convergence theorem* (note that we can do this, because $|X_n - X| \wedge 1 \leq 1$ and $\mathbb{E}[1] < \infty$) to obtain that $\lim_{n\to\infty} \mathbb{E}[|X_n - X| \wedge 1] = \mathbb{E}[\lim_{n\to\infty}(|X_n -$

$X| \wedge 1)] = 0$. Finally, if $\lim_{n \to \infty, L^p} X_n = X$, we can use the fact that for all $p \geq 1$,

$$\mathbb{E}[|X_n - X| \wedge 1] \leq \underbrace{\mathbb{E}[|X_n - X|]}_{\|X_n - X\|_1} \leq \|X_n - X\|_p \xrightarrow{n \to \infty} 0.$$

□

Proposition 2.7.1.3. *Let $(\Omega, \mathcal{A}, \mathbb{P})$ be a probability space. Let $(X_n)_{n\geq 1}$ be a sequence of random variables and let $\lim_{n \to \infty, \mathbb{P}} X_n = X$. Assume there is some $r < 1$ such that $(X_n)_{n\geq 1}$ is bounded in L^r, i.e.,*

$$\sup_n \mathbb{E}[|X_n|^r] < \infty.$$

Then for every $p \in [1, \infty)$, we get that $\lim_{n \to \infty, L^p} X_n = X$.

Proof. Note that the fact that $(X_n)_{n\geq 1}$ is bounded in L^r implies that there is some $C > 0$, such that for all $n \geq 1$ we get

$$\mathbb{E}[|X_n|^r] \leq C.$$

Using *Fatou's lemma*, we get

$$\mathbb{E}[|X|] \leq C.$$

Hence, it follows that $X \in L^r$. Now we can apply *Hölder's inequality* to obtain for $p \in [1, r)$,

$$\begin{aligned}\mathbb{E}[|X_n - X|^p] &= \mathbb{E}[|X_n - X|^p \mathbb{1}_{\{|X_n - X| \leq \varepsilon\}}] + \mathbb{E}[|X_n - X|^p \mathbb{1}_{\{|X_n - X| > \varepsilon\}}] \\ &\leq \varepsilon^p + \mathbb{E}[|X_n - X|^r]^{\frac{p}{r}} \mathbb{P}(|X_n - X| > \varepsilon)^{1 - \frac{p}{r}} \\ &\leq \varepsilon^p + 2^p C^{\frac{p}{r}} \mathbb{P}(|X_n - X| > \varepsilon) \xrightarrow{n \to \infty} 0.\end{aligned}$$

□

2.7.2. The strong law of large numbers

Theorem 2.7.2.1 (Kolmogorov's 0–1 law). *Let $(\Omega, \mathcal{A}, \mathbb{P})$ be a probability space. Let $(X_n)_{n\geq 1}$ be a sequence of independent random* *variables with values in arbitrary measure spaces. For $n \geq 1$, define the σ-algebra*

$$\mathcal{B}_n := \sigma(X_k \mid k \geq n).$$

The tail σ-algebra $\mathcal{B}_\infty$ is defined as

$$\mathcal{B}_\infty := \bigcap_{n=1}^{\infty} \mathcal{B}_n.$$

Then $\mathcal{B}_\infty$ is trivial in the sense that for all $B \in \mathcal{B}_\infty$ we get that $\mathbb{P}(B) \in \{0, 1\}$.

Remark 2.7.2.1. We can easily see that a random variable which is $\mathcal{B}_\infty$-measurable is constant a.s. and in fact its distribution function can only take the values 0 and 1.

Proof of Theorem 2.7.2.1. Define $\mathcal{D}_n := \sigma(X_k \mid k \leq n)$. We have already observed in Section 2.5.3 that $\mathcal{D}_n$ and $\mathcal{B}_{n+1}$ are independent and hence since $\mathcal{B}_\infty \subset \mathcal{B}_{n+1}$, we get that for all $n \geq 1$, $\mathcal{D}_n$ and $\mathcal{B}_\infty$ are also independent. This implies that for all $A \in \bigcup_{n=1}^{\infty} \mathcal{D}_n$ and for all $B \in \mathcal{B}_\infty$ we get

$$\mathbb{P}(A \cap B) = \mathbb{P}(A)\mathbb{P}(B).$$

Since $\bigcup_{n=1}^{\infty} \mathcal{D}_n$ is stable under finite intersection, we obtain that $\sigma\left(\bigcup_{n=1}^{\infty} \mathcal{D}_n\right)$ is independent of $\mathcal{B}_\infty$ and thus

$$\sigma\left(\bigcup_{n=1}^{\infty} \mathcal{D}_n\right) = \sigma\left(X_1, X_2, \ldots\right).$$

We also note that the fact $\mathcal{B}_\infty \subset \sigma(X_1, X_2, \ldots)$ implies that $\mathcal{B}_\infty$ is independent of itself. Thus, it follows that for all $B \in \mathcal{B}_\infty$, we get $\mathbb{P}(B) = \mathbb{P}(B \cap B) = \mathbb{P}(B)\mathbb{P}(B) = \mathbb{P}(B)^2$. Hence, $\mathbb{P}(B) = \mathbb{P}(B)^2$ and therefore $\mathbb{P}(B) \in \{0, 1\}$. $\square$

Remark 2.7.2.2. If $(X_n)_{n \geq 1}$ is a sequence of independent random variables, then

$$\limsup_n \frac{1}{n}(X_1 + \cdots + X_n) \in [-\infty, \infty]$$

is $\mathcal{B}_\infty$ measurable. It follows that $\frac{1}{n}(X_1 + \cdots + X_n)$ converges a.s. and its limit random variable is a.s. constant.

Proposition 2.7.2.2. *Let $(\Omega, \mathcal{A}, \mathbb{P})$ be a probability space. Let $(X_n)_{n\geq 1}$ be a sequence of independent random variables with the same distribution*

$$\mathbb{P}(X_n = 1) = \mathbb{P}(X_n = -1) = \frac{1}{2}$$

for all $n \geq 1$ and define $S_n := \sum_{j=1}^n X_j$. Then we get that

$$\begin{cases} \sup_{n\geq 1} S_n = \infty & a.s. \\ \inf_{n\geq 1} S_n = -\infty & a.s. \end{cases}$$

Exercise 2.7.2.3. Show that for $p \geq 1$ we get $\mathbb{P}(-p \leq \inf_n S_n \leq \sup_n S_n \leq p) = 0$.

Proof. Assume that Exercise 2.7.2.3 holds. Take $p \to \infty$ to obtain

$$\mathbb{P}(\{\inf_n S_n > -\infty\} \cap \{\sup_n S_n < \infty\}) = 0,$$

and therefore $\mathbb{P}(\{\inf_n S_n = -\infty\} \cup \{\sup_n S_n = \infty\}) = 1$. So it follows that

$$1 \leq \mathbb{P}(\inf_n S_n = -\infty) + \mathbb{P}(\sup_n S_n = \infty).$$

By symmetry, we get $\mathbb{P}(\inf_n S_n = -\infty) = \mathbb{P}(\sup_n S_n = \infty)$ and hence $\mathbb{P}(\sup_n S_n = \infty) > 0$. Now note that $\{\sup_n S_n = \infty\} \in \mathcal{B}_\infty$. Indeed, for all $k \geq 1$ we get $\{\sup_n S_n = \infty\} = \{\sup_{n\geq k}(X_k, X_{k+1} + \cdots + X_n) = \infty\} \in \mathcal{B}_k$. Since $\{\sup_n S_n = \infty\} \in \mathcal{B}_\infty$, it follows that $\mathbb{P}(\sup_n S_n = \infty) \in \{0, 1\}$, but we have just seen that $\mathbb{P}(\sup_n S_n = \infty) \geq \frac{1}{2} > 0$, which implies then $\mathbb{P}(\sup_n S_n = \infty) = 1$. □

Theorem 2.7.2.3 (Strong law of large numbers). *Let $(\Omega, \mathcal{A}, \mathbb{P})$ be a probability space. Let $(X_n)_{n\geq 1}$ be a sequence of i.i.d. random variables such that $X_i \in L^1(\Omega, \mathcal{A}, \mathbb{P})$ for all $i \in \{1, \dots, n\}$. Then*

$$\lim_{\substack{n\to\infty \\ a.s.}} \frac{1}{n}(X_1 + \cdots + X_n) = \mathbb{E}[X_1].$$

Moreover, for $\bar{X}_n := \frac{1}{n}\sum_{j=1}^n (X_j - \mathbb{E}[X_j])$ we have

$$\mathbb{P}\left(\limsup_{n\to\infty} |\bar{X}_n| = 0\right) = 1.$$

Remark 2.7.2.4. The assumption $\mathbb{E}[X_1] < \infty$ is important, but if $X_1 \geq 0$ and $\mathbb{E}[X_1] = \infty$, we can apply Theorem 2.7.2.3 to $X_1 \wedge k$ for $k > 0$, and obtain that the result also holds with $\mathbb{E}[X_1] = \infty$.

Proof of Theorem 2.7.2.3. Let $S_n := X_1 + \cdots + X_n$ with $S_0 := 0$ and take $a > \mathbb{E}[X_1]$. Define moreover $M := \sup_{n>0}(S_n - na)$. We shall show that $M < \infty$ a.s. Since we obviously have $S_n \leq na + M$, it follows immediately that $\frac{S_n}{n} \leq a$ a.s. Choosing $a \searrow \mathbb{E}[X_1]$ we obtain that $\limsup_n \frac{S_n}{n} \leq \mathbb{E}[X_1]$. Replacing $(X_n)_{n\geq 1}$ with $(-X_n)_{n\geq 1}$, we also get $\liminf_n \frac{S_n}{n} \geq \mathbb{E}[X_1]$ a.s. So it follows that

$$\liminf_n \frac{S_n}{n} = \limsup_n \frac{S_n}{n} = \mathbb{E}[X_1] \quad a.s.$$

Hence, we only need to show that $M < \infty$ a.s. We first note that $\{M < \infty\} \in \mathcal{B}_\infty$. Indeed, for all $k \geq 0$ we get that $\{M < \infty\} = \{\sup_{n\in\mathbb{N}}(S_n - an) < \infty\} = \{\sup_{n\geq k}(S_n - S_k - (n-k)a) < \infty\}$. So it follows that $\mathbb{P}(M < \infty) \in \{0, 1\}$. Now we need to show that $\mathbb{P}(M < \infty) = 1$ or equivalently $\mathbb{P}(M = \infty) < 1$. We do it by contradiction. For $k \in \mathbb{N}$, set $M_k = \sup_{0\leq n\leq k}(S_n - na)$ and $M_k' = \sup_{0\leq n\leq k}(S_{n+1} - S_n - na)$. Then M_k and M_k' have the same distribution. Indeed, $(X_1, \ldots, X_k)$ and $(X_2, \ldots, X_{k+1})$ have the same distribution and $M_k = F_k(X_1, \ldots, X_k)$ and $M_k' = F_k(X_2, \ldots, X_{k+1})$ with some map $F_k : \mathbb{R}^k \to \mathbb{R}$. Moreover, $M = \lim_{k\to\infty} \uparrow M_k$ and therefore $M' = \lim_{k\to\infty} M_k$. Since M_k and M_k' have the same distribution, M and M' also have the same distribution. Indeed, $\mathbb{P}(M' \leq X) = \lim_{k\to\infty} \downarrow \mathbb{P}(M_k' \leq X) = \lim_{k\to\infty} \downarrow \mathbb{P}(M_k \leq X) = \mathbb{P}(M \leq X)$. So M and M' have the same distribution function. Moreover, $M_{k+1} = \sup\{0, \sup_{1\leq n\leq k+1}(S_n - na)\} = \sup\{0, M_k' + X_1 - a\}$, which implies that $M_{k+1} = M_k' - \inf\{a - X_1, M_k'\}$. Now we can use the fact that M_k' and M_k are bounded to obtain

$$\begin{aligned}
\mathbb{E}[\inf\{a - X_1, M_k'\}] &= \mathbb{E}[M_k'] - \mathbb{E}[M_{k+1}] \\
&\leq |a - X_1| \\
&\leq |a| + |X_1| \in \mathcal{L}^1(\Omega, \mathcal{A}, \mathbb{P})
\end{aligned}$$

and apply *Lebesgue's dominated convergence theorem* to obtain

$$\mathbb{E}[\inf\{a - X_1, M'\}] \leq 0.$$

If we had $\mathbb{P}(M \leq \infty) = 1$, then since M' and M have the same distribution we would also have $\mathbb{P}(M' = \infty) = 1$, in which case

$\inf\{a - X_1, M'\} < a - X_1$ and $\mathbb{E}[a - X_1] > 0$ and this contradicts

$$\mathbb{E}[\inf\{a - X_1, M'\}] \leq 0. \qquad \square$$

2.8. More Convergence in Probability, L^p and Almost Surely

Proposition 2.8.0.1. *Let $(\Omega, \mathcal{A}, \mathbb{P})$ be a probability space. Let $(X_n)_{n\geq 1}$ be a sequence of random variables and assume that for all $\varepsilon > 0$ we have*

$$\sum_{n\geq 1} \mathbb{P}(|X_n - X| > \varepsilon) < \infty.$$

Then we get

$$\lim_{\substack{n\to\infty \\ a.s.}} X_n = X.$$

Proof. Take $\varepsilon_k = \frac{1}{k}$ for $k \in \mathbb{N}$ with $k \geq 1$. Now using Lemma 2.5.4.1, we get

$$\mathbb{P}\left(\limsup_n \left\{|X_n - X| > \frac{1}{k}\right\}\right) = 0,$$

which implies that $\mathbb{P}\left(\bigcup_{k\geq 1} \limsup_n \left\{|X_n - X| > \frac{1}{k}\right\}\right) = 0$ and hence

$$\mathbb{P}\left(\underbrace{\bigcap_{k\geq 1} \liminf_n \left\{|X_n - X| \leq \frac{1}{k}\right\}}_{\Omega'}\right) = 1.$$

Moreover, we have that $\mathbb{P}[\Omega'] = 1$ and for $\omega \in \Omega'$ we get that for all $k \geq 1$ there is $n_0(\omega) \in \mathbb{N} \setminus \{0\}$ such that for $n \geq n_0(\omega)$ we get that $|X_n(\omega) - X(\omega)| \leq \frac{1}{k}$, i.e., $\lim_{n\to\infty} X_n(\omega) = X(\omega)$ for $\omega \in \Omega'$. $\square$

Example 2.8.0.1. Let $(\Omega, \mathcal{A}, \mathbb{P})$ be a probability space. Let $(X_n)_{n\geq 1}$ be a sequence of random variables such that $\mathbb{P}(X_n = 0) = 1 - \frac{1}{1+n^2}$

and $\mathbb{P}(X_n = 1) = \frac{1}{1+n^2}$. Then for all $\varepsilon > 0$ we get $\mathbb{P}(|X_n| > \varepsilon) = \mathbb{P}(X_n > \varepsilon) = \frac{1}{1+n^2}$, so it follows

$$\sum_{n\geq 1} \mathbb{P}(|X_n| > \varepsilon) < \infty,$$

which implies that $\lim_{\substack{n\to\infty \\ a.s.}} X_n = 0$.

Proposition 2.8.0.2. *Let $(\Omega, \mathcal{A}, \mathbb{P})$ be a probability space. Let $(X_n)_{n\geq 1}$ be a sequence of random variables. Then*

$$\lim_{\substack{n\to\infty \\ a.s.}} X_n = X \iff \lim_{\substack{n\to\infty \\ \mathbb{P}}} \sup_{m>n} |X_m - X| = 0.$$

Exercise 2.8.0.2. Prove Proposition 2.8.0.2.

Example 2.8.0.3. Let $(Y_n)_{n\geq 1}$ be i.i.d. random variables such that $\mathbb{P}(Y_n \leq X) = 1 - \frac{1}{1+X}$ for $X \geq 0$ and $n \geq 1$. Take $X_n = \frac{Y_n}{n}$ and let $\varepsilon > 0$. Then

$$\mathbb{P}(|X_n| > \varepsilon) = \mathbb{P}(|Y_n| > n\varepsilon) = \frac{1}{1 + n\varepsilon} \xrightarrow{n\to\infty} 0,$$

and thus $\lim_{\substack{n\to\infty \\ \mathbb{P}}} X_n = 0$. Moreover, we have

$$\mathbb{P}\left(\sup_{m\geq n} |X_m| > \varepsilon\right) = 1 - \mathbb{P}\left(\sup_{m\geq n} |X_n| \leq \varepsilon\right) = 1 - \prod_{m=n}^{\infty}\left(1 - \frac{1}{1+m\varepsilon}\right),$$

but $\prod_{m=n}^{\infty}\left(1 - \frac{1}{1+m\varepsilon}\right) = 0$. Hence, $\mathbb{P}(\sup_{m\geq n} |X_n| > \varepsilon) \not\to 0$ as $n \to \infty$ and therefore $(X_n)_{n\geq 1}$ does not converge a.s. to X.

Lemma 2.8.0.3. *Let $(\Omega, \mathcal{A}, \mathbb{P})$ be a probability space. Let $(X_n)_{n\geq 1}$ be a sequence of random variables. Then $\lim_{\substack{n\to\infty \\ \mathbb{P}}} X_n = X$ if and only if for very subsequence of $(X_n)_{n\geq 1}$, there exists a further subsequence which converges a.s.*

Proof. If $\lim_{\substack{n\to\infty \\ \mathbb{P}}} X_n = X$, then any of its subsequences also converges in probability. We already know that there exists a subsequence which converges a.s. Conversely, if $\lim_{\substack{n\to\infty \\ \mathbb{P}}} X_n = X$, then there is an $\varepsilon > 0$, some $n_k \in \mathbb{N}$ and a $\nu > 0$ such that for all $k \geq 1$ we get

$$\mathbb{P}(|X_{n_k} - X| > \varepsilon) > \nu$$

and therefore we can not extract a subsequence from $(X_{n_k})_{k\geq 1}$ which would converge a.s. $\square$

Proposition 2.8.0.4. *Let $(\Omega, \mathcal{A}, \mathbb{P})$ be a probability space. Let $(X_n)_{n\geq 1}$ be a sequence of random variables and $g : \mathbb{R} \to \mathbb{R}$ a continuous map. Moreover, assume that $\lim_{\substack{n\to\infty \\ \mathbb{P}}} X_n = X$. Then*

$$\lim_{\substack{n\to\infty \\ \mathbb{P}}} g(X_n) = g(X).$$

Proof. Any subsequence $g((X_{n_k})_{k\geq 1})$ and $(X_{n_k})_{k\geq 1}$ converges in probability. So it follows that there exists a subsequence $(X_{m_k})_{k\geq 1}$ of $(X_{n_k})_{k\geq 1}$ such that

$$\lim_{\substack{n\to\infty \\ a.s.}} X_n = X \quad \text{and} \quad \lim_{\substack{k\to\infty \\ a.s.}} g(X_{m_k}) = g(X)$$

because g is continuous. Now, using Lemma 2.8.0.3, we get that

$$\lim_{\substack{n\to\infty \\ \mathbb{P}}} g(X_n) = g(X).$$

$\square$

Proposition 2.8.0.5. *Let $(\Omega, \mathcal{A}, \mathbb{P})$ be a probability space. Let $(X_n)_{n\geq 1}$ and $(Y_n)_{n\geq 1}$ be sequences of random variables such that $\lim_{\substack{n\to\infty \\ \mathbb{P}}} X_n = X$ and $\lim_{\substack{n\to\infty \\ \mathbb{P}}} Y_n = Y$. Then*

(i) $\lim_{\substack{n\to\infty \\ \mathbb{P}}} X_n + Y_n = X + Y$;
(ii) $\lim_{\substack{n\to\infty \\ \mathbb{P}}} X_n \cdot Y_n = X \cdot Y$.

Proof. (i) Let $\varepsilon > 0$. Then $|X_n - X| \leq \frac{\varepsilon}{2}$ and $|Y_n - Y| \leq \frac{\varepsilon}{2}$ implies that $|(X_n + Y_n) - (X + Y)| \leq \varepsilon$, and thus we get

$$\mathbb{P}(|X_n + Y_n - (X+Y)| > \varepsilon) \leq \mathbb{P}\left(|X_n - X| > \frac{\varepsilon}{2}\right) + \mathbb{P}\left(|Y_n - Y| > \frac{\varepsilon}{2}\right).$$

(ii) We apply Proposition 2.8.0.4 to the continuous map $g(X) := X^2$. Hence, we get

$$2X_nY_n = (X_n + Y_n)^2 - X_n^2 - Y_n^2.$$

$\square$

2.9. Convergence in Law

We denote by $C_b(\mathbb{R}^d)$ the space of bounded and continuous functions $\varphi : \mathbb{R}^d \to \mathbb{R}$. Moreover, we endow $C_b(\mathbb{R}^d)$ with the supremum norm $\|\varphi\|_\infty := \sup_{x\in\mathbb{R}^d} |\varphi(x)|$. The space $(C_b(\mathbb{R}^d), \|\cdot\|_\infty)$ forms a *Banach space*, i.e., it is a complete normed vector space. We want to introduce the notion of law convergence in terms of probability measures.

Definition 2.9.0.1 (Weak and Law convergence). (i) Let $(\mu_n)_{n\geq 1}$ be a sequence of probability measures on $\mathbb{R}^d$. We say that $(\mu_n)_{n\geq 1}$ is *converging weakly* to a probability measure μ on $\mathbb{R}^d$, and we write

$$\lim_{\substack{n\to\infty \\ w}} \mu_n = \mu,$$

if for all $\varphi \in C_b(\mathbb{R}^d)$ we have

$$\lim_{n\to\infty} \int_{\mathbb{R}^d} \varphi \mathrm{d}\mu_n = \int_{\mathbb{R}^d} \varphi \mathrm{d}\mu.$$

(ii) Let $(\Omega, \mathcal{A}, \mathbb{P})$ be a probability space. A sequence of random variables $(X_n)_{n\geq 1}$, taking values in $\mathbb{R}^d$, is said to *converge in law* to a random variable X with values in $\mathbb{R}^d$ and we write

$$\lim_{\substack{n\to\infty \\ law}} X_n = X,$$

if $\lim_{\substack{n\to\infty \\ w}} \mathbb{P}_{X_n} = \mathbb{P}_X$, or equivalently, if for all $\varphi \in C_b(\mathbb{R}^d)$ we have

$$\begin{aligned}\lim_{n\to\infty} \mathbb{E}[\varphi(X_n)] = \mathbb{E}[\varphi(X)] \iff \lim_{n\to\infty} \int_{\mathbb{R}^d} \varphi(x)\mathrm{d}\mathbb{P}_{X_n}(x) \\ = \int_{\mathbb{R}^d} \varphi(x)\mathrm{d}\mathbb{P}_X(x).\end{aligned}$$

Remark 2.9.0.2. (i) There is an abuse of language when we say that $\lim_{\substack{n\to\infty \\ law}} X_n = X$ because the random variable X is not determined in a unique way, only $\mathbb{P}_X$ is unique.

(ii) Note also that the random variables X_n and X need not be defined on the same probability space $(\Omega, \mathcal{A}, \mathbb{P})$.

(iii) The space of probability measures on $\mathbb{R}^d$ can be viewed as a subspace of $C_b(\mathbb{R}^d)^*$ (the dual space of $C_b(\mathbb{R}^d)$). The weak convergence then corresponds to convergence for the *weak**-topology.

(iv) It is enough to show that $\lim_{n\to\infty}\mathbb{E}[\varphi(X_n)] = \mathbb{E}[\varphi(X)]$ or $\lim_{n\to\infty}\int_{\mathbb{R}^d}\varphi(x)\mathrm{d}\mathbb{P}_{X_n}(x) = \int_{\mathbb{R}^d}\varphi(x)\mathrm{d}\mathbb{P}_X(x)$ is satisfied for all $\varphi \in C_c(\mathbb{R}^d)$, where $C_c(\mathbb{R}^d)$ is the space of continuous functions with compact support. That is, $\varphi \in C_c(\mathbb{R}^d)$ if $\mathrm{supp}(\varphi) := \overline{\{x \in \mathbb{R}^d \mid \varphi(x) \neq 0\}}$ is compact.

Example 2.9.0.3. If X_n and $X \in \mathbb{Z}^d$ for all $n \geq 1$, then $\lim_{\substack{n\to\infty \\ law}} X_n = X$ if and only if for all $X \in \mathbb{Z}^d$ we have

$$\lim_{n\to\infty}\mathbb{P}(X_n = x) = \mathbb{P}(X = x).$$

To see this, we use point (iv) of the remark above. Let therefore $\varphi \in C_c(\mathbb{R}^d)$. Then

$$\mathbb{E}[\varphi(X_n)] = \sum_{k\in\mathbb{Z}^d}\varphi(k)\mathbb{P}(X_n = k).$$

Since φ has compact support, i.e., $\varphi(x) = 0$ for $|x| \leq C$ for some $C \geq 0$, we get

$$\mathbb{E}[\varphi(X_n)] = \sum_{\substack{k\in\mathbb{Z}^d \\ |k|\leq C}}\varphi(k)\mathbb{P}(X_n = k).$$

Hence we have

$$\begin{aligned}\lim_{n\to\infty}\mathbb{E}[\varphi(X_n)] &= \lim_{n\to\infty}\sum_{\substack{k\in\mathbb{Z}^d \\ |k|\leq C}}\varphi(k)\mathbb{P}(X_n = k) \\ &= \sum_{\substack{k\in\mathbb{Z}^d \\ |k|\leq C}}\varphi(k)\mathbb{P}[X = k] = \mathbb{E}[\varphi(X)].\end{aligned}$$

Example 2.9.0.4. If X_n has density $\mathbb{P}_{X_n}(\mathrm{d}x) = P_n(x)\mathrm{d}x$ for all $n \geq 1$ and if we assume that

$$\lim_{\substack{n\to\infty \\ a.e.}} P_n(x) = P(x),$$

then there is a positive function $Q \geq 0$ such that $\int_{\mathbb{R}^d} Q(x)\mathrm{d}x < \infty$ and $P_n(x) \leq Q(x)$ a.e. Then an application of *Lebesgue's dominated convergence theorem* shows that

$$\int_{\mathbb{R}^d} P(x)\mathrm{d}x = 1,$$

and thus there exists a random variable X with density P such that $\lim_{\substack{n\to\infty \\ law}} X_n = X$ and for $\varphi \in C_b(\mathbb{R}^d)$ we get

$$\mathbb{E}[\varphi(X_n)] = \int_{\mathbb{R}^d} \varphi(x)P_n(x)\mathrm{d}x,$$

and $|\varphi(x)P_n(x)| \leq \underbrace{\|\varphi\|_\infty q(x)}_{\in\mathcal{L}^1(\mathbb{R}^d)}$. So with the dominated convergence theorem we get

$$\lim_{n\to\infty} \int_{\mathbb{R}^d} \varphi(x)P_n(x)\mathrm{d}x = \int_{\mathbb{R}^d} \varphi(x)P(x)\mathrm{d}x = \mathbb{E}[\varphi(X)].$$

Example 2.9.0.5. Let $X_n \sim \mathcal{N}(0,\sigma_n^2)$ such that $\lim_{n\to\infty}\sigma_n = 0$. Then $\lim_{\substack{n\to\infty \\ law}} X_n = 0$ and

$$\mathbb{E}[\varphi(X_n)] = \int_{\mathbb{R}} \varphi(x)\mathrm{e}^{-\frac{x^2}{2\sigma_n^2}} \frac{1}{\sigma_n\sqrt{2\pi}}\mathrm{d}x.$$

Now using that $u = \frac{x}{\sigma_n}$, we get $\mathrm{d}x = \sigma_n\mathrm{d}u$ and hence we have

$$\mathbb{E}[\varphi(X_n)] = \int_{\mathbb{R}} \varphi(x)\mathrm{e}^{-\frac{x^2}{2\sigma_n^2}} \frac{1}{\sigma_n\sqrt{2\pi}}\mathrm{d}x = \int_{\mathbb{R}} \varphi(\sigma_n u)\mathrm{e}^{-\frac{u^2}{2}} \frac{1}{\sqrt{2\pi}}\mathrm{d}u.$$

Moreover, we have $|\varphi(\sigma_n u)\mathrm{e}^{-\frac{u^2}{2}}| \leq \underbrace{\|\varphi\|_\infty \mathrm{e}^{-\frac{u^2}{2}}}_{\in\mathcal{L}^1(\mathbb{R})}$. Hence, we get

$$\lim_{n\to\infty} \mathbb{E}[\varphi(X_n)] = \int_{\mathbb{R}} \varphi(0)\mathrm{e}^{-\frac{u^2}{2}} \frac{1}{\sqrt{2\pi}}\mathrm{d}u.$$

Proposition 2.9.0.1. *Let $(\Omega,\mathcal{A},\mathbb{P})$ be a probability space. Let $(X_n)_{n\geq 1}$ be a sequence of random variables and assume that $\lim_{\substack{n\to\infty \\ \mathbb{P}}} X_n = X$. Then $\lim_{\substack{n\to\infty \\ law}} X_n = X$.*

Proof. We first note that if $\lim_{\substack{n\to\infty \\ a.s.}} X_n = X$ then $\lim_{n\to\infty} \mathbb{E}[\varphi(X_n)] = \mathbb{E}[\varphi(X)]$ for every $\varphi \in C_b(\mathbb{R}^d)$. Let us now assume that $(X_n)_{n\geq 1}$ does not converge in law to X. Then there is a $\varphi \in C_b(\mathbb{R}^d)$ such that $\mathbb{E}[\varphi(X_n)]$ does not converge to $\mathbb{E}[\varphi(X)]$. We can hence extract a subsequence $(X_{n_k})_{k\geq 1}$ from $(X_n)_{n\geq 1}$ and find an $\varepsilon > 0$ such that

$$|\mathbb{E}[\varphi(X_{n_k})] - \mathbb{E}[\varphi(X)]| > \varepsilon.$$

But this contradicts the fact that we can extract a further subsequence $(X_{n_{k_l}})_{l\geq 1}$ from $(X_{n_k})_{k\geq 1}$ such that

$$\lim_{\substack{l\to\infty \\ a.s.}} X_{n_{k_l}} = X.$$

□

Let $(\Omega, \mathcal{A}, \mathbb{P})$ be a probability space. Let $(X_n)_{n\geq 1}$ be a sequence of random variables, A natural question would be to ask whether, under these condition, we have a $B \in \mathcal{B}(\mathbb{R})$ such that $\lim_{n\to\infty} \mathbb{P}(X_n \in B) = \mathbb{P}(X \in B)$. If we take $B = \{0\}$ and use the previous example, we would get

$$\lim_{n\to\infty} \underbrace{\mathbb{P}(X_n = 0)}_{=0} \neq \underbrace{\mathbb{P}(X = 0)}_{=1},$$

which shows that the answer to the question is negative.

Proposition 2.9.0.2. *Let $(\mu_n)_{n\geq 1}$ be a sequence of probability measures on $\mathbb{R}^d$ and μ be a probability measure on $\mathbb{R}^d$. Then the following conditions are equivalent:*

(i) $\lim_{\substack{n\to\infty \\ w}} \mu_n = \mu$.

(ii) *For all open subsets $G \subset \mathbb{R}^d$, we have*

$$\limsup_n \mu_n(G) \geq \mu(G).$$

(iii) *For all closed subsets $F \subset \mathbb{R}^d$, we have*

$$\limsup_n \mu_n(F) \leq \mu(F).$$

(iv) *For all Borel measurable sets $B \in \mathcal{B}(\mathbb{R}^d)$ with $\mu(\partial B) = 0$, we have*

$$\lim_{n\to\infty} \mu_n(B) = \mu(B).$$

Proof. We immediately note that (ii) $\Leftrightarrow$ (iii) by taking complements. First we show (i) $\Rightarrow$ (ii): Let G be an open subset of $\mathbb{R}^d$. Define $\varphi_p(x) := p(\mathsf{d}(x, G^{\mathsf{C}}) \wedge 1)$. Then φ_p is continuous, bounded, $0 \le \varphi_p(x) \le \mathbb{1}_G(x)$ for all $x \in \mathbb{R}^d$ and $\varphi_p \uparrow \mathbb{1}_G$ (recall that $\mathsf{d}(x, F) := \inf_{y \in F} d(x, y))$ as $p \to \infty$. Moreover, F is closed if and only if $\mathsf{d}(x, F) = 0$. We also get that $\varphi_p(x) = 0$ on G^{C} and $0 \le \varphi_p(x) \le 1 \le \mathbb{1}_G(x)$ for all $x \in \mathbb{R}^d$. Therefore we get

$$\begin{aligned}\liminf_n \mu_n(G) &\geq \sup_p \left(\liminf_n F \int_{\mathbb{R}^d} \varphi_p \mathrm{d}\mu_n\right) \\ &= \sup_p \int_{\mathbb{R}^d} \varphi_p \mathrm{d}\mu = \int \mathbb{1}_G \mathrm{d}\mu = \mu(G).\end{aligned}$$

Next we show that (ii) and (iii) $\Rightarrow$ (iv): For Borel measurable sets $B \in \mathcal{B}(\mathbb{R}^d)$ with $\mathring{B} \subset B \subset \bar{B}$ we get

$$\limsup_n \mu_n(B) \leq \limsup_n \mu_n(\bar{B}) \leq \mu(\bar{B}),$$

$$\liminf_n \mu_n(B) \geq \liminf_n \mu_n(\mathring{B}) \geq \mu(\mathring{B}).$$

Therefore, it follows that

$$\mu(\mathring{B}) \leq \liminf_n \mu_n(B) \leq \limsup_n \mu_n(B) \leq \mu(B).$$

Moreover, if $\mu(\partial B) = 0$, we get that $\mu(\bar{B}) = \mu(\mathring{B}) = \mu(B)$ and thus $\lim_{n\to\infty} \mu_n(B) = \mu(B)$. Next we show (iv) $\Rightarrow$ (i): Let therefore $\varphi \in C_b(\mathbb{R}^d)$. We can always use that $\varphi = \varphi^+ - \varphi^-$ and so, without loss of generality, we may assume that $\varphi \ge 0$. Let $\varphi \ge 0$ and $K \ge 0$ be such that $0 \le \varphi \le K$. Then

$$\int_{\mathbb{R}^d} \varphi(x) \mathrm{d}\mu(x) = \int_{\mathbb{R}^d} \underbrace{\left(\int_0^K \mathbb{1}_{\{t \leq \varphi(x)\}} \mathrm{d}t\right)}_{K \wedge \varphi(x) = \varphi(x)} \mathrm{d}\mu(x) = \int_0^K \mu(E_t^\varphi) \mathrm{d}t,$$

where $E_t^\varphi := \{x \in \mathbb{R}^d \mid \varphi(x) \geq t\}$. Similarly, we have

$$\int_{\mathbb{R}^d} \varphi(x) \mathrm{d}\mu_n(x) = \int_0^K \mu_n(E_t^\varphi) \mathrm{d}t.$$

Now we can note that $\partial E_t^\varphi \subset \{x \in \mathbb{R}^d \mid \varphi(x) = t\}$. Moreover, there are at most countably many values of t for which

$\mu\left(\{x \in \mathbb{R}^d \mid \varphi(x) = t\}\right) > 0$. Indeed, for an integer $K \geq 1$ we get that $\mu\left(\{x \in \mathbb{R}^d \mid \varphi(x) = t\}\right) \geq \frac{1}{K}$. This can happen for at most K distinct values of t. Thus, we have

$$\lim_{n\to\infty} \mu_n(E_t^\varphi) = \mu(E_t^\varphi)\mathrm{d}t \quad a.e.,$$

which implies that

$$\begin{aligned}\lim_{n\to\infty} \int_{\mathbb{R}^d} \varphi(x)\mathrm{d}\mu_n(x) &= \int_0^K \mu_n(E_t^\varphi)\mathrm{d}t \xrightarrow{n\to\infty} \int_0^K \mu(E_t^\varphi)\mathrm{d}t \\ &= \int_{\mathbb{R}^d} \varphi(x)\mathrm{d}\mu(x).\end{aligned}$$

□

Remark 2.9.0.6. Consider the case of $d = 1$. Let $(X_n)_{n\geq 1}$ be a sequence of random variables with values in $\mathbb{R}$ and let X be a random variable with values in $\mathbb{R}$. Using Proposition 2.9.0.2, one can show that

$$\lim_{\substack{n\to\infty \\ law}} X_n = X \iff \lim_{n\to\infty} F_{X_n}(t) = F_X(t).$$

Proposition 2.9.0.3. *Let $(\mu_n)_{n\geq 1}$ and μ be probability measures on $\mathbb{R}^d$. Let $H \subset C_b(\mathbb{R}^d)$ such that $\bar{H} \supset C_c(\mathbb{R}^d)$. Then the following conditions are equivalent:*

(i) $\lim_{\substack{n\to\infty \\ w}} \mu_n = \mu$.

(ii) *For all $\varphi \in C_c(\mathbb{R}^d)$, we have*

$$\lim_{n\to\infty} \int_{\mathbb{R}^d} \varphi \mathrm{d}\mu_n = \int_{\mathbb{R}^d} \varphi \mathrm{d}\mu.$$

(iii) *For all $\varphi \in H$, we have*

$$\lim_{n\to\infty} \int_{\mathbb{R}^d} \varphi \mathrm{d}\mu_n = \int_{\mathbb{R}^d} \varphi \mathrm{d}\mu.$$

Proof. It is obvious that (i) ⇒ (ii) and (i)⇒ (iii). Therefore, we first show (ii)⇒ (i): Consider $\varphi \in C_b(\mathbb{R}^d)$ and let $(f_k)_{k\geq 1} \in C_c(\mathbb{R}^d)$

with $0 \leq f_k \leq 1$ and $f_k \uparrow 1$ as $k \to \infty$. Then for all $k \geq 1$ we get that $\varphi f_k \in C_c(\mathbb{R}^d)$ and hence

$$\lim_{n\to\infty} \int_{\mathbb{R}^d} \varphi f_k \mathrm{d}\mu_n = \int_{\mathbb{R}^d} \varphi f_k \mathrm{d}\mu.$$

Moreover, we have

$$\left| \int_{\mathbb{R}^d} \varphi \mathrm{d}\mu - \int_{\mathbb{R}^d} \varphi f_k \mathrm{d}\mu \right| \leq \sup_{x\in\mathbb{R}^d} |\varphi(x)| \left(1 - \int_{\mathbb{R}^d} f_k \mathrm{d}\mu\right)$$

and also

$$\left| \int_{\mathbb{R}^d} \varphi \mathrm{d}\mu_n - \int_{\mathbb{R}^d} \varphi f_k \mathrm{d}\mu_n \right| \leq \sup_{x\in\mathbb{R}^d} |\varphi(x)| \left(1 - \int_{\mathbb{R}^d} f_k \mathrm{d}\mu_n\right).$$

Hence, for all $k \geq 1$ we get

$$\begin{aligned}
\limsup_n \left| \int_{\mathbb{R}^d} \varphi \mathrm{d}\mu - \int_{\mathbb{R}^d} \varphi \mathrm{d}\mu_n \right| &\leq \sup_{x\in\mathbb{R}^d} |\varphi(x)| \limsup_n \left\{ \left(1 - \int_{\mathbb{R}^d} f_k \mathrm{d}\mu_n\right) \right. \\
&\qquad \left. + \left(1 - \int_{\mathbb{R}^d} f_k \mathrm{d}\mu\right) \right\} \\
&= 2 \sup_{x\in\mathbb{R}^d} |\varphi(x)| \left(1 - \int_{\mathbb{R}^d} f_k \mathrm{d}\mu\right) \xrightarrow{k\to\infty} 0.
\end{aligned}$$

Now we show (iii) $\Rightarrow$ (ii): Let therefore $\varphi \in C_c(\mathbb{R}^d)$. Then there is a sequence $(\varphi_k)_{k\geq 1} \subset H$ such that $\|\varphi - \varphi_k\|_\infty \leq \frac{1}{k}$ for all $k \geq 1$. This implies that

$$\begin{aligned}
&\limsup_n \left| \int_{\mathbb{R}^d} \varphi \mathrm{d}\mu_n - \int_{\mathbb{R}^d} \varphi \mathrm{d}\mu \right| \\
&\quad \leq \limsup_n \left\{ \left| \int_{\mathbb{R}^d} \varphi \mathrm{d}\mu_n - \int_{\mathbb{R}^d} \varphi_k \mathrm{d}\mu_n \right| + \underbrace{\left| \int_{\mathbb{R}^d} \varphi_k \mathrm{d}\mu_n - \int_{\mathbb{R}^d} \varphi_k \mathrm{d}\mu \right|}_{\xrightarrow{n\to\infty} 0} \right. \\
&\quad \left. + \left| \int_{\mathbb{R}^d} \varphi_k \mathrm{d}\mu - \int_{\mathbb{R}^d} \varphi \mathrm{d}\mu \right| \right\} \leq \frac{2}{k}.
\end{aligned}$$

The claim follows now for $k \to \infty$. □

Theorem 2.9.0.4 (Lévy). *Let $(\Omega, \mathcal{A}, \mathbb{P})$ be a probability space. Let $(\mu_n)_{n\geq 1}$ be a sequence of probability measures on $\mathbb{R}^d$ associated to a sequence of real random variables $(X_n)_{n\geq 1}$. Moreover, let $\hat{\mu}_n(\xi) = \int_{\mathbb{R}^d} \mathrm{e}^{i\xi x}\mathrm{d}\mu_n(x)$ and $\Phi_X(\xi) = \mathbb{E}[\mathrm{e}^{i\xi x}]$. Then for all $\xi \in \mathbb{R}^d$ we get*

$$\lim_{\substack{n\to\infty \\ w}} \mu_n = \mu \Longleftrightarrow \lim_{n\to\infty} \hat{\mu}_n(\xi) = \hat{\mu}(\xi).$$

Equivalently, for all $\xi \in \mathbb{R}^d$ we get

$$\lim_{\substack{n\to\infty \\ law}} X_n = X \Longleftrightarrow \lim_{n\to\infty} \Phi_{X_n}(\xi) = \Phi_X(\xi).$$

Proof. It is obvious that $\lim_{\substack{n\to\infty \\ w}} \mu_n = \mu$ implies that $\lim_{n\to\infty} \hat{\mu}_n(\xi) = \hat{\mu}(\xi)$. Therefore $\mathrm{e}^{i\xi X}$ is continuous and bounded. For notational conventions we deal with the case $d = 1$. Consider $f \in C_c(\mathbb{R})$. For $\sigma > 0$ we also note $g_\sigma(x) = \frac{1}{\sigma\sqrt{2\pi}}\mathrm{e}^{-\frac{x^2}{2\sigma^2}}$.

Exercise 2.9.0.7. Show that $g_\sigma * f \xrightarrow{\sigma\to 0} f$ uniformly on $\mathbb{R}$.

Exercise 2.9.0.8. Show that if ν is a probability measure, then

$$\begin{aligned}\int_{\mathbb{R}} g_\sigma * f\mathrm{d}\nu &= \int_{\mathbb{R}} f(x)(g_\sigma * \nu)(x)\mathrm{d}x \\ &= \int_{\mathbb{R}} f(x)\frac{1}{\sigma\sqrt{2\pi}}\int_{\mathbb{R}} \mathrm{e}^{i\xi x} g_{\frac{1}{\sigma}}(\xi)\hat{\nu}(\xi)\mathrm{d}\xi\mathrm{d}x.\end{aligned}$$

Since $\lim_{n\to\infty} \hat{\mu}_n(\xi) = \hat{\mu}(\xi)$, we get by the dominated convergence theorem that

$$\int_{\mathbb{R}} \mathrm{e}^{i\xi x} g_{\frac{1}{\sigma}}(\xi)\hat{\mu}_n(\xi)\mathrm{d}\xi \xrightarrow{n\to\infty} \int_{\mathbb{R}} \mathrm{e}^{i\xi x} g_{\frac{1}{\sigma}}(\xi)\hat{\mu}(\xi)\mathrm{d}\xi.$$

These quantities are bounded by 1, and hence we can apply *Lebesgue's dominated convergence theorem* to obtain

$$\int_{\mathbb{R}} g_\sigma * f\mathrm{d}\mu_n \xrightarrow{n\to\infty} \int_{\mathbb{R}} g_\sigma * f\mathrm{d}\mu.$$

Let now $H := \{\varphi = g_\sigma * f \mid \sigma > 0, f \in C_c(\mathbb{R})\} \subset C_b(\mathbb{R})$. Since $f \in C_c(\mathbb{R})$, we get that $\|g_\sigma * f - f\|_\infty \xrightarrow{n\to\infty} 0$ and thus $\bar{H} \supset C_c(\mathbb{R})$. The result now follows from Proposition 2.9.0.3. □

Theorem 2.9.0.5 (Lèvy). *Let $(\mu_n)_{n\geq 1}$ be a sequence of probability measures on $\mathbb{R}^d$ with characteristic functions $(\Phi_n)_{n\geq 1}$. If Φ_n converges pointwise to a function Φ which is continuous at 0, then*

$$\lim_{\substack{n\to\infty \\ w}} \mu_n = \mu$$

for some probability measure μ on $\mathbb{R}^d$.

Example 2.9.0.9. Let $(X_n)_{n\geq 1}$ be a sequence of Poisson random variables with parameter λ (see Example 1.2.2.1). Moreover, consider the sequence $Z_n := \frac{X_n - 1}{\sqrt{n}}$. Then we have

$$\begin{aligned}\mathbb{E}\left[\mathrm{e}^{iuZ_n}\right] &= \mathbb{E}\left[\mathrm{e}^{iu\left(\frac{X_n-u}{\sqrt{n}}\right)}\right] = \mathrm{e}^{-i\frac{u}{\sqrt{n}}}\mathbb{E}\left[\mathrm{e}^{iu\frac{X_n}{\sqrt{n}}}\right] \\ &= \mathrm{e}^{-i\frac{u}{\sqrt{n}}}\mathrm{e}^{\lambda\left(\mathrm{e}^{i\frac{u}{\sqrt{n}}}-1\right)} = \mathrm{e}^{-i\frac{u}{\sqrt{n}}}\mathrm{e}^{\lambda\left(i\frac{u}{\sqrt{n}}-\frac{u^2}{2n}+O\left(\frac{1}{n}\right)\right)} \xrightarrow{n\to\infty} \mathrm{e}^{-\frac{u^2}{2}}.\end{aligned}$$

Since $\mathbb{E}\left[\mathrm{e}^{iu\mathcal{N}(0,1)}\right] = \mathrm{e}^{-\frac{u^2}{2}}$, we deduce that $\lim_{\substack{n\to\infty \\ law}} Z_n = \lim_{\substack{n\to\infty \\ law}} \frac{X_n-u}{\sqrt{n}} = \mathcal{N}(0,1)$.

Theorem 2.9.0.6. *Let $(\Omega, \mathcal{A}, \mathbb{P})$ be a probability space. Let $(X_n)_{n\geq 1}$ be a sequence of real random variables and X a real random variable and assume that $\lim_{\substack{n\to\infty \\ law}} X_n = X$ and that X is a.s. equal to a constant $a \in \mathbb{R}$. Then*

$$\lim_{\substack{n\to\infty \\ \mathbb{P}}} X_n = X.$$

Proof. Let $f(x) := |x-a| \wedge 1$. Then f is a continuous and bounded map and therefore $\lim_{n\to\infty}\mathbb{E}[f(X_n)] = \mathbb{E}[f(X)] = 0$, i.e., $\lim_{n\to\infty}\mathbb{E}[|X_n - a| \wedge 1] = 0$ which implies that $\lim_{\substack{n\to\infty \\ \mathbb{P}}} X_n = X$. □

Theorem 2.9.0.7. *Let $(\Omega, \mathcal{A}, \mathbb{P})$ be a probability space. Let $(X_n)_{n\geq 1}$ be a sequence of $\mathbb{R}^d$-valued random variables and X be random variable in $\mathbb{R}^d$. Assume that X_n has density f_n for all $n \geq 1$ and X has density f. Moreover, assume that $\lim_{n\to\infty} f_n(x) = f(x)$, a.e. Then*

$$\lim_{\substack{n\to\infty \\ law}} X_n = X.$$

Proof. We need to show that $\mathbb{E}[h(X_n)] \xrightarrow{n\to\infty} \mathbb{E}[h(X)]$, where $h : \mathbb{R}^d \to \mathbb{R}$ is a bounded and measurable map and

$$\mathbb{E}[h(X_n)] = \int_{\mathbb{R}^d} h(x) f_n(x) \mathrm{d}x,$$
$$\mathbb{E}[h(X)] = \int_{\mathbb{R}^d} h(x) f(x) \mathrm{d}x.$$

Let $h : \mathbb{R}^d \to \mathbb{R}$ be a bounded and measurable map. Moreover, let $\alpha := \sup_{x\in\mathbb{R}^d} |h(x)|$. Set $h_1(x) = h(x) + \alpha \geq 0$ and $h_2(x) = \alpha - h(x) \geq 0$. So it follows that $h_1 f_n \geq 0$ and $h_2 f_n \geq 0$. Moreover, we also get that

$$h_1 f_n \xrightarrow{n\to\infty} h_1 f \quad a.e.,$$
$$h_2 f_n \xrightarrow{n\to\infty} h_2 f \quad a.e.,$$

Using *Fatou's lemma*, we get

$$\mathbb{E}[h_1(X)] = \int_{\mathbb{R}^d} h_1(x) f(x) \mathrm{d}x \leq \liminf_n \int_{\mathbb{R}^d} h_1(x) f_n(x) \mathrm{d}x$$
$$= \liminf_n \mathbb{E}[h_1(X_n)].$$

Similarly, we get $\mathbb{E}[h_2(X)] \leq \liminf_n \mathbb{E}[h_2(X_n)]$. Now substitute $h_1(x) = h(x)+\alpha$ and $h_2(x) = \alpha - h(x)$, where we use $\liminf_n(-a_n) = \liminf_n(a_n)$ to obtain

$$\limsup_n \mathbb{E}[h(X_n)] \leq \mathbb{E}[h(X)] \leq \liminf_n \mathbb{E}[h(X_n)],$$

which implies that

$$\liminf_n \mathbb{E}[h(X_n)] = \limsup_n \mathbb{E}[h(X_n)] = \mathbb{E}[h(X)]. \qquad \square$$

2.10. The Central Limit Theorem (Real Case)

Theorem 2.10.0.1 (Central limit theorem (CLT)). *Let $(\Omega, \mathcal{A}, \mathbb{P})$ be a probability space. Let $(X_n)_{n\geq 1}$ be a sequence of i.i.d. random variables with values in $\mathbb{R}$. We assume that $\mathbb{E}[X_k^2] < \infty$*

(i.e., $X_k \in L^2(\Omega, \mathcal{A}, \mathbb{P})$) and let $\sigma^2 = \mathbb{V}(X_k)$ for all $k \in \{1, \dots, n\}$. Then for all $k \in \{1, \dots, n\}$ we get

$$\sqrt{n}\left(\frac{1}{n}\sum_{i=0}^{n} X_i - \mathbb{E}[X_k]\right) \xrightarrow[law]{n\to\infty} \mathcal{N}(0, \sigma^2).$$

Equivalently, for all $a, b \in \bar{\mathbb{R}} := \mathbb{R} \cup \{-\infty, \infty\}$ with $a < b$ and for all $k \in \{1, \dots, n\}$ we get

$$\lim_{n\to\infty} \mathbb{P}\left(\sum_{i=0}^{n} X_i \in \left[n\mathbb{E}[X_k] + a\sqrt{n}, n\mathbb{E}[X_k]+b\sqrt{n}\right]\right)$$
$$= \frac{1}{\sigma\sqrt{2\pi}}\int_a^b \mathrm{e}^{-\frac{x^2}{2\sigma^2}}\mathrm{d}x.$$

Remark 2.10.0.1. If $\mathbb{E}[X_k] = 0$ for all $k \in \{1, \dots, n\}$, then

$$\lim_{n\to\infty} \mathbb{P}\left[\frac{\sum_{i=1}^{n} X_i - n\mathbb{E}[X_k]}{\sqrt{n}} \in [a, b]\right] = \frac{1}{\sigma\sqrt{2\pi}}\int_a^b e^{-\frac{x^2}{2\sigma^2}}\mathrm{d}x.$$

Proof of Theorem 2.10.0.1. Without loss of generality we may assume that $\mathbb{E}[X_k] = 0$, for all $k \in \{1, \dots, n\}$. Define now a sequence $Z_n := \frac{\sum_{i=1}^{n} X_i}{\sqrt{n}}$. Then, we can obtain

$$\Phi_{Z_n}(\xi) = \mathbb{E}\left[\mathrm{e}^{i\xi Z_n}\right] = \mathbb{E}\left[\mathrm{e}^{i\xi\left(\frac{\sum_{i=1}^{n} X_i}{\sqrt{n}}\right)}\right]$$
$$= \prod_{j=1}^{n} \mathbb{E}\left[\mathrm{e}^{i\xi\frac{X_j}{\sqrt{n}}}\right] = \mathbb{E}\left[\mathrm{e}^{i\xi\frac{X_k}{\sqrt{n}}}\right]^n = \Phi_{X_k}^n(\xi).$$

We have already seen that

$$\Phi_{X_k}(\xi) = 1 + n\xi\mathbb{E}[X_k] - \frac{1}{2}\xi^2\mathbb{E}[X_k^2] + O(\xi^2) = 1 - \frac{\sigma^2}{2}\xi^2 + O(\xi^2).$$

Finally, for fixed ξ, we have

$$\Phi_{X_k}\left(\frac{\xi}{\sqrt{n}}\right) = 1 - \frac{\sigma^2\xi^2}{2n} + O\left(\frac{1}{n}\right)$$

$$\lim_{n\to\infty} \Phi_{X_k}^n(\xi) = \lim_{n\to\infty}\left(1 - \frac{\sigma^2\xi^2}{2} + O\left(\frac{1}{n}\right)\right)^n = \mathbb{E}\left[\mathrm{e}^{i\xi\mathcal{N}(0,\sigma^2)}\right]. \qquad \square$$

Theorem 2.10.0.2. *Let $(X_n)_{n\geq 1}$ be independent random variables but not necessarily i.i.d. We assume that $\mathbb{E}[X_j] = 0$ and that*

$\mathbb{E}[X_j^2] = \sigma_j^2 < \infty$, for all $j \in \{1, \ldots, n\}$. Assume further that $\sup_n \mathbb{E}[|X_n|^{2+\delta}] < \infty$ for some $\delta > 0$, and that $\sum_{j=1}^{\infty} \sigma_j^2 < \infty$. Then

$$\frac{\sum_{j=1}^{n} X_j}{\sqrt{\sum_{j=1}^{n} \sigma_j^2}} \xrightarrow[law]{n\to\infty} \mathcal{N}(0,1).$$

Example 2.10.0.2. Let $(X_n)_{n\geq 1}$ be i.i.d. random variables with $\mathbb{P}(X_n = 1) = p$ and $\mathbb{P}(X_n = 0) = 1 - p$. Then $S_n = \sum_{i=1}^{n} X_i$ is a binomial random variable $\mathcal{B}(p, n)$ (see Section 2.2.1.3). We have $\mathbb{E}[S_n] = np$ and $\mathbb{V}(S_n) = np(1-p)$. Now using the strong law of large numbers (Theorem 2.7.2.3), we get $\frac{S_n}{n} \xrightarrow[a.s.]{n\to\infty} p$ and using the central limit theorem (Theorem 2.10.0.1), we get

$$\frac{S_n - np}{\sqrt{np(1-p)}} \xrightarrow[law]{n\to\infty} \mathcal{N}(0,1).$$

Example 2.10.0.3 (The Erdős-Kac theorem). Let $\mathcal{P}$ be the set of prime numbers. For $p \in \mathcal{P}$, define $\mathcal{B}_p$ as $\mathbb{P}(\mathcal{B}_p = 1) = \frac{1}{p}$ and $\mathbb{P}(\mathcal{B}_p = 0) = 1 - \frac{1}{p}$. We take the $(\mathcal{B}_p)_{p\in\mathcal{P}}$ to be independent and

$$W_n := \sum_{\substack{\mathcal{P}\subset\mathbb{N} \\ p\in\mathcal{P}}} \mathcal{B}_p,$$

the probabilistic model for the total numbers of distinct prime divisors of $n := W(0)$. It is a simple exercise to check that $(W_n)_{n\geq 1}$ satisfies the assumption of Theorem 2.10.0.1 and using the fact that $\sum_{\substack{p\leq n \\ p\in\mathcal{P}}} \frac{1}{p} \sim \log\log n$, we obtain

$$\frac{W_n - \log\log n}{\sqrt{\log\log n}} \xrightarrow[law]{n\to\infty} \mathcal{N}(0,1).$$

Theorem 2.10.0.3 (Erdős-Kac). *Let N_n be a random variable with uniformly distribution in $\{1, \ldots, n\}$. Then*

$$\frac{W(N_n) - \log\log n}{\sqrt{\log\log n}} \xrightarrow[law]{n\to\infty} \mathcal{N}(0,1),$$

where $W(n) := \sum_{\substack{p\leq n \\ p\in\mathcal{P}}} \mathbb{1}_{p|n}$.

Example 2.10.0.4. Suppose that $(X_n)_{n\geq 1}$ are i.i.d. random variables with distribution function $F(x) = \mathbb{P}(X_1 \leq x)$. Let $Y_n(x) := \mathbb{1}_{X_n \leq x}$, where $(Y_n)_{n\geq 1}$ are i.i.d. Define $F_n(x) := \frac{1}{n}\sum_{k=1}^{n} Y_k(x) = \frac{1}{n}\sum_{k=1}^{n} \mathbb{1}_{X_k \leq x}$. F_n is called the *empirical distribution function.* Using the strong law of large numbers (Theorem 2.7.2.3), we get $\lim_{\substack{n\to\infty \\ a.s.}} F_n(x) = \mathbb{E}[Y_1(x)]$ and

$$\mathbb{E}[Y_1(x)] = \mathbb{E}[\mathbb{1}_{X_1 \leq x}] = \mathbb{P}(X_1 \leq x) = F(x).$$

In fact, there is a theorem (the *Gliwenko–Cantelli theorem*) which says that

$$\sup_{x\in\mathbb{R}} |F_n(x) - F(x)| \xrightarrow[a.s.]{n\to\infty} 0.$$

Next, we note that

$$\begin{aligned}
\sqrt{n}(F_n(x) - F(x)) &= \sqrt{n}\left(\frac{1}{n}\sum_{k=1}^{n} Y_k(x) - \mathbb{E}[Y_1(x)]\right) \\
&= \frac{1}{\sqrt{n}}\left(\sum_{k=1}^{n} Y_k(x) - n\mathbb{E}[Y_1(x)]\right).
\end{aligned}$$

Using the central limit theorem (Theorem 2.10.0.1), we get

$$\sqrt{n}(F_n(x) - F(x)) \xrightarrow[law]{n\to\infty} \mathcal{N}(0, \sigma^2(x)),$$

where $\sigma^2(x) = \mathbb{V}(Y_1) = \mathbb{E}[Y_1^2(x)] = \mathbb{E}[Y_1(x)] = F(x) - F^2(x) = F(x)(1 - F(x))$. Hence, we have

$$\sqrt{n}(F_n(x) - F(x)) \xrightarrow[law]{n\to\infty} \mathcal{N}(0, F(x)(1 - F(x))).$$

Theorem 2.10.0.4 (Berry–Esseen). *Let $(X_n)_{n\geq 1}$ be i.i.d. random variables and suppose that $\mathbb{E}[|X_k|^3] < \infty$, $\forall k \in \{1, \ldots, n\}$. Define*

$$G_n(x) := \mathbb{P}\left(\frac{\sum_{i=1}^{n} X_i - n\mathbb{E}[X_k]}{\sigma\sqrt{n}} \leq x\right), \quad \forall k \in \{1, \ldots, n\},$$

where $\sigma^2 = \mathbb{E}[X_k^2]$ $\forall k \in \{1, \ldots, n\}$ and $\Phi(x) = \mathbb{P}[\mathcal{N}(0,1) \leq x] = \int_{-\infty}^{x} e^{-\frac{u^2}{2}} \frac{1}{\sqrt{2\pi}} \mathrm{d}u$. Then

$$\sup_{x\in\mathbb{R}} |G_n(x) - \Phi(x)| \leq C\frac{\mathbb{E}[|X_k|^3]}{\sigma^3\sqrt{n}}, \quad \forall k \in \{1, \ldots, n\}$$

where C is a universal constant.

2.11. Gaussian Vectors and Multidimensional CLT

2.11.1. Gaussian vectors

Definition 2.11.1.1 (Gaussian random vector). An $\mathbb{R}^n$-valued random variable $X = (X_1, \ldots, X_n)$ is called a *Gaussian random vector* if every linear combination $\sum_{j=1}^n \lambda_j X_j$, with $\lambda_j \in \mathbb{R}$, is a Gaussian random variable (possibly degenerated $\mathcal{N}(\mu, 0) = \mu$ a.s.).

Theorem 2.11.1.1. *X is an $\mathbb{R}^n$-valued Gaussian random variable if and only if its characteristic function has the form*

$$\Phi_X(\xi) = \exp\left(i\langle \xi, \mu\rangle - \frac{1}{2}\langle \xi, \mathbf{\Sigma}_X \xi\rangle\right), \tag{2.11.1.1}$$

where $\mu \in \mathbb{R}^n$ and $\mathbf{\Sigma}_X$ is a symmetric, non-negative and semi-definite *matrix of size $n \times n$. $\mathbf{\Sigma}_X$ is then the covariance matrix of X and μ is the mean vector, i.e.,* $\mathbb{E}[X_j] = \mu_j$.

Proof. Suppose that (2.11.1.1) holds. Let $Y := \sum_{j=1}^n a_j X_j = \langle a, X\rangle$. For $v \in \mathbb{R}$, $\Phi_Y(v) = \Phi_X(va) = \exp\left(iv\langle a, \mu\rangle - \frac{v^2}{2}\langle a, \mathbf{\Sigma}_X a\rangle\right) \Rightarrow Y \sim \mathcal{N}\left(\langle a, \mu\rangle, \langle a, \mathbf{\Sigma}_X a\rangle\right) \Rightarrow X$ is a Gaussian vector. Conversely, assume that X is a Gaussian vector and let $Y := \sum_{j=1}^n a_j X_j = \langle a, X\rangle$. Moreover, let $\omega := \mathbb{CV}(X)$ and note that $\mathbb{E}[Y] = \langle \sigma, \mu\rangle$ and $\mathbb{V}(Y) = \sigma^2(Y) = \langle a, \mathbf{\Sigma}_X a\rangle$. Since Y is a Gaussian random variable, we get

$$\Phi_Y(v) = \exp\left(iv\langle a, \mu\rangle - \frac{v^2}{2}\langle a, \omega a\rangle\right).$$

Now, it is easy to see that $\Phi_X(v) = \Phi_Y(1) = \exp\left(i\langle v, \mu\rangle - \frac{1}{2}\ \langle v, \mathbf{\Sigma}_X v\rangle\right)$. □

Remark 2.11.1.2. We will also write $X \sim \mathcal{N}(\mu, \mathbf{\Sigma}_X)$.

Example 2.11.1.3. Let $X_1, \ldots, X_n$ be independent Gaussian random variables with $X_k \sim \mathcal{N}(\mu_k, \sigma_k^2)$ for all $1 \leq k \leq n$.

Then $X = (X_1, \ldots, X_n)$ is a Gaussian vector. Indeed, we have

$$\begin{aligned}\Phi_X(v_1, \ldots, v_n) &= \mathbb{E}\left[e^{i(v_1X_1+\cdots+v_nX_n)}\right] \\ &= \prod_{j=1}^{n} \mathbb{E}\left[e^{iv_jX_j}\right] \\ &= e^{i\langle v,\mu\rangle} e^{-\frac{1}{2}\langle v,\mathbf{\Sigma}_X v\rangle},\end{aligned}$$

where $\mu = (\mu_1, \ldots, \mu_n)$, $\mathbf{\Sigma}_X = \mathrm{diag}(\sigma_1^2, \ldots, \sigma_n^2)$.

Corollary 2.11.1.2. *Let X be an $\mathbb{R}^n$-valued Gaussian vector. The components X_j of X are independent if and only if $\mathbf{\Sigma}_X$ is a diagonal matrix.*

Proof. Suppose $\mathbf{\Sigma}_X = \mathrm{diag}(\sigma_1^2, \ldots, \sigma_n^2)$, then (2.11.1.1) shows that

$$\Phi_X(v_1, \ldots, v_n) = \prod_{j=1}^{n} \Phi_{X_j}(v_j),$$

where $X_j \sim \mathcal{N}(\mu_j, \sigma_j^2)$. The result follows from the uniqueness of the random variables. □

Theorem 2.11.1.3. *Let X be an $\mathbb{R}^n$-valued Gaussian vector with mean μ. Then there exists independent Gaussian random variables $Y_1, \ldots, Y_n$ with*

$$Y_j \sim \mathcal{N}(0, \lambda_j), \qquad \lambda_j \geq 0, \qquad 1 \leq j \leq n$$

and an orthogonal matrix $A \in \mathrm{O}(n)$ such that

$$X = \mu + AY.$$

Remark 2.11.1.4. It is possible that $\lambda_j = 0$. In that case, it is also possible to get $Y_j = 0$ a.s.

Proof of Theorem 2.11.1.3. Note first that there is an $A \in \mathrm{O}(n)$, such that $\mathbf{\Sigma}_X = A\Lambda A^*$, with $\Lambda := \mathrm{diag}(\lambda_1, \ldots, \lambda_n)$ such that $\lambda_j \leq 0$ for $1 \leq j \leq n$. Set $Y := A^*(X - \mu)$. Then one can check that Y is indeed Gaussian. So we get that $\mathbb{CV}(Y) = A^*\mathbf{\Sigma}_X A = \Lambda$, which implies that $Y_1, \ldots, Y_n$ are independent by Corollary 2.11.1.2 because $\mathbb{CV}(Y)$ is diagonal. □

Corollary 2.11.1.4. *An $\mathbb{R}^n$-valued Gaussian vector X has density on $\mathbb{R}^n$ if and only if*

$$\det \mathbf{\Sigma}_X \neq 0.$$

Remark 2.11.1.5. If $\det \mathbf{\Sigma}_X \neq 0$, then $f_X(x) = \frac{1}{\sqrt{2\pi \det \mathbf{\Sigma}_X}} e^{-\frac{1}{2}\langle x-\mu, \mathbf{\Sigma}_X^{-1}(\lambda-\mu)\rangle}$.

Theorem 2.11.1.5. *Let X be an $\mathbb{R}^n$-valued Gaussian random variable and let Y be an $\mathbb{R}^m$-valued Gaussian random variable. If X and Y are independent, then $Z = (X, Y)$ is an $\mathbb{R}^{n+m}$-valued Gaussian vector.*

Proof. Let $u = (w, v)$, $w \in \mathbb{R}^n$ and $v \in \mathbb{R}^m$. Take $\mathbf{\Sigma} = \begin{pmatrix} \mathbf{\Sigma}_X & 0 \\ 0 & \mathbf{\Sigma}_Y \end{pmatrix}$. Now we get

$$\begin{aligned}
\Phi_Z(u) &= \Phi_X(w)\Phi_Y(v) \\
&= \exp\left(i\langle w, \mu^X\rangle - \frac{1}{2}\langle w, \mathbf{\Sigma}_X w\rangle\right) + \exp\left(i\langle v, \mu^Y\rangle - \frac{1}{2}\langle v, \mathbf{\Sigma}_Y v\rangle\right) \\
&= \exp\left(i\langle (w, v), (\mu^X, \mu^Y)\rangle - \frac{1}{2}\langle u, \mathbf{\Sigma} u\rangle\right),
\end{aligned}$$

implying that Z is a Gaussian vector. □

Theorem 2.11.1.6. *Let X be an $\mathbb{R}^n$-valued Gaussian vector. Two components X_j and X_k of X are independent if and only if $\mathbb{CV}(X_j, X_k) = 0$.*

Proof. Consider $Y := (Y_1, Y_2)$, with $Y_1 := X_j$ and $Y_2 ;= X_k$. If Y is a Gaussian vector, then $\mathbb{CV}(Y_1, Y_2) = 0$, which implies that Y_1 and Y_2 are independent. □

Remark 2.11.1.6. For a random variable $Y \sim \mathcal{N}(0, 1)$ and for a positive constant $a > 0$, define $Z := Y\mathbb{1}_{|Y|\leq a} - Y\mathbb{1}_{|Y|>a}$. Then $Z \sim \mathcal{N}(0, 1)$, $Y + Z = 2Y\mathbb{1}_{|Y|\leq a}$ is *not* Gaussian because it is a bounded random variable and it is not constant. Therefore, (Y, Z) is *not* a Gaussian vector.

2.11.2. Multidimensional CLT

Theorem 2.11.2.1 (Multidimensional CLT). *Let $(\Omega, \mathcal{A}, \mathbb{P})$ be a probability space. Let $(X_n)_{n\geq 1}$ be a sequence of i.i.d. random* *variables with values in* $\mathbb{R}^d$*, i.e., each* X_n *has the form* $X_n = (X_{n_1}, \dots, X_{n_d})$*. We assume that* $\mathbb{E}[X_{n_j}^2] < \infty$ *(i.e.,* $X_{n_j} \in L^2(\Omega, \mathcal{A}, \mathbb{P})$*) and let*

$$\mathbf{\Sigma}_X := (\mathbb{CV}(X_{n_i}, X_{n_j}))_{1 \le i,j \le d}.$$

Then for all $j \in \{1, \dots, d\}$ *we get*

$$\sqrt{n}\left(\frac{1}{n}\sum_{i=0}^{n} X_i - \mathbb{E}[X_i]\right) \xrightarrow[law]{n\to\infty} \mathcal{N}(0, \mathbf{\Sigma}_X).$$

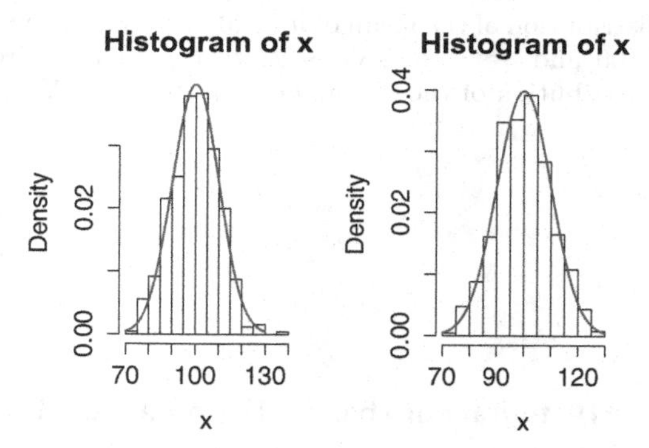

Fig. 2.6 For the illustration of the CLT: the distribution (density) function of a Gaussian distributed random variable X with $\mu = 100$ and $\sigma = 10$ ($X \sim \mathcal{N}(\mu, \sigma^2)$).

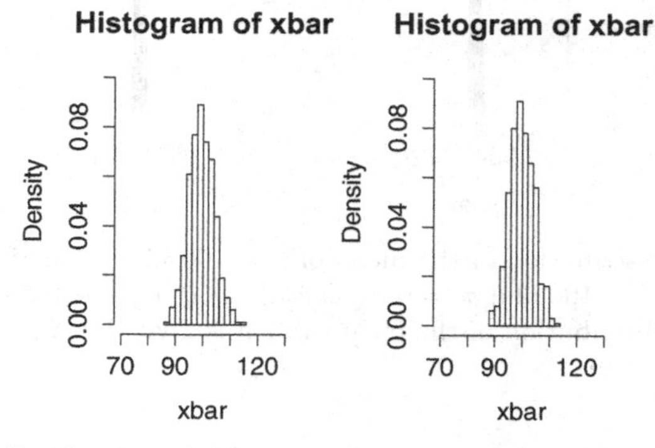

Fig. 2.7 The distribution of the mean of $n = 5$ i.i.d. Gaussian random variables (X_n) with $\mu = 100$ and $\sigma^2 = 10$ ($X_k \sim \mathcal{N}(\mu, \sigma^2)$). The figure illustrates the behavior of the distribution of the mean, which is given by $\bar{X}_n = \frac{1}{n}\sum_{k=1}^{n} X_k$.

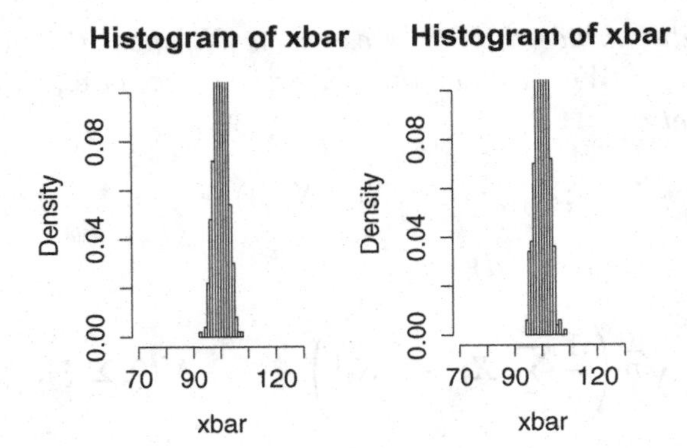

Fig. 2.8 The distribution of the mean of $n = 20$ i.i.d. Gaussian random variables (X_n) with $\mu = 100$ and $\sigma^2 = 10$ ($X_k \sim \mathcal{N}(\mu, \sigma^2)$). The figure illustrates the behavior of the distribution of the mean, which is given by $\bar{X}_n = \frac{1}{n}\sum_{k=1}^{n} X_k$.

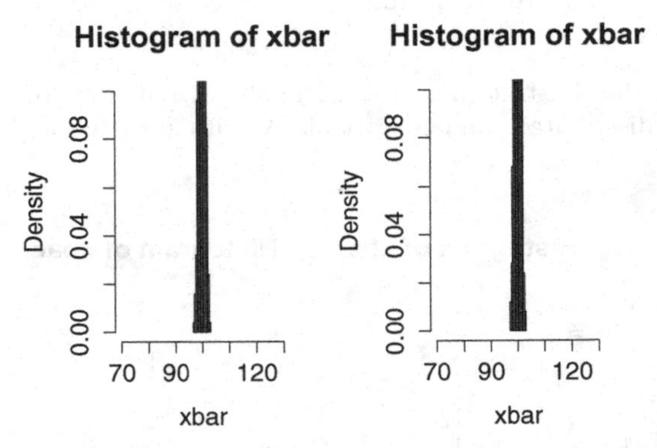

Fig. 2.9 The distribution of the mean of $n = 100$ i.i.d. Gaussian random variables (X_n) with $\mu = 100$ and $\sigma^2 = 10$ ($X_k \sim \mathcal{N}(\mu, \sigma^2)$). The figure illustrates the behavior of the distribution of the mean, which is given by $\bar{X}_n = \frac{1}{n}\sum_{k=1}^{n} X_k$.

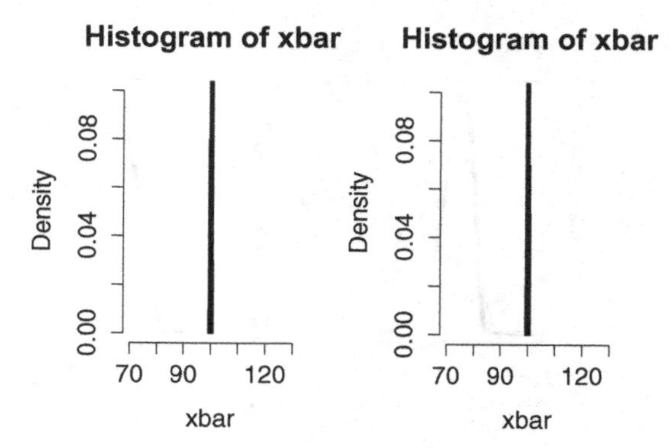

Fig. 2.10 The distribution of the mean of $n = 1000$ i.i.d. Gaussian random variables (X_n) with $\mu = 100$ and $\sigma^2 = 10$ $(X_k \sim \mathcal{N}(\mu, \sigma^2))$. The figure illustrates the behavior of the distribution of the mean, which is given by $\bar{X}_n = \frac{1}{n}\sum_{k=1}^{n} X_k$. Clearly, we see that it is converging to μ for $n \to \infty$.

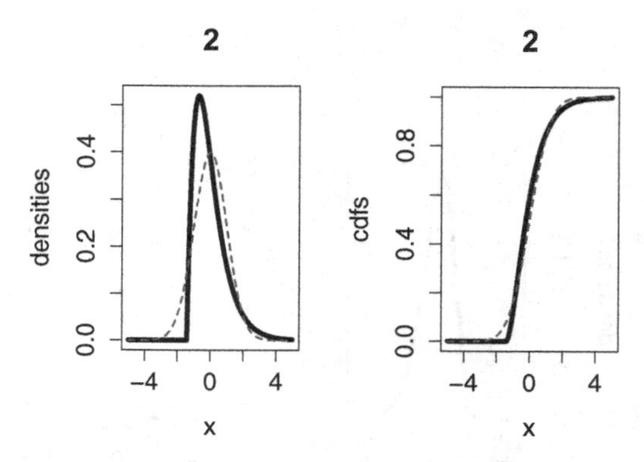

Fig. 2.11 An illustration of the CLT, where the random variables (X_n) are i.i.d. exponentially distributed. Also with the cumulative distribution functions. Here we have $n = 2$ exponentially distributed random variables $\left(X_k \sim \lambda e^{-\lambda}\right)$. On the left side, the thick curve represents the density function of the random variables and the thin curve represents the density of a Gaussian random variable Y with $\mu = 0$ $(Y \sim \mathcal{N}(0, \sigma^2))$. On the right side, the thick curve represents the cumulative distribution function of the random variables X_k and the thin curve represents the cumulative distribution function of a Gaussian random variable Y.

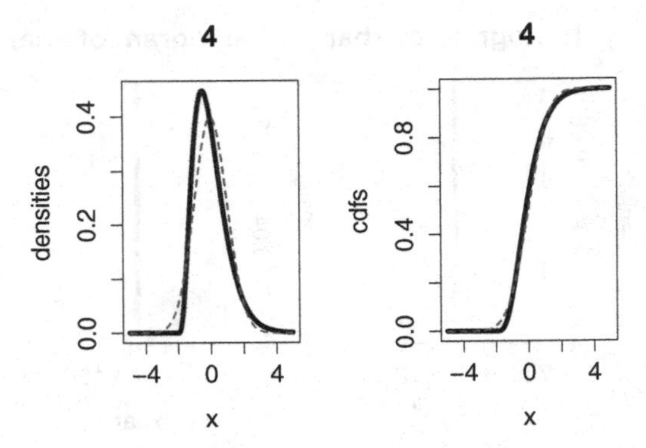

Fig. 2.12 An illustration of the CLT, where the random variables (X_n) are i.i.d. exponentially distributed. Also with the cumulative distribution functions. Here we have $n = 4$ exponentially distributed random variables $\left(X_k \sim \lambda e^{-\lambda}\right)$. On the left side, the thick curve represents the density function of the random variables and the thin curve represents the density of a Gaussian random variable Y with $\mu = 0$ $(Y \sim \mathcal{N}(0, \sigma^2))$. On the right side, the thick curve represents the cumulative distribution function of the random variables X_k and the thin curve represents the cumulative distribution function of a Gaussian random variable Y.

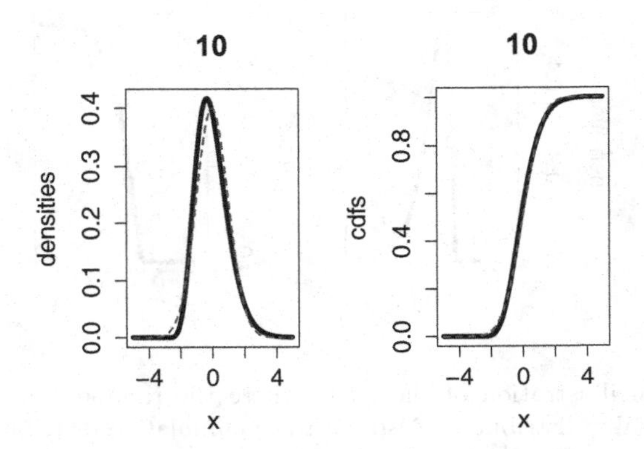

Fig. 2.13 An illustration of the CLT, where the random variables (X_n) are i.i.d. exponentially distributed. Also with the cumulative distribution functions. Here we have $n = 10$ exponentially distributed random variables $\left(X_k \sim \lambda e^{-\lambda}\right)$. On the left side, the thick curve represents the density function of the random variables and the thin curve represents the density of a Gaussian random variable Y with $\mu = 0$ $(Y \sim \mathcal{N}(0, \sigma^2))$. On the right side, the thick curve represents the cumulative distribution function of the random variables X_k and the thin curve represents the cumulative distribution function of a Gaussian random variable Y.

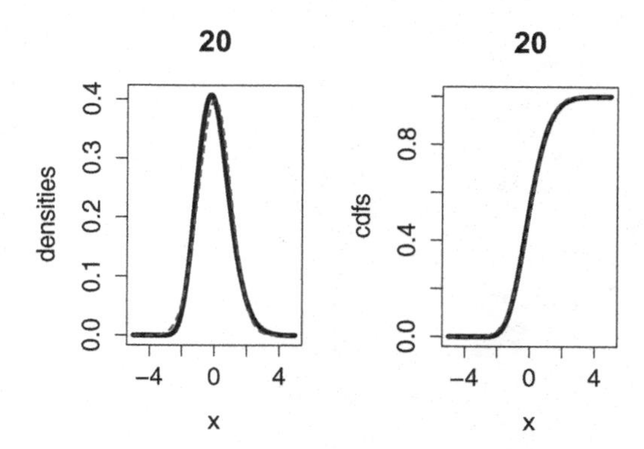

Fig. 2.14 An illustration of the CLT, where the random variables (X_n) are i.i.d. exponentially distributed. Also with the cumulative distribution functions. Here we have $n = 20$ exponentially distributed random variables $\left(X_k \sim \lambda e^{-\lambda}\right)$. On the left side, the thick curve represents the density function of the random variables and the thin curve represents the density of a Gaussian random variable Y with $\mu = 0$ ($Y \sim \mathcal{N}(0, \sigma^2)$). On the right side, the thick curve represents the cumulative distribution function of the random variables X_k and the thin curve represents the cumulative distribution function of a Gaussian random variable Y. Now we can see that both, the density and the cumulative distribution function, are converging to a Gaussian density and a Gaussian cumulative distribution function for $n \to \infty$.

Part II

Conditional Expectations, Martingales and Markov Chains

Introduction

If we consider a probability space $(\Omega, \mathcal{F}, \mathbb{P})$ with a sequence of i.i.d. random variables $(X_n)_{n\geq 1}$, we can look at the expectation $\mathbb{E}[|X_1|] = \cdots = \mathbb{E}[|X_n|] < \infty$. Now *limit theorems* play a central role, as we have seen in Part I. For example, we have seen the *strong law of large numbers* (Theorem 2.7.2.3), that is

$$\frac{X_1 + \cdots + X_n}{n} \xrightarrow{n\to\infty} \mathbb{E}[X_1] \quad \text{a.s.}$$

Note that $\mathbb{E}[X_1] = \cdots = \mathbb{E}[X_n]$ since $(X_n)_{n\geq 1}$ are i.i.d. random variables. Recall that another very important limit theorem is the *central limit theorem (CLT)* (Theorem 2.10.0.1), that is

$$\sqrt{n}\left(\frac{X_1 + \cdots + X_n}{n}\right) \xrightarrow[\text{law}]{n\to\infty} \mathcal{N}(0,1).$$

This means that the distribution of the sum of the random variables over $\sqrt{n}$ converges to a standard Gaussian distribution. Note that it does not matter what the distribution of the X_i is. The way we have proved this in Part I was by using the approach of the characteristic function. Take

$$\mathbb{E}\left[\mathrm{e}^{it\left(\frac{X_1+\cdots+X_n}{\sqrt{n}}\right)}\right] = \prod_{i=1}^{n} \mathbb{E}\left[\mathrm{e}^{it\frac{X_1}{\sqrt{n}}}\right] = \left(\mathbb{E}\left[\mathrm{e}^{it\frac{X_1}{\sqrt{n}}}\right]\right)^n \xrightarrow{n\to\infty} \mathrm{e}^{-\frac{t^2}{2}}.$$

Since we know that the characteristic function of a standard Gaussian is $\mathrm{e}^{-\frac{t^2}{2}}$ (see Section 2.4), we get the claim.

Now the more interesting question is what kind of dependence structure one can put on a family of random variables $(Z_n)_{n\geq 1}$?

This question will be discussed in detail in this part and will lead to two very important notions in probability theory:

- the notion of *martingales*;
- the notion of *Markov chains* (ergodicity).

The additional notion that we get is that n is a representative for *time*, i.e., one can imagine stochastic processes changing with time (time evolution).

Hence, we consider tuples of the form $(Z_n, \mathcal{F}_n)_{n\geq 1}$, where $\mathcal{F}_n$ is a σ-algebra for all $n \geq 1$. It is often important to write the tuple down and to emphasize the $\mathcal{F}_n$'s. Assume that the space of events in infinite time steps is known and denote it by $\mathcal{F}$. Then one will see that $\mathcal{F}_n \subset \mathcal{F}$ and moreover $\mathcal{F}_n \subset \mathcal{F}_{n+1}$. This is a very important fact for the study of stochastic processes and is known as *filtration.*

Another very important thing is the notion of *conditional expectation* and *conditional distribution*, which we will cover as the first chapter of this part.

Chapter 3

Conditional Expectations

3.1. $L^2(\Omega, \mathcal{F}, \mathbb{P})$ as a Hilbert Space and Orthogonal Projections

In this section, we will always work on a probability space $(\Omega, \mathcal{F}, \mathbb{P})$. Consider the space $L^2(\Omega, \mathcal{F}, \mathbb{P})$, which is given by

$$L^2(\Omega, \mathcal{F}, \mathbb{P}) := \{X : \Omega \to \mathbb{R} \text{ a random variable} \mid \mathbb{E}[X^2] < \infty\}.$$

More precisely, it consists only of equivalence classes, i.e., X and Y are identified if $X = Y$ a.s. We also know that we have a norm on this space given by

$$\|X\|_2 = \mathbb{E}[X^2]^{1/2}.$$

Remark 3.1.0.1. Recall that on $\mathcal{L}^2(\Omega, \mathcal{F}, \mathbb{P})$ this would only be a *semi-norm* rather than a norm.

We can also define an inner product on $L^2(\Omega, \mathcal{F}, \mathbb{P})$ by

$$\langle X, Y\rangle := \mathbb{E}[XY].$$

One can easily check that this satisfies the conditions of an inner product. It remains to show that $|\langle X, Y\rangle| < \infty$. By the Cauchy–Schwarz inequality, we get

$$|\langle X, Y\rangle| \leq \|X\|_2\|Y\|_2 = \mathbb{E}[X^2]^{1/2}\mathbb{E}[Y^2]^{1/2},$$

and since we assume that the second moment of our random variables exists, this is finite. Moreover, one can see that

$$\sqrt{\langle X, X\rangle} = \mathbb{E}[X^2]^{1/2} = \|X\|_2.$$

Now we know that $L^2(\Omega, \mathcal{F}, \mathbb{P})$ is a Banach space and therefore a complete, normed vector space, i.e., every Cauchy sequence has a limit inside the space with respect to the norm. Since we have also an inner product on $L^2(\Omega, \mathcal{F}, \mathbb{P})$, it is also a *Hilbert space*. We shall recall what a Hilbert space is.

Definition 3.1.0.2 (Hilbert space). An inner product space $(\mathcal{H}, \langle \cdot, \cdot \rangle)$ is called a *Hilbert space*, if it is complete and the norm is derived from the inner product. If it is not complete it is called a *pre-Hilbert space.*

Remark 3.1.0.3. It is in fact important that we have written out $L^2(\Omega, \mathcal{F}, \mathbb{P})$, for instance if $\mathcal{G}$ is a σ-algebra and $\mathcal{G} \subset \mathcal{F}$, then $L^2(\Omega, \mathcal{G}, \mathbb{P}) \subset L^2(\Omega, \mathcal{F}, \mathbb{P})$. When we have several σ-algebras, we write explicitly the dependence of them, by writing for instance $L^2(\Omega, \mathcal{G}, \mathbb{P}), L^2(\Omega, \mathcal{F}, \mathbb{P})$, etc.

Example 3.1.0.4. Take $\mathcal{H} = \mathbb{R}^n$. This is indeed a Hilbert space with the Euclidean inner product, i.e., if $X, Y \in \mathbb{R}^n$ then

$$\langle X, Y \rangle = \sum_{i=1}^{n} X_i Y_i.$$

Be aware that this is an example of a finite-dimensional Hilbert space, but $L^2(\mathbb{P})$ is a infinite-dimensional Hilbert space.

Example 3.1.0.5. Take $\mathcal{H} = \ell^2(\mathbb{N})$, the space of *square summable sequences* a subspace of $\ell^\infty(\mathbb{N})$ which is the space of convergent sequences, i.e.,

$$\ell^2(\mathbb{N}) := \left\{ x = (x_n)_{n \geq 1} \in \ell^\infty(\mathbb{N}) \,\middle|\, \sum_{n=1}^{\infty} |x_n|^2 < \infty \right\}.$$

This is also an example of an infinite-dimensional Hilbert space. It is clearly related to the $L^p(\mu)$ spaces, where we use the counting measure.

Example 3.1.0.6 (Sobolev spaces). *Sobolev spaces*, usually denoted by H^s or $W^{s,2}$, are Hilbert spaces. These are a special kind of a function space in which differentiation may be performed, but that support the structure of an inner product. Because differentiation is permitted, Sobolev spaces are a convenient setting for the theory of partial differential equations. They also form the basis of the theory of direct methods in the calculus of variations. For s a non-negative integer and $\Omega \subset \mathbb{R}^n$, the Sobolev space $H^s(\Omega)$ contains L^2-functions whose weak derivatives of order up to s are also in L^2. The inner product in $H^s(\Omega)$ is

$$\langle f, g\rangle_{H^s(\Omega)} := \int_\Omega f(x)\bar{g}(x)\mathrm{d}\mu(x) + \int_\Omega Df(x)\cdot D\bar{g}(x)\mathrm{d}\mu(x) + \cdots + \int_\Omega D^s f(x)\cdot D^s\bar{g}(x)\mathrm{d}\mu(x),$$

where the dot indicates the dot product in the Euclidean space of partial derivatives of each order. In fact, Sobolev spaces can also be defined when s is not an integer.

Example 3.1.0.7 (Hardy spaces). The *Hardy spaces* are function spaces, arising in complex analysis and harmonic analysis, whose elements are certain holomorphic functions in a complex domain. Let U denote the unit disc in the complex plane. Then the Hardy space $\mathcal{H}^2(U)$ is defined as the space of holomorphic functions f on U such that the means

$$M_r(f) = \frac{1}{2\pi}\int_0^{2\pi} |f(re^{i\theta})|^2 \mathrm{d}\theta$$

remains bounded for $r < 1$. The norm on this Hardy space is defined by

$$\|f\|_{\mathcal{H}^2(U)} = \lim_{r\to 1}\sqrt{M_r(f)}.$$

Hardy spaces in the disc are related to *Fourier series.* A function f is in $\mathcal{H}^2(U)$ if and only if $f(z) = \sum_{n=0}^\infty a_n z^n$, where $\sum_{n=0}^\infty |a_n|^2 < \infty$. Thus $\mathcal{H}^2(U)$ consists of those functions that are L^2 on the circle and whose negative frequency Fourier coefficients vanish.

Example 3.1.0.8 (Bergman spaces). The *Bergman spaces* are another family of Hilbert spaces of holomorphic functions. Let D be a bounded open set in the complex plane (or a higher-dimensional complex space) and let $L^{2,h}(D)$ be the space of holomorphic functions f in D that are also in $L^2(D)$ in the sense that

$$\|f\|^2 = \int_D |f(z)|^2 \mathrm{d}\mu(z) < \infty,$$

where the integral is taken with respect to the Lebesgue measure in D. Clearly $L^{2,h}(D)$ is a subspace of $L^2(D)$; in fact, it is a closed subspace and so a Hilbert space in its own right. This is a consequence of the estimate, valid on compact subsets K of D, that

$$\sup_{z\in K} |f(z)| \leq C_K \|f\|_2,$$

which in turn follows from Cauchy's integral formula. Thus convergence of a sequence of holomorphic functions in $L^2(D)$ implies also compact convergence and so the function is also holomorphic. Another consequence of this inequality is that the linear functional that evaluates a function f at a point of D is actually continuous on $L^{2,h}(D)$. The *Riesz representation theorem* implies that the evaluation functional can be represented as an element of $L^{2,h}(D)$. Thus, for every $z \in D$, there is a function $\eta_z \in L^{2,h}(D)$ such that

$$f(z) = \int_D f(\zeta)\overline{\eta_z(\zeta)}\mathrm{d}\mu(\zeta),$$

for all $f \in L^{2,h}(D)$. The integrand $K(\zeta, z) = \overline{\eta_z(\zeta)}$ is known as the *Bergman kernel* of D. This integral kernel satisfies a reproducing property

$$f(z) = \int_D f(\zeta)K(\zeta, z)\mathrm{d}\mu(\zeta).$$

A Bergman space is an example of a reproducing kernel Hilbert space, which is a Hilbert space of functions along with a kernel $K(\zeta, z)$ that verifies a reproducing property analogous to this one. The Hardy space $\mathcal{H}^2(D)$ also admits a reproducing kernel, known as the *Szegő kernel.* Reproducing kernels are common in other areas of mathematics as well. For instance, in harmonic analysis the Poisson kernel is a reproducing kernel for the Hilbert space of square integrable harmonic functions in the unit ball. That the latter is a Hilbert space at all is a consequence of the mean value theorem for harmonic functions.

Remark 3.1.0.9. The notion of Hilbert spaces allows us to do basic geometry on them. Even if our space is an infinite-dimensional vector space, we can still make sense of geometrical meanings, for example orthogonality, only by using the inner product on the space.

Definition 3.1.0.10 (Orthogonal). Two elements X, Y in a Hilbert space $(\mathcal{H}, \langle\cdot,\cdot\rangle)$ are said to be *orthogonal* if

$$\langle X, Y\rangle = 0.$$

Remark 3.1.0.11. For a real-valued Hilbert space $\mathcal{H}$ we get the following identity. For every $X, Y \in \mathcal{H}$

$$\begin{aligned}\|X+Y\|^2 &= \langle X+Y, X+Y\rangle = \langle X,X\rangle + \langle X,Y\rangle + \langle Y,X\rangle + \langle Y,Y\rangle\\ &= \|X\|^2 + \|Y\|^2 + 2\langle X,Y\rangle,\end{aligned}$$

and if X and Y are orthogonal, i.e., $\langle X,Y\rangle = 0$, we get the usual Pythagorean relation

$$\|X+Y\|^2 = \|X\|^2 + \|Y\|^2.$$

Theorem 3.1.0.1. *Let $(X_n)_{n\geq 1}$ and $(Y_n)_{n\geq 1}$ be two converging sequences in a Hilbert space $\mathcal{H}$ such that $X_n \xrightarrow{n\to\infty} X$ and $Y_n \xrightarrow{n\to\infty} Y$. Then*

$$\langle X_n, Y_n\rangle \xrightarrow{n\to\infty} \langle X, Y\rangle.$$

(In particular, $X_n = Y_n$ gives us that $\|X_n\| \xrightarrow{n\to\infty} \|X\|$).

Proof. We can look at the difference, which is given by

$$\begin{aligned}|\langle X,Y\rangle - \langle X_n,Y_n\rangle| &= |\langle X-X_n, Y\rangle + \langle X_n, Y\rangle - \langle X_n, Y_n\rangle|\\ &= |\langle X-X_n, Y\rangle + \langle X_n, Y-Y_n\rangle|\\ &\leq |\langle X-X_n, Y\rangle| + |X_n, Y-Y_n\rangle|\\ &\leq \|X-X_n\|\|Y\| + \|X_n\|\|Y-Y_n\|,\end{aligned}$$

where we have used the Cauchy–Schwarz inequality. Now since for n large enough there is some $\varepsilon > 0$ such that $\|X - X_n\| < \varepsilon$ and $\|Y - Y_n\| < \varepsilon$ by assumption, and the fact that $\|X_n\|$ is bounded independently of n, we get the claim. □

Lemma 3.1.0.2 (Parallelogram identity). *Let $\mathcal{H}$ be a Hilbert space. For all $X, Y \in \mathcal{H}$ we get*

$$\|X+Y\|^2 + \|X-Y\|^2 = 2(\|X\|^2 + \|Y\|^2).$$

Moreover, if a norm satisfies the parallelogram identity, it can be derived from an inner product.

Exercise 3.1.0.12. Prove[1] Lemma 3.1.0.2.

Definition 3.1.0.13 (Closed linear subset). Let $\mathcal{H}$ be a Hilbert space and let $\mathcal{L} \subset \mathcal{H}$ be a linear subset. $\mathcal{L}$ is called *closed* if for every sequence $(X_n)_{n\geq 1}$ in $\mathcal{L}$ with $X_n \xrightarrow{n\to\infty} X$ we get that $X \in \mathcal{L}$.

Theorem 3.1.0.3. *Let $\mathcal{H}$ be a Hilbert space and let $\Gamma \subset \mathcal{H}$ be a subset. Let $\Gamma^\perp$ denote the set of all elements of $\mathcal{H}$ which are orthogonal to Γ, i.e.,*

$$\Gamma^\perp = \{X \in \mathcal{H} \mid \langle X, \gamma \rangle = 0, \forall \gamma \in \Gamma\}.$$

Then $\Gamma^\perp$ is a closed subspace of $\mathcal{H}$. We call $\Gamma^\perp$ the orthogonal complement *of Γ.*

Proof. Let $\alpha, \beta \in \mathbb{R}$ and $X, X' \in \Gamma^\perp$. It is clear that for all $Y \in \Gamma$

$$\langle \alpha X + \beta X', Y \rangle = 0.$$

Hence $\Gamma^\perp$ is a linear subspace of $\mathcal{H}$. Next we want to check whether it is closed. Take a sequence $(X_n)_{n\geq 1}$ in $\Gamma^\perp$ such that $X_n \xrightarrow{n\to\infty} X$ with $X \in \mathcal{H}$. Now for all $Y \in \Gamma$ we get $\langle X_n, Y \rangle = 0$ and $\langle X_n, Y \rangle \xrightarrow{n\to\infty} \langle X, Y \rangle$ because of Theorem 3.1.0.1. Hence $\langle X, Y \rangle = 0$ and therefore $X \in \Gamma^\perp$ and the claim follows. □

Definition 3.1.0.14 (Distance to a closed subspace). Let $\mathcal{H}$ be a Hilbert space and let $X \in \mathcal{H}$. Moreover let $\mathcal{L} \subset \mathcal{H}$ be a closed subspace. The *distance* of X to $\mathcal{L}$ is given by

$$\mathsf{d}(X, \mathcal{L}) := \inf_{Y \in \mathcal{L}} \|X - Y\| = \inf\{\|X - Y\| \mid Y \in \mathcal{L}\}.$$

Remark 3.1.0.15. Since $\mathcal{L}$ is closed, $X \in \mathcal{L}$ if and only if $\mathsf{d}(X, \mathcal{L}) = 0$.

3.1.1. Convex sets in uniformly convex spaces

While the emphasis in this section is on Hilbert spaces, it is useful to isolate a more abstract property which is precisely what is needed for several proofs.

[1] Use the fact $\|X + Y\|^2 = \langle X + Y, X + Y \rangle$.

Definition 3.1.1.1 (Uniformly convex vector space). A normed vector space $(V, \|\cdot\|)$ is called *uniformly convex* if for $X, Y \in V$

$$\|X\|, \|Y\| \leq 1 \Rightarrow \left\|\frac{X+Y}{2}\right\| \leq 1 - \psi(\|X-Y\|),$$

where $\psi : [0,2] \to [0,1]$ is a monotonically increasing function with $\psi(r) > 0$ for all $r > 0$.

Lemma 3.1.1.1. *A Hilbert space $(\mathcal{H}, \langle\cdot,\cdot\rangle)$ is uniformly convex.*

Proof. For $X, Y \in \mathcal{H}$ with $\|X\|, \|Y\| \leq 1$ then by *parallelogram identity* (Lemma 3.1.0.2) we have

$$\begin{aligned}\left\|\frac{X+Y}{2}\right\| &= \sqrt{\frac{1}{2}\|X\|^2 + \frac{1}{2}\|Y\|^2 - \frac{1}{2}\|X-Y\|^2} \\ &\leq \sqrt{1 - \frac{1}{2}\|X-Y\|^2} = 1 - \psi(\|X-Y\|)\end{aligned}$$

as required, with $\psi(r) = 1 - \sqrt{1 - \frac{1}{2}r^2}$. □

Heuristically, we can think of Definition 3.1.1.1 as having the following geometrical meaning. If vectors X and Y have norm (length) one, then their mid-point $\frac{X+Y}{2}$ has much smaller norm unless X and Y are very close together. This accords closely with the geometrical intuition from finite-dimensional spaces with Euclidean distance. The following theorem, whose conclusion is illustrated in the figure, will have many important consequences for the study of Hilbert spaces.

Theorem 3.1.1.2 (Unique approximation of a closed convex set). *Let $(V, \|\cdot\|)$ be a Banach space with a uniformly convex norm, let $K \subset V$ be a closed convex subset and assume that $v_0 \in V$. Then there exists a unique element $w \in K$ that is closest to v_0 in the sense that w is the only element of K with*

$$\|w - v_0\| = \mathsf{d}(v_0, K) = \inf_{k \in K} \|k - v_0\|.$$

Proof. By translating both the set K and the point v_0 by $-v_0$, we may assume without loss of generality that $v_0 = 0$. We define

$$s = \inf_{k \in K} \|k - v_0\| = \inf_{k \in K} \|k\|.$$

If $s = 0$, then we must have $0 \in K$ since K is closed and the only choice is then $w = v_0 = 0$ (the uniqueness of w is a consequence of the strict positivity of the norm). So assume that $s > 0$. By multiplying by the scalar $\frac{1}{s}$ we have found a point $w \in K$ with norm 1, then its uniqueness is an immediate consequence of the uniform convexity: if $w_1, w_2 \in K$ have $\|w_1\| = \|w_2\| = 1$, then $\frac{w_1+w_2}{2} \in K$ because K is convex. Also, $\left\|\frac{w_1+w_2}{2}\right\| = 1$ by the triangle inequality and since $s = 1$. By uniform convexity this implies that $w_1 = w_2$. Turning to the existence, let us first sketch the argument. Choose a sequence (k_n) in K with $\|k_n\| \to 1$ as $n \to \infty$. Then the midpoints $\frac{k_n+k_m}{2}$ also lie in K, since K is convex. However, this shows that the midpoint must have norm greater than or equal to 1, since $s = 1$. Therefore, k_n and k_m must be close together by uniform convexity. Making this precise, we will see that (k_n) is a Cauchy sequence. Since V is complete and K is closed, this will give a point $w \in K$ with $\|w\| = 1 = s$ as required. To make this more precise, we apply uniform convexity to the normalized vectors

$$x_n = \frac{1}{s_n} k_n,$$

where $s_n = \|k_n\|$. The midpoint of x_n and x_m can now be expressed as

$$\frac{x_m + x_n}{2} = \frac{1}{2s_m} k_m + \frac{1}{2s_n} k_n = \frac{1}{2s_m} + \frac{1}{2s_n}(ak_m + bk_n)$$

with

$$a = \frac{\frac{1}{2s_m}}{\frac{1}{2s_m} + \frac{1}{2s_n}} \geq 0,$$

$$b = \frac{\frac{1}{2s_n}}{\frac{1}{2s_m} + \frac{1}{2s_n}} \geq 0$$

and $a + b = 1$. Therefore, $ak_m + bk_n \in K$ by convexity and so

$$\left\|\frac{x_m + x_n}{2}\right\| = \left(\frac{1}{2s_m} + \frac{1}{2s_n}\right) \|ak_m + bk_n\| \geq \frac{1}{2s_m} + \frac{1}{2s_n}.$$

Let ψ be as in Definition 3.1.1.1 and fix $\varepsilon > 0$. Choose $N = N(\varepsilon)$ large enough to ensure that $m \geq N$ implies that

$$\frac{1}{s_m} \geq 1 - \psi(\varepsilon).$$

Then $m, n \geq N$ implies that

$$\frac{1}{2s_m} + \frac{1}{s_n} \geq 1 - \psi(\varepsilon),$$

which together with the definition of uniform convexity gives

$$1 - \psi(\|x_m - x_n\|) \geq \left\| \frac{x_m + x_n}{2} \right\| \geq 1 - \psi(\varepsilon).$$

By monotonicity of the function ψ this implies for all $m, n \geq N$ that $\|x_m - x_n\| \leq \varepsilon$, showing that (x_n) is a Cauchy sequence. As V is assumed to be complete, we deduce that (x_n) converges to some $x \in V$. Since $s_n \to 1$ and $k_n \to s_n x_n$ as $n \to \infty$ it follows that $\lim_{n\to\infty} k_n = x$. As K is closed the limit x belongs to K and by contradiction is an (and hence is the unique) element closest to $v_0 = 0$. □

Remark 3.1.1.2. This unique approximation is clearly true for Hilbert spaces, since they are uniformly convex spaces.

Corollary 3.1.1.3 (Orthogonal decomposition). *Let $\mathcal{H}$ be a Hilbert space and let $\mathcal{L} \subset \mathcal{H}$ be a closed subspace. Then $\mathcal{L}^\perp$ is a closed subspace with*

$$\mathcal{H} = \mathcal{L} \oplus \mathcal{L}^\perp,$$

meaning that every element $H \in \mathcal{H}$ can be written in the form

$$H = Y + Z$$

with $Y \in \mathcal{L}$ and $Z \in \mathcal{L}^\perp$ and Y and Z are unique with these properties. Moreover, $Y = (Y^\perp)^\perp$ and

$$\|H\|^2 = \|Y\|^2 + \|Z\|^2$$

if $H = Y + Z$ with $Y \in \mathcal{L}$ and $Z \in \mathcal{L}^\perp$.

Proof. As $H \mapsto \langle H, Y\rangle$ is a (continuous linear) functional for each $Y \in \mathcal{L}$, the set $\mathcal{L}^\perp$ is an intersection of closed subspaces and hence is a closed subspace. Using positivity of the inner product, it is easy to see that $\mathcal{L} \cap \mathcal{L}^\perp = \{0\}$ and from this the uniqueness of the decomposition

$$H = Y + Z$$

with $Y \in \mathcal{L}$ and $Z \in \mathcal{L}^\perp$ follows at once. So it remains to show the existence of this decomposition. Fix $H \in \mathcal{H}$ and apply the theorem of unique approximation with $K = \mathcal{L}$ to find a point $Y \in \mathcal{L}$ that is closest to h. Let $Z = H - Y$, so that for any $v \in \mathcal{L}$ and any scalar t we have

$$\|Z\|^2 \leq \|H - \underbrace{(tv + Y)}_{\in \mathcal{L}}\|^2 = \|Z - tv\|^2 = \|Z\|^2 - 2t\langle v, Z\rangle + |t|^2\|Z\|^2.$$

However, this shows that $t\langle v, Z\rangle = 0$ for all scalars t and $v \in \mathcal{L}$ and so $\langle v, Z\rangle = 0$ for all $v \in \mathcal{L}$. Thus $Z \in \mathcal{L}^\perp$ and hence

$$\|H\|^2 = \langle H, H\rangle = \langle Y + Z, Y + Z\rangle = \|Y\|^2 + \|Z\|^2.$$

It is clear from the definitions that $\mathcal{L} \subset (\mathcal{L}^\perp)^\perp$. If $v \in (\mathcal{L}^\perp)^\perp$ then

$$v = Y + Z$$

for some $Y \in \mathcal{L}$ and $Z \in \mathcal{L}^\perp$ by the first part of the proof. However,

$$0 = \langle v, Z\rangle = \|Z\|^2$$

implies that $v = Y$ and so $\mathcal{L} = (\mathcal{L}^\perp)^\perp$. □

3.1.2. Orthogonal projection

Let again $\mathcal{H}$ be a Hilbert space. The projection of an element $X \in \mathcal{H}$ onto a closed subspace $\mathcal{L} \subset \mathcal{H}$ is the unique point $Y \in \mathcal{L}$ such that

$$\mathsf{d}(X, \mathcal{L}) = \|X - Y\|.$$

We denote this projection by

$$\Pi : \mathcal{H} \to \mathcal{L}$$

$$X \mapsto \Pi X = Y$$

Theorem 3.1.2.1. *Let $\mathcal{H}$ be a Hilbert space and let $\mathcal{L} \subset \mathcal{H}$ be a closed subspace. Then the projection operator Π of $\mathcal{H}$ onto $\mathcal{L}$ satisfies*

(i) $\Pi^2 = \Pi$;

(ii) $\begin{cases} \Pi X = X, & X \in \mathcal{L}, \\ \Pi X = 0, & X \in \mathcal{L}^\perp, \end{cases}$;

(iii) $(X - \Pi X) \perp \mathcal{L}$ *for all* $X \in \mathcal{H}$.

Proof. (i) Is clear. The first statement of (ii) is clear from (i). For the second statement of (ii), if $X \in \mathcal{L}^\perp$, then for $Y \in \mathcal{L}$ we get

$$\|X - Y\|^2 = \|X\|^2 + \|Y\|^2.$$

This is going to be minimized if $Y = 0$. Hence, $\Pi X = 0$.

(iii) If $Y \in \mathcal{L}$ we get

$$\|X - \underbrace{\Pi X}_{\in \mathcal{L}}\|^2 \leq \|X - \underbrace{\Pi X - Y}_{\in \mathcal{L}}\|^2 = \|Y\|^2 + \|X - \Pi X\|^2 - 2\langle X - \Pi X, Y\rangle.$$

Therefore

$$2\langle X - \Pi X, Y\rangle \leq \|Y\|^2 \tag{3.1.2.1}$$

for all $Y \in \mathcal{L}$. Now since $\mathcal{L}$ is a linear space we get that for all $\alpha > 0$

$$\alpha Y \in \mathcal{L}.$$

So in particular, (3.1.2.1) is true when Y is replaced by αY. Therefore, we get

$$2\langle X - \Pi X, Y\rangle \leq \alpha\|Y\|^2.$$

Now let $\alpha \to 0$. Then we obtain

$$\langle X - \Pi X, Y\rangle \leq 0$$

for all $Y \in \mathcal{L}$. Since $-Y \in \mathcal{L}$ we get that

$$-\langle X - \Pi X, Y\rangle \leq 0.$$

But this means $\langle X - \Pi X, Y\rangle = 0$ and the claim follows. □

Corollary 3.1.2.2. *Let* $\mathcal{H}$ *be a Hilbert space and let* $\mathcal{L} \subset \mathcal{H}$ *be a closed subspace. Moreover let* Π *be the projection operator of* $\mathcal{H}$ *onto* $\mathcal{L}$. *Then*

$$X = (X - \Pi X) + \Pi X$$

is the unique representation of X *as the sum of an element of* $\mathcal{L}$ *and an element of* $\mathcal{L}^\perp$.

Proof. This is a direct consequence of Corollary 3.1.1.3. □

Remark 3.1.2.1. The uniqueness of the projection operator implies that for $X_1 \in \mathcal{L}$ and $X_2 \in \mathcal{L}^\perp$ we get

$$\Pi(X_1 + X_2) = X_1.$$

Corollary 3.1.2.3. *Let $\mathcal{H}$ be a Hilbert space and let $\mathcal{L} \subset \mathcal{H}$ be a closed subspace. Moreover, let Π be the projection operator of $\mathcal{H}$ onto $\mathcal{L}$. Then we have the following conditions:*

(i) $\langle \Pi X, Y\rangle = \langle X, \Pi Y\rangle$ *for all* $X, Y \in \mathcal{H}$.
(ii) $\Pi(\alpha X + \beta Y) = \alpha \Pi X + \beta \Pi Y$ *for all* $\alpha, \beta \in \mathbb{R}$ *and* $X, Y \in \mathcal{H}$.

Proof. (i) Let $X, Y \in \mathcal{H}$, $X = X_1 + X_2$ with $X_1 \in \mathcal{L}$ and $X_2 \in \mathcal{L}^\perp$ and $Y = Y_1 + Y_2$ with $Y_1 \in \mathcal{L}$ and $Y_2 \in \mathcal{L}^\perp$. Then we get

$$\langle \Pi X, Y\rangle = \langle \Pi(X_1 + X_2), Y_1 + Y_2\rangle = \langle X_1, Y_1 + Y_2\rangle = \langle X_1, Y_1\rangle,$$
$$\langle X, \Pi Y\rangle = \langle X_1 + X_2, \Pi(Y_1 + Y_2)\rangle = \langle X_1 + X_2, Y_1\rangle = \langle X_1, Y_1\rangle.$$

Therefore, they are the same.

(ii) Take $\alpha, \beta \in \mathbb{R}$ and look at

$$\alpha X + \beta Y = \underbrace{(\alpha X_1 + \beta Y_1)}_{\in \mathcal{L}} + \underbrace{(\alpha X_2 + \beta Y_2)}_{\in \mathcal{L}^\perp}.$$

Hence, we get

$$\Pi(\alpha X + \beta Y) = \alpha \Pi(X_1 + X_2) + \beta \Pi(Y_1 + Y_2) = \alpha \Pi X + \beta \Pi Y. \quad \square$$

3.2. Conditional Expectation

3.2.1. Review of conditional probability

Let $(\Omega, \mathcal{F}, \mathbb{P})$ be a probability space and let $A, B \in \mathcal{F}$ such that $\mathbb{P}(B) > 0$. Then the conditional probability (see Section 2.5.2) of A given B is defined as

$$\mathbb{P}(A \mid B) = \frac{\mathbb{P}(A \cap B)}{\mathbb{P}(B)}.$$

The important fact here is that the map $\mathcal{F} \to [0, 1]$, $A \mapsto \mathbb{P}(A \mid B)$ defines a new probability measure on $\mathcal{F}$ called the *conditional probability* given B. There are several facts, which we need to recall:

(i) If $A_1, \ldots, A_n \in \mathcal{F}$ and if $\mathbb{P}\left(\bigcap_{k=1}^n A_k\right) > 0$, then

$$\mathbb{P}\left(\bigcap_{k=1}^{n} A_k\right) = \prod_{j=1}^{n} \mathbb{P}\left(A_j \,\middle|\, \bigcap_{k=1}^{j-1} A_k\right).$$

(ii) Let $(E_n)_{n\geq 1}$ be a measurable partition of Ω, i.e., for all $n \geq 1$ we have that $E_n \in \mathcal{F}$ and for $n \neq m$ we get $E_n \cap E_m = \emptyset$ and $\bigcup_{n\geq 1} E_n = \Omega$. Then, for all $A \in \mathcal{F}$, we get

$$\mathbb{P}(A) = \sum_{n\geq 1} \mathbb{P}(A \mid E_n)\mathbb{P}(E_n).$$

(iii) (Bayes' theorem, Theorem 2.5.2.4) Let $(E_n)_{n\geq 1}$ be a measurable partition of Ω and $A \in \mathcal{F}$ with $\mathbb{P}(A) > 0$. Then

$$\mathbb{P}(E_n \mid A) = \frac{\mathbb{P}(A \mid E_n)\mathbb{P}(E_n)}{\sum_{m\geq 1} \mathbb{P}(A \mid E_m)\mathbb{P}(E_m)}.$$

Remark 3.2.1.1. We can reformulate the definition of the conditional probability (Definition 2.5.2.1) to obtain

$$\mathbb{P}(A \mid B)\mathbb{P}(B) = \mathbb{P}(A \cap B), \tag{3.2.1.1}$$

$$\mathbb{P}(B \mid A)\mathbb{P}(A) = \mathbb{P}(A \cap B). \tag{3.2.1.2}$$

Therefore, one can prove statements (i), (ii) and (iii) by using equations[2] (3.2.1.1) and (3.2.1.2).

3.2.2. Discrete construction of the conditional expectation

Let X and Y be two random variables on a probability space $(\Omega, \mathcal{F}, \mathbb{P})$. Let Y take values in $\mathbb{R}$ and X take values in a countable discrete set $\{x_1, x_2, \ldots, x_n, \ldots\}$. The goal is to describe the expectation of the random variable Y by knowing the observed random

[2] One should also note that if A and B are two independent events, then $\mathbb{P}(A \mid B) = \frac{\mathbb{P}(A\cap B)}{\mathbb{P}(B)} = \frac{\mathbb{P}(A)\mathbb{P}(B)}{\mathbb{P}(B)} = \mathbb{P}(A)$.

variable X. For instance, let $X = x_j \in \{x_1, x_2, \ldots, x_n, \ldots\}$. Therefore, we look at a set $\{\omega \in \Omega \mid X(\omega) = x_j\}$ rather than looking at whole Ω. For $\Lambda \in \mathcal{F}$, we thus define

$$\mathbb{Q}(\Lambda) := \mathbb{P}(\Lambda \mid \{X = x_j\}),$$

a new probability measure $\mathbb{Q}$, with $\mathbb{P}(X = x_j) > 0$. Therefore, it makes more sense to compute

$$\mathbb{E}_{\mathbb{Q}}[Y] = \int_{\Omega} Y(\omega)\mathrm{d}\mathbb{Q}(\omega) = \int_{\{\omega \in \Omega \mid X(\omega) = x_j\}} Y(\omega)\mathrm{d}\mathbb{P}(\omega)$$

rather than

$$\mathbb{E}_{\mathbb{P}}[Y] = \int_{\Omega} Y(\omega)\mathrm{d}\mathbb{P}(\omega) = \int_{\mathbb{R}} y\mathrm{d}\mathbb{P}_Y(y).$$

Definition 3.2.2.1 (Conditional expectation (X discrete, Y real-valued, single-valued case)). Let $(\Omega, \mathcal{F}, \mathbb{P})$ be a probability space. Let $X : \Omega \to \{x_1, x_2, \ldots, x_n, \ldots\}$ be a random variable taking values in a discrete set and let Y be a real-valued random variable on that space. If $\mathbb{P}(X = x_j) > 0$, we can define the conditional expectation of Y given $\{X = x_j\}$ to be

$$\mathbb{E}[Y \mid X = x_j] = \mathbb{E}_{\mathbb{Q}}[Y],$$

where $\mathbb{Q}$ is the probability measure on $\mathcal{F}$ defined by

$$\mathbb{Q}(\Lambda) = \mathbb{P}(\Lambda \mid X = x_j),$$

for $\Lambda \in \mathcal{F}$, provided that $\mathbb{E}_{\mathbb{Q}}[|Y|] < \infty$.

Theorem 3.2.2.1 (Conditional expectation (X discrete, Y discrete, single-valued case)). *Let $(\Omega, \mathcal{F}, \mathbb{P})$ be a probability space. Let X be a random variable on that space with values in $\{x_1, x_2, \ldots, x_n, \ldots\}$ and let Y also be a random variable with values in $\{y_1, y_2, \ldots, y_n, \ldots\}$. If $\mathbb{P}(X = x_j) > 0$, we can write the conditional expectation of Y given $\{X = x_j\}$ as*

$$\mathbb{E}[Y \mid X = x_j] = \sum_{k=1}^{\infty} y_k \mathbb{P}(Y = y_k \mid X = x_j)$$

provided that the series is absolutely convergent.

Proof. Apply the definitions above to obtain

$$\begin{aligned}\mathbb{E}[Y \mid X = x_j] = \mathbb{E}_{\mathbb{Q}}[Y] &= \sum_{k=1}^{\infty} y_k \mathbb{Q}(Y = y_k) \\ &= \sum_{k=1}^{\infty} y_k \mathbb{P}(Y = y_k \mid X = x_j). \qquad \square\end{aligned}$$

Now let again X be a random variable with values in $\{x_1, x_2, \ldots, x_n, \ldots\}$ and Y be a real-valued random variable. The next step is to define $\mathbb{E}[Y \mid X]$ as a function $f(X)$. Therefore, we introduce the function

$$\begin{aligned} f &: \{x_1, x_2, \ldots, x_n, \ldots\} \to \mathbb{R}, \\ f(x) &:= \begin{cases} \mathbb{E}[Y \mid X = x], & \mathbb{P}(X = x) > 0, \\ \text{any value in } \mathbb{R}, & \mathbb{P}(X = x) = 0. \end{cases} \end{aligned} \tag{3.2.2.1}$$

Remark 3.2.2.2. It does not matter which value we assign to f for $\mathbb{P}(X = x) = 0$, since it does not affect the expectation because it is defined on a null set. For convention we want to assign to it the value 0.

Definition 3.2.2.3 (Conditional expectation (X discrete, Y real-valued, complete case)). Let $(\Omega, \mathcal{F}, \mathbb{P})$ be a probability space. Let X be a countably valued random variable and let Y be a real-valued random variable. The conditional expectation of Y given X is defined by

$$\mathbb{E}[Y \mid X] = f(X),$$

with f as in (3.2.2.1), provided that for all j if $\mathbb{Q}_j(\Lambda) = \mathbb{P}(\Lambda \mid X = x_j)$, with $\mathbb{P}(X = x_j) > 0$, we get $\mathbb{E}_{\mathbb{Q}_j}[|Y|] < \infty$.

Remark 3.2.2.4. The above definition does not define $\mathbb{E}[Y \mid X]$ everywhere but rather almost everywhere, since on each set $\{X = x\}$, where $\mathbb{P}(X = x) = 0$, its value is arbitrary.

Example 3.2.2.5. Let $X \sim \Pi(\lambda)$ be a Poisson distributed random variable (see Example 1.2.2.1) with parameter λ. Let us consider a tossing game, where we say that when $X = n$, we do n independent tossing of a coin where each time one obtains 1 with probability

$p \in [0,1]$ and 0 with probability $1-p$. Define also S to be the random variable giving the total number of 1 obtained in the game. Therefore, if $X = n$ is given, we get that S is binomially distributed with parameters (p, n) (see Section 2.2.1.3). We want to compute

(i) $\mathbb{E}[S \mid X]$,
(ii) $\mathbb{E}[X \mid S]$.

Remark 3.2.2.6. It is more natural to ask for the expectation of the amount of 1 obtained for the whole game by knowing how many games were played. The reverse is a bit more difficult. Logically, we may also notice that it definitely does not make sense to say $S \geq X$, because we cannot obtain more wins in a game than the amount of games that were played.

(i) First we compute $\mathbb{E}[S \mid X = n]$: If $X = n$, we know that S is binomially distributed with parameters (p, n) ($S \sim \mathcal{B}(p, n)$) and therefore we already know[3]

$$\mathbb{E}[S \mid X = n] = pn.$$

Now we need to identify the function f defined as in (3.2.2.1) by

$$f : \mathbb{N} \to \mathbb{R},$$
$$n \mapsto pn.$$

Therefore, we get by definition

$$\mathbb{E}[S \mid X] = pX.$$

(ii) Next we want to compute $\mathbb{E}[X \mid S = k]$: For $n \geq k$ we have

$$\begin{aligned}\mathbb{P}(X = n \mid S = k) &= \frac{\mathbb{P}(S = k \mid X = n)\mathbb{P}(X = n)}{\mathbb{P}(S = k)} \\ &= \frac{\binom{n}{k} p^k (1-p)^{n-k} \mathrm{e}^{-\lambda} \frac{\lambda^n}{n!}}{\sum_{m=k}^{\infty} \binom{m}{k} p^k (1-p)^{m-k} \mathrm{e}^{-\lambda} \frac{\lambda^m}{m!}},\end{aligned}$$

[3]Recall that if $X \sim \mathcal{B}(p, n)$ then $\mathbb{E}[X] = pn$ (see Section 2.2.1.3).

since $\{S = k\} = \bigsqcup_{m\geq k}\{S = k, X = m\}$. By some algebraic computation we obtain that

$$\frac{\binom{n}{k}p^k(1-p)^{n-k}\mathrm{e}^{-\lambda}\frac{\lambda^n}{n!}}{\sum_{m=k}^{\infty}\binom{m}{k}p^k(1-p)^{m-k}\mathrm{e}^{-\lambda}\frac{\lambda^m}{m!}} = \frac{(\lambda(1-p))^{n-k}\mathrm{e}^{-\lambda(1-p)}}{(n-k)!}.$$

Hence, we get that

$$\mathbb{E}[X \mid S = k] = \sum_{n\geq k} n\mathbb{P}(X = n \mid S = k) = k + \lambda(1-p).$$

Therefore, we get $\mathbb{E}[X \mid S] = S + \lambda(1-p)$.

3.2.3. Continuous construction of the conditional expectation

Now we want to define $\mathbb{E}[Y \mid X]$, where X is no longer assumed to be countably valued. Therefore, we want to recall some facts. Recall first the following definition.

Definition 3.2.3.1 (σ-algebra generated by a random variable). Let $(\Omega, \mathcal{F}, \mathbb{P})$ be a probability space. Let $X : (\Omega, \mathcal{F}, \mathbb{P}) \to (\mathbb{R}^n, \mathcal{B}(\mathbb{R}^n), \lambda)$ be a random variable on that space. The σ-algebra generated by X is given by

$$\sigma(X) := X^{-1}(\mathcal{B}(\mathbb{R}^n)) = \{A \in \Omega \mid A = X^{-1}(B), B \in \mathcal{B}(\mathbb{R}^n)\}.$$

Theorem 3.2.3.1. *Let $(\Omega, \mathcal{F}, \mathbb{P})$ be a probability space. Let X: $(\Omega, \mathcal{F}, \mathbb{P}) \to (\mathbb{R}^n, \mathcal{B}(\mathbb{R}^n), \lambda)$ be a random variable on that space and let Y be a real-valued random variable on that space. Y is measurable with respect to $\sigma(X)$ if and only if there exists a Borel measurable function $f : \mathbb{R}^n \to \mathbb{R}$ such that*

$$Y = f(X).$$

Remark 3.2.3.2. We want to make use of the fact that for the Hilbert space $L^2(\Omega, \mathcal{F}, \mathbb{P})$ we get that $L^2(\Omega, \sigma(X), \mathbb{P}) \subset L^2(\Omega, \mathcal{F}, \mathbb{P})$ is a complete subspace, since $\sigma(X) \subset \mathcal{F}$ is a sub σ-algebra. This allows us to use the orthogonal projections and to interpret the conditional expectation as such a projection.

Definition 3.2.3.3 (Conditional expectation (As a projection onto a closed subspace)). Let $(\Omega, \mathcal{F}, \mathbb{P})$ be a probability space. Let $Y \in L^2(\Omega, \mathcal{F}, \mathbb{P})$. Then the conditional expectation of Y given X is the unique element $\hat{Y} \in L^2(\Omega, \sigma(X), \mathbb{P})$ such that for all $Z \in L^2(\Omega, \sigma(X), \mathbb{P})$

$$\mathbb{E}[YZ] = \mathbb{E}[\hat{Y}Z]. \tag{3.2.3.1}$$

This is due to the fact that if $Y - \hat{Y} \in L^2(\Omega, \mathcal{F}, \mathbb{P})$ then for all $Z \in L^2(\Omega, \sigma(X), \mathbb{P})$ we get $\langle Y - \hat{Y}, Z \rangle = 0$. We also write $\mathbb{E}[Y \mid X]$ for $\hat{Y}$.

Remark 3.2.3.4. Note that $\hat{Y}$ is the orthogonal projection of Y onto $L^2(\Omega, \sigma(X), \mathbb{P})$.

Remark 3.2.3.5. Since X takes values in $\mathbb{R}^n$, there exists a Borel measurable function $f : \mathbb{R}^n \to \mathbb{R}$ such that

$$\mathbb{E}[Y \mid X] = f(X)$$

with $\mathbb{E}[f^2(X)] < \infty$. We can also rewrite (3.2.3.1) as follows: for all Borel measurable $g : \mathbb{R}^n \to \mathbb{R}$, such that $\mathbb{E}[g^2(X)] < \infty$, we get

$$\mathbb{E}[Yg(X)] = \mathbb{E}[f(X)g(X)].$$

Now let $\mathcal{G} \subset \mathcal{F}$ be a sub σ-algebra of $\mathcal{F}$ and consider the space $L^2(\Omega, \mathcal{G}, \mathbb{P}) \subset L^2(\Omega, \mathcal{F}, \mathbb{P})$. It is clear that $L^2(\Omega, \mathcal{G}, \mathbb{P})$ is a Hilbert space and thus we can project to it.

Definition 3.2.3.6 (Conditional expectation (Projection case)). Let $(\Omega, \mathcal{F}, \mathbb{P})$ be a probability space. Let $Y \in L^2(\Omega, \mathcal{F}, \mathbb{P})$ and let $\mathcal{G} \subset \mathcal{F}$ be a sub σ-algebra of $\mathcal{F}$. Then the conditional expectation of Y given $\mathcal{G}$ is defined as the unique element $\mathbb{E}[Y \mid \mathcal{G}] \in L^2(\Omega, \mathcal{G}, \mathbb{P})$ such that for all $Z \in L^2(\Omega, \mathcal{G}, \mathbb{P})$

$$\mathbb{E}[YZ] = \mathbb{E}[\mathbb{E}[Y \mid \mathcal{G}]Z]. \tag{3.2.3.2}$$

Remark 3.2.3.7. In (3.2.3.1) or (3.2.2.1), it is enough[4] to restrict the test random variable Z to the class of random variables of the form

$$Z = \mathbb{1}_A, \quad A \in \mathcal{G}.$$

[4]We can always consider linear combinations of $\mathbb{1}_A$ and then apply some density theorems to it.

Remark 3.2.3.8. The conditional expectation is in L^2, so it is only defined almost surely and not everywhere in a unique way. So in particular, any statement like $\mathbb{E}[Y \mid \mathcal{G}] \geq 0$ or $\mathbb{E}[Y \mid \mathcal{G}] = Z$ has to be understood with an implicit *almost surely.*

Theorem 3.2.3.2. *Let $(\Omega, \mathcal{F}, \mathbb{P})$ be a probability space. Let $Y \in L^2(\Omega, \mathcal{F}, \mathbb{P})$ and let $\mathcal{G} \subset \mathcal{F}$ be a sub σ-algebra of $\mathcal{F}$.*

(i) *If $Y \geq 0$, then $\mathbb{E}[Y \mid \mathcal{G}] \geq 0$.*
(ii) $\mathbb{E}[\mathbb{E}[Y \mid \mathcal{G}]] = \mathbb{E}[Y]$.
(iii) *The map $Y \mapsto \mathbb{E}[Y \mid \mathcal{G}]$ is linear.*

Proof. (i) Take $Z = \mathbb{1}_{\{\mathbb{E}[Y|\mathcal{G}]<0\}}$ to obtain

$$\underbrace{\mathbb{E}[YZ]}_{\geq 0} = \underbrace{\mathbb{E}[\mathbb{E}[Y \mid \mathcal{G}]Z]}_{\leq 0}.$$

This implies that $\mathbb{P}(\mathbb{E}[Y \mid \mathcal{G}] < 0) = 0$.
(ii) Take $Z = \mathbb{1}_\Omega$ and plug into (3.2.3.2).
(iii) Note that linearity comes from the orthogonal projection operator. Nevertheless, we can also do it directly by taking $Y, Y' \in L^2(\Omega, \mathcal{F}, \mathbb{P})$, $\alpha, \beta \in \mathbb{R}$ and $Z \in L^2(\Omega, \mathcal{G}, \mathbb{P})$ to obtain

$$\begin{aligned}\mathbb{E}[(\alpha Y + \beta Y')Z] &= \mathbb{E}[YZ] + \beta\mathbb{E}[Y'Z] = \alpha\mathbb{E}[\mathbb{E}[Y \mid \mathcal{G}]Z] \\ &\quad + \beta\mathbb{E}[\mathbb{E}[Y' \mid \mathcal{G}]Z] = \mathbb{E}[(\alpha\mathbb{E}[Y \mid \mathcal{G}] + \beta\mathbb{E}[Y' \mid \mathcal{G}])Z].\end{aligned}$$

Now we can conclude, by using the uniqueness property, that

$$\mathbb{E}[\alpha Y + \beta Y' \mid \mathcal{G}] = \alpha\mathbb{E}[Y \mid \mathcal{G}] + \beta\mathbb{E}[Y' \mid \mathcal{G}]. \qquad \square$$

Now we want to extend the definition of the conditional expectation to random variables in $L^1(\Omega, \mathcal{F}, \mathbb{P})$ or to $L^+(\Omega, \mathcal{F}, \mathbb{P})$, which is the space of non-negative random variables allowing the value ∞.

Lemma 3.2.3.3. *Let $(\Omega, \mathcal{F}, \mathbb{P})$ be a probability space. Let $Y \in L^+(\Omega, \mathcal{F}, \mathbb{P})$ and let $\mathcal{G} \subset \mathcal{F}$ be a sub σ-algebra of $\mathcal{F}$. Then, there exists a unique element $\mathbb{E}[Y \mid \mathcal{G}] \in L^+(\Omega, \mathcal{G}, \mathbb{P})$ such that for all $X \in L^+(\Omega, \mathcal{G}, \mathbb{P})$ we get*

$$\mathbb{E}[YX] = \mathbb{E}[\mathbb{E}[Y \mid \mathcal{G}]X]. \tag{3.2.3.3}$$

Moreover, this conditional expectation agrees with the previous definition when $Y \in L^2(\Omega, \mathcal{F}, \mathbb{P})$. In particular, if $0 \leq Y \leq Y'$, then

$$\mathbb{E}[Y \mid \mathcal{G}] \leq \mathbb{E}[Y' \mid \mathcal{G}].$$

Proof. If $Y \leq 0$ and $Y \in L^2(\Omega, \mathcal{F}, \mathbb{P})$, then we define $\mathbb{E}[Y \mid \mathcal{G}]$ as before. If $X \in L^+(\Omega, \mathcal{G}, \mathbb{P})$, we get that $X_n = X \wedge n$, is in $L^2(\Omega, \mathcal{G}, \mathbb{P})$ and is positive with $X_n \uparrow X$ for $n \to \infty$. Using the *monotone convergence theorem*, we get

$$\begin{aligned}\mathbb{E}[YX] &= \mathbb{E}\left[Y \lim_{n\to\infty} X_n\right] = \lim_{n\to\infty} \mathbb{E}[YX_n] = \lim_{n\to\infty} \mathbb{E}[\mathbb{E}[Y \mid \mathcal{G}]X_n] \\ &= \mathbb{E}\left[\mathbb{E}[Y \mid \mathcal{G}] \lim_{n\to\infty} X\right] = \mathbb{E}[\mathbb{E}[Y \mid \mathcal{G}]X].\end{aligned}$$

This shows that (3.2.3.3) is true whenever $Y \in L^2(\Omega, \mathcal{F}, \mathbb{P})$ with $Y \geq 0$ and $X \in L^+(\Omega, \mathcal{G}, \mathbb{P})$. Now let $Y \in L^1(\Omega, \mathcal{F}, \mathbb{P})$ and define $Y_m = Y \wedge m$. Hence, we get $Y_m \in L^2(\Omega, \mathcal{F}, \mathbb{P})$ and $Y_m \uparrow Y$ as $n \to \infty$. Note that each $\mathbb{E}[Y_m \mid \mathcal{G}]$ is well-defined,[5] positive and increasing. We define

$$\mathbb{E}[Y \mid \mathcal{G}] = \lim_{n\to\infty} \mathbb{E}[Y_m \mid \mathcal{G}].$$

Several applications of the *monotone convergence theorem* will give us for $X \in L^+(\Omega, \mathcal{G}, \mathbb{P})$

$$\mathbb{E}[YX] = \lim_{m\to\infty} \mathbb{E}[Y_m X] = \lim_{m\to\infty} \mathbb{E}[\mathbb{E}[Y_m \mid \mathcal{G}]X] = \mathbb{E}[\mathbb{E}[Y \mid \mathcal{G}]X].$$

Furthermore, if $0 \leq Y \leq Y'$, then $Y \wedge m \leq Y' \wedge m$ and therefore

$$\mathbb{E}[Y \mid \mathcal{G}] \leq \mathbb{E}[Y' \mid \mathcal{G}].$$

Now we need to show uniqueness.[6] Let U and V be two versions of $\mathbb{E}[Y \mid \mathcal{G}]$. Define

$$\Lambda_n := \{U < V \leq n\} \in \mathcal{G}$$

and assume $\mathbb{P}(\Lambda_n) > 0$. We then have

$$\mathbb{E}[Y\mathbb{1}_{\Lambda_n}] = \underbrace{\mathbb{E}[U\mathbb{1}_{\Lambda_n}] = \mathbb{E}[V\mathbb{1}_{\Lambda_n}]}_{\mathbb{E}[(U-V)\mathbb{1}_{\Lambda_n}]=0}.$$

[5]This is true since for $Y \in L^2$ and $U \in L^2$ we get $Y \geq U \to Y - U \geq 0 \to \mathbb{E}[Y \mid \mathcal{G}] \geq \mathbb{E}[U \mid \mathcal{G}]$.

[6]Note that for any $W \in L^+$, the set E on which $W = \infty$ is a null set. Indeed, suppose this is not true, then $\mathbb{E}[W] \geq \mathbb{E}[\infty \mathbb{1}_E] = \infty\mathbb{P}(E)$. But since $\mathbb{P}(E) > 0$ this cannot happen.

This contradicts the fact that $\mathbb{P}(\Lambda_n) > 0$. Moreover, $\{U < V\} = \bigcup_{n\geq 1} \Lambda_n$ and therefore

$$\mathbb{P}(U < V) = 0,$$

and similarly $\mathbb{P}(V < U) = 0$. This implies that

$$\mathbb{P}(U = V) = 1. \qquad \square$$

Theorem 3.2.3.4. *Let $(\Omega, \mathcal{F}, \mathbb{P})$ be a probability space. Let $Y \in L^1(\Omega, \mathcal{F}, \mathbb{P})$ and let $\mathcal{G} \subset \mathcal{F}$ be a sub σ-algebra of $\mathcal{F}$. Then there exists a unique element $\mathbb{E}[Y \mid \mathcal{G}] \in L^1(\Omega, \mathcal{G}, \mathbb{P})$ such that for every X bounded and $\mathcal{G}$-measurable we have*

$$\mathbb{E}[YX] = \mathbb{E}[\mathbb{E}[Y \mid \mathcal{G}]X]. \tag{3.2.3.4}$$

Moreover, this conditional expectation agrees with the definition for the L^2 and satisfies the following:

(i) *If $Y \geq 0$, then $\mathbb{E}[Y \mid \mathcal{G}] \geq 0$.*
(ii) *The map $Y \mapsto \mathbb{E}[Y \mid \mathcal{G}]$ is linear.*

Proof. We will only prove the existence, since the rest is exactly the same as before. Write $Y = Y^+ - Y^-$ with $Y^+, Y^- \in L^1(\Omega, \mathcal{F}, \mathbb{P})$ and $Y^+, Y^- \geq 0$. So, we get that $\mathbb{E}[Y^+ \mid \mathcal{G}]$ and $\mathbb{E}[Y^- \mid \mathcal{G}]$ are well-defined. Now we define

$$\mathbb{E}[Y \mid \mathcal{G}] := \mathbb{E}[Y^+ \mid \mathcal{G}] - \mathbb{E}[Y^- \mid \mathcal{G}].$$

This is well-defined because

$$\mathbb{E}[\mathbb{E}[Y^\pm \mid \mathcal{G}]] = \mathbb{E}[Y^\pm] < \infty,$$

whenever we consider $X = \mathbb{1}_\Omega$ in Lemma 3.2.3.3 and therefore $\mathbb{E}[Y^+ \mid \mathcal{G}]$ and $\mathbb{E}[Y^- \mid \mathcal{G}] \in L^1(\Omega, \mathcal{G}, \mathbb{P})$. In fact, for all X bounded and $\mathcal{G}$-measurable we can also write $X = X^+ - X^-$ and it follows from Lemma 3.2.3.3 that

$$\mathbb{E}[\mathbb{E}[Y^\pm \mid \mathcal{G}]X] = \mathbb{E}[Y^\pm X].$$

This implies that $\mathbb{E}[Y \mid \mathcal{G}]$ satisfies (3.2.3.4). $\square$

Corollary 3.2.3.5. *Let $(\Omega, \mathcal{F}, \mathbb{P})$ be a probability space and let $X \in L^1(\Omega, \mathcal{F}, \mathbb{P})$ be a random variable on that space. Then*

$$\mathbb{E}[\mathbb{E}[X \mid \mathcal{G}]] = \mathbb{E}[X].$$

Proof. Take equation (3.2.3.2) and set $Z = \mathbb{1}_\Omega$. □

Corollary 3.2.3.6. *Let $(\Omega, \mathcal{F}, \mathbb{P})$ be a probability space. Let $X \in L^1(\Omega, \mathcal{F}, \mathbb{P})$ be a random variable on that space. Then*

$$|\mathbb{E}[X \mid \mathcal{G}]| \leq \mathbb{E}[|X| \mid \mathcal{G}].$$

In particular, we have

$$\mathbb{E}[|\mathbb{E}[X \mid \mathcal{G}]|] \leq \mathbb{E}[|X|].$$

Proof. We can always write $X = X^+ - X^-$ and also $|X| = X^+ + X^-$. Therefore, we get

$$\begin{aligned} |\mathbb{E}[X \mid \mathcal{G}]| &= |\mathbb{E}[X^+ \mid \mathcal{G}] - \mathbb{E}[X^- \mid \mathcal{G}]| \leq \mathbb{E}[X^+ \mid \mathcal{G}] + \mathbb{E}[X^- \mid \mathcal{G}] \\ &= \mathbb{E}[X^+ + X^- \mid \mathcal{G}] = \mathbb{E}[|X| \mid \mathcal{G}]. \end{aligned}$$ □

Proposition 3.2.3.7. *Let $(\Omega, \mathcal{F}, \mathbb{P})$ be a probability space. Let $Y \in L^1(\Omega, \mathcal{F}, \mathbb{P})$ be a random variable on that space and assume that Y is independent of the sub-σ-algebra $\mathcal{G} \subset \mathcal{F}$, i.e., $\sigma(Y)$ is independent of $\mathcal{G}$. Then*

$$\mathbb{E}[Y \mid \mathcal{G}] = \mathbb{E}[Y].$$

Proof. Let Z be a bounded and $\mathcal{G}$-measurable random variable and therefore Y and Z are independent. Hence, we get

$$\mathbb{E}[YZ] = \mathbb{E}[Y]\mathbb{E}[Z] = \mathbb{E}[\mathbb{E}[Y]Z].$$

Since $\mathbb{E}[Y]$ is constant, this implies that $\mathbb{E}[Y] \in L^1(\Omega, \mathcal{G}, \mathbb{P})$ and it satisfies (3.2.3.2). Therefore, by uniqueness, we get that $\mathbb{E}[Y \mid \mathcal{G}] = \mathbb{E}[Y]$. □

Theorem 3.2.3.8. *Let $(\Omega, \mathcal{F}, \mathbb{P})$ be a probability space. Let X and Y be two random variables on that space and let $\mathcal{G} \subset \mathcal{F}$ be a sub-σ-algebra of $\mathcal{F}$. Assume further that at least one of these two conditions holds:*

(i) *X, Y and XY are in $L^1(\Omega, \mathcal{F}, \mathbb{P})$ with X being $\mathcal{G}$-measurable;*
(ii) *$X \geq 0$, $Y \geq 0$ with X being $\mathcal{G}$-measurable.*

Then

$$\mathbb{E}[XY \mid \mathcal{G}] = \mathbb{E}[Y \mid \mathcal{G}]X.$$

In particular, if X is a positive random variable or in $L^1(\Omega, \mathcal{G}, \mathbb{P})$ and $\mathcal{G}$-measurable, then

$$\mathbb{E}[X \mid \mathcal{G}] = X.$$

Proof. (ii) Assume first that $X, Y \leq 0$. Let Z be a positive and $\mathcal{G}$-measurable random variable. Then we can obtain that

$$\mathbb{E}[(XY)Z] = \mathbb{E}[Y(XZ)] = \mathbb{E}[\mathbb{E}[Y \mid \mathcal{G}]XZ] = \mathbb{E}[(\mathbb{E}[Y \mid \mathcal{G}]X)Z].$$

Note that $\mathbb{E}[(\mathbb{E}[Y|\mathcal{G}]X)Z]$ is a positive random variable and $\mathcal{G}$-measurable. Hence $\mathbb{E}[XY \mid \mathcal{G}] = X\mathbb{E}[Y \mid \mathcal{G}]$.

(i) We can write $X = X^+ + X^-$ and use (ii). We leave this as an exercise. □

Remark 3.2.3.9. Next we want to show that the classical limit theorems from measure theory also make sense in terms of the conditional expectation.[7]

Theorem 3.2.3.9 (Limit theorems for the conditional expectation). *Let $(\Omega, \mathcal{F}, \mathbb{P})$ be a probability space. Let $(Y_n)_{n\geq 1}$ be a sequence of random variables on that space and let $\mathcal{G} \subset \mathcal{F}$ be a sub-σ-algebra of $\mathcal{F}$. Then, the following assertions hold:*

---

[7]Recall the classical limit theorems for integrals (see also Appendix B).

- *Monotone convergence theorem*: Let $(f_n)_{n\geq 1}$ be an increasing sequence of positive and measurable functions and let $f = \lim_{n\to\infty} \uparrow f_n$. Then $\int f d\mu = \lim_{n\to\infty} f_n d\mu$.
- *Fatou's lemma*: Let $(f_n)_{n\geq 1}$ be a sequence of measurable and positive functions. Then $\int \liminf_n f_n d\mu \leq \liminf_n \int f_n d\mu$.
- *Lebesgue's dominated convergence theorem*: Let $(f_n)_{n\geq 1}$ be a sequence of integrable functions with $|f_n| \leq g$ for all n with g integrable. Denote $f := \lim_{n\to\infty} f_n$. Then $\lim_{n\to\infty} \int f_n d\mu = \int f d\mu$.

(i) (Monotone convergence theorem) *Assume that* $(Y_n)_{n\geq 1}$ *is a sequence of positive random variables for all* n *such that* $\lim_{n\to\infty} \uparrow Y_n = Y$ *a.s. Then*

$$\lim_{n\to\infty} \mathbb{E}[Y_n \mid \mathcal{G}] = \mathbb{E}[Y \mid \mathcal{G}].$$

(ii) (Fatou's lemma) *Assume that* $(Y_n)_{n\geq 1}$ *is a sequence of positive random variables for all* n*. Then*

$$\mathbb{E}[\liminf_n Y_n \mid \mathcal{G}] = \liminf_n \mathbb{E}[Y_n \mid \mathcal{G}.$$

(iii) (Lebesgue's dominated convergence theorem) *Assume that* $Y_n \xrightarrow{n\to\infty} Y$ *a.s. and that there exists* $Z \in L^1(\Omega, \mathcal{F}, \mathbb{P})$ *such that* $|Y_n| \leq Z$ *for all* n*. Then*

$$\lim_{n\to\infty} \mathbb{E}[Y_n \mid \mathcal{G}] = \mathbb{E}[Y \mid \mathcal{G}].$$

Proof. We will only prove (i), since (ii) and (iii) are proved in a similar way (we leave them as an exercise). Since $(Y_n)_{n\geq 1}$ is an increasing sequence, it follows that

$$\mathbb{E}[Y_{n+1} \mid \mathcal{G}] \geq \mathbb{E}[Y_n \mid \mathcal{G}].$$

Hence, we can deduce that $\lim_{n\to\infty} \uparrow \mathbb{E}[Y_n \mid \mathcal{G}]$ exists and we denote it by Y'. Moreover, note that Y' is $\mathcal{G}$-measurable, since it is a limit of $\mathcal{G}$-measurable random variables. Let X be a positive and $\mathcal{G}$-measurable random variable and note that then

$$\begin{aligned}\mathbb{E}[Y'X] &= \mathbb{E}[\lim_{n\to\infty} \mathbb{E}[Y_n \mid \mathcal{G}]X] = \lim_{n\to\infty} \uparrow \mathbb{E}[\mathbb{E}[Y_n \mid \mathcal{G}]X] \\ &= \lim_{n\to\infty} \mathbb{E}[Y_n X] = \mathbb{E}[YX],\end{aligned}$$

where we have used the *monotone convergence theorem* twice and Equation (3.2.3.2). Therefore, we get

$$\lim_{n\to\infty} \mathbb{E}[Y_n \mid \mathcal{G}] = \mathbb{E}[Y \mid \mathcal{G}]. \qquad \square$$

Theorem 3.2.3.10 (Jensen's inequality). *Let* $(\Omega, \mathcal{F}, \mathbb{P})$ *be a probability space. Let* $\varphi : \mathbb{R} \to \mathbb{R}$ *be a real, convex function. Let* $X \in L^1(\Omega, \mathcal{F}, \mathbb{P})$ *such that* $\varphi(X) \in L^1(\Omega, \mathcal{F}, \mathbb{P})$*. Then*

$$\varphi(\mathbb{E}[X \mid \mathcal{G}]) \leq \mathbb{E}[\varphi(X) \mid \mathcal{G}]$$

for all sub-σ*-algebras* $\mathcal{G} \subset \mathcal{F}$.

Exercise 3.2.3.10. Prove Theorem 3.2.3.10.

Example 3.2.3.11. Let $(\Omega, \mathcal{F}, \mathbb{P})$ be a probability space. Let $\varphi(X) = X^2$ and let $X \in L^2(\Omega, \mathcal{F}, \mathbb{P})$. Then

$$(\mathbb{E}[X \mid \mathcal{G}])^2 \leq \mathbb{E}[X^2 \mid \mathcal{G}]$$

for all sub-σ-algebras $\mathcal{G} \subset \mathcal{F}$.

Theorem 3.2.3.11 (Tower property). *Let $(\Omega, \mathcal{F}, \mathbb{P})$ be a probability space. Let $X \in L^1(\Omega, \mathcal{F}, \mathbb{P})$ be a positive random variable on that space. Let $\mathcal{C} \subset \mathcal{G} \subset \mathcal{F}$ be a tower of sub-σ-algebras of $\mathcal{F}$. Then, we get*

$$\mathbb{E}[\mathbb{E}[X \mid \mathcal{G}]\mathcal{C}] = \mathbb{E}[X \mid \mathcal{C}].$$

Proof. Let Z be a bounded and $\mathcal{C}$-measurable random variable. Then we obtain

$$\mathbb{E}[XZ] = \mathbb{E}[\mathbb{E}[X \mid \mathcal{C}]Z].$$

But Z is also $\mathcal{G}$-measurable and thus we get

$$\mathbb{E}[XZ] = \mathbb{E}[\mathbb{E}[X \mid \mathcal{G}]Z].$$

Therefore, for all Z bounded and $\mathcal{C}$-measurable random variables, we get

$$\mathbb{E}[\mathbb{E}[X \mid \mathcal{G}]Z] = \mathbb{E}[\mathbb{E}[X \mid \mathcal{C}]Z]$$

and thus

$$\mathbb{E}[\mathbb{E}[X \mid \mathcal{G}]\mathcal{C}] = \mathbb{E}[X \mid \mathcal{C}].$$ □

3.3. The Radon–Nikodym Approach for the Conditional Expectation

Remark 3.3.0.1. Before stating the *Radon–Nikodym theorem*, we recall some definitions from measure theory. Let $(\Omega, \mathcal{B})$ be a measurable space. A measure ν is *absolutely continuous* with respect to another measure μ, written $\nu \ll \mu$ if there exists some measurable

$f \geq 0$ with $\mathrm{d}\nu = f\mathrm{d}\mu$, that is if there is a finite measurable function $f \geq 0$ with

$$\nu(B) = \int_B f\mathrm{d}\mu$$

for all $B \in \mathcal{B}$. Two measures μ and ν are *singular* with respect to each other if there exists disjoint measurable sets $A_1, A_2 \subset \Omega$ with $\Omega = A_1 \sqcup A_2$ and with $\nu(A_1) = 0 = \mu(A_2)$. Finally, recall that a measure μ is σ-finite if there is a decomposition of Ω into measurable sets,

$$\Omega = \bigsqcup_{i=1}^{\infty} A_i$$

with $\mu(A_i) < \infty$.

Theorem 3.3.0.1 (Radon–Nikodym). *Let μ and ν be two σ-finite measures on a measurable space $(\Omega, \mathcal{B})$. Then ν can be decomposed as*

$$\nu = \nu_{\text{abs}} + \nu_{\text{sing}}$$

into the sum of two σ-finite measure with $\nu_{\text{abs}} \ll \mu$ being absolutely continuous with respect to μ, and with ν_{sing} and μ being singular to each other (which will be written $\nu_{\text{sing}} \perp \mu$).

Remark 3.3.0.2. Theorem 3.3.0.1 implies that there exists another, more practical way of checking whether a given σ-finite measure ν is absolutely continuous with respect to another σ-finite measure μ. If $\mu(N) = 0$ implies that $\nu(N) = 0$ for every measurable $N \subset \Omega$, then $\nu = \nu_{\text{abs}}$ is absolutely continuous. We also note that the density function f with $f\mathrm{d}\mu = \mathrm{d}\nu$ is called the *Radon–Nikodym derivative* and is often written $f = \frac{\mathrm{d}\nu}{\mathrm{d}\mu}$.

Remark 3.3.0.3. To prove Theorem 3.3.0.1, we need a lemma which gives us a nice relationship between a Hilbert space and its dual space. Actually, we can identify a Hilbert space $\mathcal{H}$ with its dual space $\mathcal{H}^*$.

Lemma 3.3.0.2 (Riesz-representation for Hilbert spaces). *For a Hilbert space $\mathcal{H}$, the map sending $h \in \mathcal{H}$ to $\phi(h) \in \mathcal{H}^*$, defined by*

$$\phi(h)(x) := \langle x, h \rangle,$$

is a linear (respectively, sesqui-linear in the complex case) isometric isomorphism between $\mathcal{H}$ and its dual space $\mathcal{H}^*$.

Proof of Theorem 3.3.0.1. Suppose that μ and ν are both finite measures (the general case can be reduced to this case by using the assumption that μ and ν are both σ-finite). We define a new measure $m = \mu + \nu$ and will work with the real Hilbert space $\mathcal{H} = L^2(\Omega, m)$. On this Hilbert space we define a linear functional ϕ by

$$\phi(g) = \int g \mathrm{d}\nu$$

for $g \in \mathcal{H}$. For g a simple function on Ω, this is clearly well-defined and satisfies

$$|\phi(g)| = \left| \int g \mathrm{d}\nu \right| \leq \int |g| \mathrm{d}\nu \leq \int |g| \mathrm{d}m \leq \|g\|_{\mathcal{H}} \|\mathbb{1}\|_{\mathcal{H}},$$

where we have used the fact that $m = \mu + \nu$, that μ is a positive measure and the Cauchy–Schwarz inequality on $\mathcal{H}$. Since the simple functions are dense in $\mathcal{H}$, the functional extends to a functional on all of $\mathcal{H}$. By the Riesz-representation for Hilbert spaces (Lemma 3.3.0.2) there is some $k \in \mathcal{H}$ such that

$$\int g \mathrm{d}\nu = \phi(g) = \int gk \mathrm{d}m. \tag{3.3.0.1}$$

We claim that k takes values in $[0, 1]$ almost surely with respect to m. Indeed, for any $B \in \mathcal{B}$ we have

$$0 \leq \nu(B) \leq m(B),$$

so (using $g = \mathbb{1}_B$),

$$0 \leq \int_B k \mathrm{d}m \leq m(B).$$

Using the choices

$$B = \{\omega \in \Omega \mid k(\omega) < 0\},$$

and

$$B = \{\omega \in \Omega \mid k(\omega) > 1\},$$

implies the claim that k takes m-almost surely values in $[0, 1]$. Since $m = \mu + \nu$, we can reformulate (3.3.0.1) as

$$\int g(1 - k)\mathrm{d}\nu = \int gk\mathrm{d}\mu. \tag{3.3.0.2}$$

This holds by construction for all simple functions g, and hence for all non-negative measurable functions by the *monotone convergence theorem*. Now define ν_{sing} to be $\nu \mid_A$, where

$$A := \{\omega \in \Omega \mid k(\omega) = 1\}.$$

By definition, $\nu_{\text{sing}}(\Omega \setminus A) = 0$ and by (3.3.0.2) applied with $g = \mathbb{1}_A$ we also have $\mu(A) = 0$. Therefore

$$\nu_{\text{sing}} \perp \mu.$$

We also define

$$\nu_{\text{abs}} = \nu \mid_{\Omega \setminus A} = \nu_{\{\omega \in \Omega \mid k(\omega) < 1\}}$$

so that $\nu = \nu_{\text{sing}} + \nu_{\text{abs}}$. Define the function $f = \frac{k}{1-k} \geq 0$ on $\Omega \setminus A$ and let $g \geq 0$ be measurable. Then by (3.3.0.2), we have

$$\int_{\Omega \setminus A} gf\mathrm{d}\mu = \int_{\Omega \setminus A} \frac{g}{1-k} k\mathrm{d}\mu = \int_{\Omega \setminus A} \frac{g}{1-k}(1-k)\mathrm{d}\nu = \int_{\Omega \setminus A} g\mathrm{d}\nu_{\text{abs}},$$

which shows that $\mathrm{d}\nu_{\text{abs}} = f\mathrm{d}\mu$ and so $\nu_{\text{abs}} \ll \mu$. □

Theorem 3.3.0.3. *Let $(\Omega, \mathcal{F}, \mathbb{P})$ be a probability space. Let $\mathcal{G} \subset \mathcal{F}$ be a sub-σ-algebra of $\mathcal{F}$ and let $X \in L^1(\Omega, \mathcal{F}, \mathbb{P})$ be a random variable Then, there exists a unique random variable in $L^1(\Omega, \mathcal{G}, \mathbb{P})$, denoted by $\mathbb{E}[X \mid \mathcal{G}]$, such that for all $B \in \mathcal{G}$*

$$\mathbb{E}[X\mathbb{1}_B] = \mathbb{E}[\mathbb{E}[X \mid \mathcal{G}]\mathbb{1}_B].$$

More generally, for every bounded and $\mathcal{G}$-measurable random variable Z, we get

$$\mathbb{E}[XZ] = \mathbb{E}[\mathbb{E}[X \mid \mathcal{G}]Z]$$

and if $X \geq 0$, then $\mathbb{E}[X \mid \mathcal{G}] \geq 0$.

Proof. We have already shown uniqueness. To show existence, assume first that X is positive. Define a new measure $\mathbb{Q}$ on $(\Omega, \mathcal{G})$ by

$$\mathbb{Q}(A) := \mathbb{E}[X\mathbb{1}_A] = \int_A X(\omega)\mathrm{d}\mathbb{P}(\omega)$$

for all $A \in \mathcal{G}$. Now consider the measure $\mathbb{P}$ restricted to $\mathcal{G}$. Then we get

$$\mathbb{Q} \ll \mathbb{P}$$

on $\mathcal{G}$. Theorem 3.3.0.1 implies that there exists a positive and $\mathcal{G}$-measurable random variable $\tilde{X}$ such that

$$\mathbb{Q}(A) = \mathbb{E}[\tilde{X}\mathbb{1}_A]$$

for all $A \in \mathcal{G}$. For $A \in \mathcal{G}$, we get

$$\mathbb{E}[X\mathbb{1}_A] = \mathbb{E}[\tilde{X}\mathbb{1}_A].$$

Now taking $A = \Omega$, we get $\mathbb{E}[X] = \mathbb{E}[\tilde{X}]$. Therefore, we have that $\tilde{X} \in L^1(\Omega, \mathcal{G}, \mathbb{P})$ and hence we see that $\tilde{X} = \mathbb{E}[X \mid \mathcal{G}]$. For the general case, we can just write $X = X^+ + X^-$ and do the same as before (we leave this as an exercise). □

3.4. More Properties of the Conditional Expectation

Theorem 3.4.0.1. *Let $(\Omega, \mathcal{F}, \mathbb{P})$ be a probability space. Let $\mathcal{G}_1 \subset \mathcal{F}$ and $\mathcal{G}_2 \subset \mathcal{F}$ be two sub-σ-algebras of $\mathcal{F}$. Then $\mathcal{G}_1$ and $\mathcal{G}_2$ are independent if and only if for every positive and $\mathcal{G}_2$-measurable random variable X (or for $X \in L^1(\Omega, \mathcal{G}_2, \mathbb{P})$ or $X = \mathbb{1}_A$ for $A \in \mathcal{G}_2$) we have*

$$\mathbb{E}[X \mid \mathcal{G}_2] = \mathbb{E}[X].$$

Proof. We only need to prove that the statement in the bracket implies that $\mathcal{G}_1$ and $\mathcal{G}_2$ are independent. Assume that for all $A \in \mathcal{G}_2$ we have that

$$\mathbb{E}[\mathbb{1}_A \mid \mathcal{G}_1] = \mathbb{P}(A),$$

and moreover for all $B \in \mathcal{G}_1$ we have that

$$\mathbb{E}[\mathbb{1}_B\mathbb{1}_A] = \mathbb{E}[\mathbb{E}[\mathbb{1}_A \mid \mathcal{G}_1]\mathbb{1}_B].$$

Note that $\mathbb{E}[\mathbb{1}_A \mid \mathcal{G}_1] = \mathbb{P}(A)$ and therefore $\mathbb{E}[\mathbb{1}_B\mathbb{1}_A] = \mathbb{P}(A \cap B) = \mathbb{P}(A)\mathbb{E}[\mathbb{1}_B] = \mathbb{P}(A)\mathbb{P}(B)$ and hence the claim follows. □

Remark 3.4.0.1. Let Z and Y be two real-valued random variables. Then Z and Y are independent if and only if for any Borel measurable function h, such that $\mathbb{E}[|h(Z)|] < \infty$, we get $\mathbb{E}[h(Z) \mid Y] = \mathbb{E}[h(Z)]$. To see this, we can apply Theorem 3.4.0.1 with $\mathcal{G}_2 = \sigma(Z)$ and note that all random variables in $L^1(\Omega, \mathcal{G}_2, \mathbb{P})$ are of the form $h(Z)$ with $\mathbb{E}[|h(Z)|] < \infty$. In particular, if $Z \in L^1(\Omega, \mathcal{F}, \mathbb{P})$, we get $\mathbb{E}[Z \mid Y] = \mathbb{E}[Z]$. Be aware that the latter equation does not imply that Y and Z are independent. For example, take $Z \sim \mathcal{N}(0,1)$ and $Y = |Z|$. Now for all h with $\mathbb{E}[|h(|Z|)|] < \infty$ we get $\mathbb{E}[Zh(|Z|)] = 0$. Thus, $\mathbb{E}[Z \mid |Z|] = 0$, but Z and $|Z|$ are not independent.

Theorem 3.4.0.2. *Let $(\Omega, \mathcal{F}, \mathbb{P})$ be a probability space. Let X and Y be two random variables on that space with values in the measurable spaces E and F, respectively. Assume that X is independent of the sub-σ-algebra $\mathcal{G} \subset \mathcal{F}$ and that Y is $\mathcal{G}$-measurable. Then, for every measurable map $g : E \times F \to \mathbb{R}_{\geq 0}$, we have*

$$\mathbb{E}[g(X,Y) \mid \mathcal{G}] = \int_E g(x,y) \mathrm{d}\mathbb{P}_X(x),$$

where $\mathbb{P}_X$ is the law of X and the right-hand side has to be understood as a function $\varphi(Y)$ with

$$\varphi : y \mapsto \int_E g(x,y) \mathrm{d}\mathbb{P}_X(x).$$

Proof. We need to show that for all $\mathcal{G}$-measurable random variable Z we get that

$$\mathbb{E}[g(X,Y)Z] = \mathbb{E}[\phi(Y)Z].$$

Let us denote by $\mathbb{P}_{(X,Y,Z)}$ the distribution of (X, Y, Z) on $E \times F \times \mathbb{R}_{\geq 0}$. Since X is independent of $\mathcal{G}$, we have $\mathbb{P}_{(X,Y,Z)} = \mathbb{P}_X \otimes \mathbb{P}_{(Y,Z)}$. Thus, we have

$$\begin{aligned}
\mathbb{E}[g(X,Y)Z] &= \int_{E \times F \times \mathbb{R}_{\geq 0}} g(x,y) z \mathrm{d}\mathbb{P}_{(X,Y,Z)}(x,y,z) \\
&= \int_{F \times \mathbb{R}_{\geq 0}} z \left(\int_E g(x,y) \mathrm{d}\mathbb{P}_X(x) \right) \mathrm{d}\mathbb{P}_{(Y,Z)}(y,z) \\
&= \int_{F \times \mathbb{R}_{\geq 0}} z\varphi(y) \mathrm{d}\mathbb{P}_{(Y,Z)}(y,z) \\
&= \mathbb{E}[Z\varphi(Y)].
\end{aligned}$$

□

3.4.1. Important examples

We need to take a look at two important examples.

Example 3.4.1.1 (Variables with densities). Let $(X, Y) \in \mathbb{R}^m \times \mathbb{R}^n$. Assume that (X, Y) has density $P(x, y)$, i.e., for all Borel measurable maps $h : \mathbb{R}^m \times \mathbb{R}^n \to \mathbb{R}_{\geq 0}$ we have

$$\mathbb{E}[h(X, Y)] = \int_{\mathbb{R}^m \times \mathbb{R}^n} h(x, y) P(x, y) \mathrm{d}x \mathrm{d}y.$$

The density of Y is given by

$$Q(y) = \int_{\mathbb{R}^m} P(x, y) \mathrm{d}x.$$

We want to compute $\mathbb{E}[h(X) \mid Y]$ for some measurable map $h : \mathbb{R}^m \to \mathbb{R}_{\geq 0}$. Note that we have

$$\begin{aligned}
\mathbb{E}[h(X)g(Y) &= \int_{\mathbb{R}^m \times \mathbb{R}^n} h(x)g(y)P(x, y)\mathrm{d}x\mathrm{d}y \\
&= \int_{\mathbb{R}^n} \left(\int_{\mathbb{R}^m} h(x)P(x, y)\mathrm{d}x \right) g(y)\mathrm{d}y \\
&= \int_{\mathbb{R}^n} \frac{1}{Q(y)} \left(\int_{\mathbb{R}^m} h(x)P(x, y)\mathrm{d}x \right) g(y)Q(y)\mathbb{1}_{\{Q(y)>0\}}\mathrm{d}y \\
&= \int_{\mathbb{R}^n} \varphi(y)g(y)Q(y)\mathbb{1}_{\{Q(y)>0\}}\mathrm{d}y = \mathbb{E}[\varphi(Y)g(y)],
\end{aligned}$$

where

$$\varphi(y) = \begin{cases} \dfrac{1}{Q(y)} \int_{\mathbb{R}^n} h(x)Q(x, y)\mathrm{d}x, & Q(y) > 0, \\ h(0), & Q(y) = 0. \end{cases}$$

Proposition 3.4.1.1. *For $Y \in \mathbb{R}^n$, let $\nu(y, \mathrm{d}x)$ be the probability measure on $\mathbb{R}^n$ defined by*

$$\nu(x, \mathrm{d}y) = \begin{cases} \dfrac{1}{Q(y)} P(x, y) & Q(y) > 0, \\ \delta_0(\mathrm{d}x) & Q(y) = 0. \end{cases}$$

Then, for all measurable maps $h : \mathbb{R}^m \to \mathbb{R}_{\geq 0}$, we get

$$\mathbb{E}[h(X) \mid Y] = \int_{\mathbb{R}^m} h(x)\nu(Y, \mathrm{d}x),$$

where the right-hand side has to be understood as $\varphi(Y)$, where

$$\varphi(Y) = \int_{\mathbb{R}^m} h(x)\nu(Y, \mathrm{d}x).$$

Remark 3.4.1.2. In the literature, one also abusively writes

$$\mathbb{E}[h(X) \mid Y = y] = \int_{\mathbb{R}^m} h(x)\nu(y, \mathrm{d}x),$$

and $\nu(y, \mathrm{d}x)$ is called the *conditional distribution* of X given $Y = y$ (even though in general we have $\mathbb{P}(Y = y) = 0$).

Example 3.4.1.3 (The Gaussian case). Let $(\Omega, \mathcal{F}, \mathbb{P})$ be a probability space. Let $X, Y_1, \ldots, Y_p \in L^2(\Omega, \mathcal{F}, \mathbb{P})$. We have seen that

$$\mathbb{E}[X \mid Y_1, \ldots, Y_p]$$

is the orthogonal projection of X on $L^2(\Omega, \sigma(Y_1, \ldots, Y_p), \mathbb{P})$. Since this conditional expectation is $\sigma(Y_1, \ldots, Y_p)$-measurable, it is of the form $\varphi(Y_1, \ldots, Y_p)$. In general, $L^2(\Omega, \sigma(Y_1, \ldots, Y_p), \mathbb{P})$ is of infinite dimension, so it is bad to obtain φ explicitly. We have also seen that $\varphi(Y_1, \ldots, Y_p)$ is the best approximation of X in the $L^2(\Omega, \sigma(Y_1, \ldots, Y_p), \mathbb{P})$-sense by an element of $L^2(\Omega, \sigma(Y_1, \ldots, Y_p), \mathbb{P})$. Moreover, it is well known that the best L^2-approximation of X by an affine function of $\mathbb{1}, Y_1, \ldots, Y_p$ is the best orthogonal projection of X on the vector space $\{\mathbb{1}, Y_1, \ldots, Y_p\}$, i.e.,

$$\mathbb{E}[(X - (\alpha_0 + \alpha_1 Y_1 + \cdots + \alpha_p Y_p)^2)].$$

In general, this is different from the orthogonal projection on $L^2(\Omega, \sigma(Y_1, \ldots, Y_p), \mathbb{P})$, except in the Gaussian case.

3.5. Basic Facts on Gaussian Vectors

Recall that a random vector $Z = (Z_1, \dots, Z_n)$ is said to be *Gaussian*, if for all $\lambda_1, \dots, \lambda_n \in \mathbb{R}$

$$\lambda_1 Z_1 + \cdots + \lambda_n Z_n$$

is Gaussian. Moreover, Z is called *centered*, if $\mathbb{E}[Z_j] = 0$ for all $1 \leq j \leq n$. Let Z be a Gaussian vector. Then, for all $\xi \in \mathbb{R}^n$, we get

$$\mathbb{E}\left[e^{i\langle \xi, Z\rangle}\right] = \exp\left(-\frac{1}{2}\xi^t C_Z \xi\right),$$

where $C_Z := (C_{ij})$ with $C_{ij} = \mathbb{E}[Z_i Z_j]$. If $\mathbb{CV}(Z_i, Z_j) = 0$, then Z_i and Z_j are independent. More generally, we can consider the Gaussian vectors

$$(\underbrace{X_1, \dots, X_{i_1}}_{Y_1}, \underbrace{X_{i_1+1}, \dots, X_{i_2}}_{Y_2}, \dots, \underbrace{X_{i_{n-1}+1}, \dots, X_{i_n}}_{Y_n}).$$

Then Y_1 and Y_2 are independent if and only if $\mathbb{CV}(X_j, X_n) = 0$, where $1 \leq j \leq i_1$ and $i_1 + 1 \leq k \leq i_2$. If $Z_1, \dots, Z_n$ are independent Gaussian random variables, we have that

$$Z = (Z_1, \dots, Z_n)$$

is a Gaussian vector. If Z is a Gaussian vector and A a real $m \times n$ matrix, we get that AZ is again a Gaussian vector.

Theorem 3.5.0.1. *Let $(\Omega, \mathcal{F}, \mathbb{P})$ be a probability space. Let $X \in L^1(\Omega, \mathcal{F}, \mathbb{P})$ and $Y_1, \dots, Y_p \in L^1(\Omega, \mathcal{F}, \mathbb{P})$ and let $(X, Y_1, \dots, Y_p)$ be a centered Gaussian vector. Then*

$$\mathbb{E}[X \mid Y_1, \dots, Y_p]$$

is the orthogonal projection of X onto the vector space

$$\operatorname{span}(Y_1, \dots, Y_p).$$

Consequently, there exists real numbers $\lambda_1, \dots, \lambda_p$ such that

$$\mathbb{E}[X \mid Y_1, \dots, Y_p] = \lambda_1 Y_1 + \cdots + \lambda_p Y_p.$$

Remark 3.5.0.1. For a Borel measurable map $h : \mathbb{R} \to \mathbb{R}_{\geq 0}$ we get

$$\mathbb{E}[h(X) \mid Y_1, \dots, Y_p] = \int_{\mathbb{R}} h(x) Q_{\sum_{j=1}^n \lambda_j Y_j, \sigma^2}(x) \mathrm{d}x,$$

where $\sigma^2 := \mathbb{E}[X - \sum_{j=1}^n \lambda_j Y_j]$ and

$$Q_{m,\sigma^2}(x) := \frac{1}{\sigma\sqrt{2\pi}} \exp\left(-\frac{(x-m)^2}{2\sigma^2}\right).$$

Exercise 3.5.0.2. Prove[8] Remark 3.5.0.1.

Proof of Theorem 3.5.0.1. Let $\tilde{X} = \lambda_1 Y_1 + \cdots + \lambda_p Y_p$ be the orthogonal projection of X onto $\mathrm{span}(Y_1, \dots, Y_p)$, meaning that for all $1 \leq j \leq p$ we have

$$\mathbb{E}[(X - \tilde{X})Y_j] = 0.$$

Note that this condition gives us explicitly the λ_j's. We obtain thus that $(Y_1, \dots, Y_p, (X - \tilde{X}))$ is a Gaussian vector. Moreover, we get $\mathbb{E}[(X - \tilde{X})Y_j] = \mathbb{CV}(X - \tilde{X}, Y_j) = 0$ and thus $X - \tilde{X}$ is independent of $Y_1, \dots, Y_p$. Hence, we have

$$\begin{aligned}\mathbb{E}[X \mid Y_1, \dots, Y_p] &= \mathbb{E}[X - \tilde{X} + \tilde{X} \mid Y_1, \dots, Y_p] \\ &= \mathbb{E}[X - \tilde{X} \mid Y_1, \dots, Y_p] + \mathbb{E}[\tilde{X} \mid Y_1, \dots, Y_p] \\ &= \mathbb{E}[X - \tilde{X}] + \tilde{X} = \tilde{X}.\end{aligned}$$

□

3.6. Transition Kernel and Conditional Distribution

Definition 3.6.0.1 (Transition kernel). Let $(E, \mathcal{E})$ and $(F, \mathcal{F})$ be two measurable spaces. A *transition kernel* from E to F is a map

$$\nu : E \times \mathcal{F} \to [0, 1],$$

such that

(i) $\nu(x, \cdot)$ is a probability measure on $\mathcal{F}$ for all $x \in E$;
(ii) $x \mapsto \nu(x, A)$ is $\mathcal{E}$-measurable for all $A \in \mathcal{F}$.

---

[8] The proof of Remark 3.5.0.1 is similar to the proof of Theorem 3.5.0.1.

Example 3.6.0.2. Let ρ be a σ-finite measure on $\mathcal{F}$ and let $f\colon E \times F \to \mathbb{R}_{\geq 0}$ be a map such that

$$\int_F f(x,y)\mathrm{d}\rho(y) = 1.$$

Then, we get that

$$\nu(x,A) = \int_A f(x,y)\mathrm{d}\rho(y),$$

is a transition kernel. An example for such a function f would be

$$f(x,y) = \frac{1}{\sigma\sqrt{2\pi}}\exp\left(-\frac{(x-y)^2}{2\sigma^2}\right).$$

Proposition 3.6.0.1. *The following assertions hold:*

(i) *Let h be a non-negative (respectively, bounded) Borel measurable function on a measurable space $(F,\mathcal{F})$. Then*

$$\varphi(x) = \int_F h(y)\nu(x,\mathrm{d}y)$$

is a non-negative (respectively, bounded) measurable function on a measurable space $(E,\mathcal{E})$.

(ii) *If ρ is a probability measure on a measurable space $(E,\mathcal{E})$, then*

$$\mu(A) = \int_E \nu(x,A)\mathrm{d}\rho(x)$$

is a probability measure on a measurable space $(F,\mathcal{F})$ for all $A \in \mathcal{F}$.

Definition 3.6.0.3 (Conditional distribution). Let X and Y be two random variables with values in a measurable space $(E,\mathcal{E})$. The *conditional distribution* of Y given X is any transition kernel ν from E to F such that for all non-negative (respectively, bounded), measurable maps h on a measurable space $(F,\mathcal{F})$ one has

$$\mathbb{E}[h(Y) \mid X] = \int_F h(y)\nu(X,\mathrm{d}y) \quad \text{a.s.},$$

where the last equality should be understood as a map $\varphi(X)$ given by

$$\varphi : x \mapsto \int_F h(y)\nu(x,\mathrm{d}y).$$

Remark 3.6.0.4. If ν is the conditional distribution of Y given X, we get for all $A \in F$

$$\mathbb{P}(Y \in A \mid X) = \nu(X, A) \quad \text{a.s.},$$

where we have set $h = \mathbb{1}_A$ in Definition 3.6.0.3. If ν' is another such conditional distribution, we get

$$\nu(X, A) = \nu'(X, A) \quad \text{a.s.}$$

This implies that

$$\nu(x, A) = \nu'(x, A) \mathrm{d}\mathbb{P}_X(x) \quad \text{a.s.}$$

Theorem 3.6.0.2. *Assume that $(E, \mathcal{E})$ and $(F, \mathcal{F})$ are two complete, separable, metric, measurable spaces endowed with their Borel σ-algebras. Then, the conditional distribution of Y given X, exists and is almost surely unique.*

Chapter 4

Martingales

4.1. Discrete Time Martingales

Recall that the strong law of large numbers tells us, if $(X_n)_{n\geq 1}$ is i.i.d., $\mathbb{E}[|X_i|] < \infty$ and $\mathbb{E}[X_i] = \mu$, then

$$\frac{1}{n}S_n \xrightarrow[\text{a.s.}]{n\to\infty} \mu,$$

with $S_n = \sum_{i=1}^{n} X_i$. We saw that the 0–1 law of Kolmogorov (Theorem 2.7.2.1) implied that in this case the limit, if it exists, is constant. It is of course of interest to have a framework in which the sequence of random variables converges a.s. to another random variable. This can be achieved in the framework of *martingales*. In this chapter, we shall consider a probability space $(\Omega, \mathcal{F}, \mathbb{P})$ as well as an increasing family $(\mathcal{F}_n)_{n\geq 0}$ of sub-σ-algebras of $\mathcal{F}$, i.e., $\mathcal{F}_n \subset \mathcal{F}_{n+1} \subset \mathcal{F}$. Such a sequence is called a *filtration*. The space $(\Omega, \mathcal{F}, (\mathcal{F}_n)_{n\geq 0}, \mathbb{P})$ is called a *filtered probability space*. We shall also consider a sequence $(X_n)_{n\geq 0}$ of random variables. Such a sequence is generally called a *stochastic process* (n is thought of as time). If for every $n \geq 0$, X_n is $\mathcal{F}_n$-measurable, we say that $(X_n)_{n\geq 0}$ is *adapted* (to the filtration $(\mathcal{F}_n)_{n\geq 0}$). One can think of $\mathcal{F}_n$ as the information at time n and the filtration $(\mathcal{F}_n)_{n\geq 0}$ as the flow of information in time.

Remark 4.1.0.1. Let us start with a stochastic process $(X_n)_{n\geq 0}$. We define

$$\mathcal{F}_n := \sigma(X_0, \ldots, X_n) = \sigma(X_k \mid 0 \leq k \leq n).$$

By construction, $\mathcal{F}_n \subset \mathcal{F}_{n+1}$ and $(X_n)_{n\geq 0}$ is $(\mathcal{F}_n)_{n\geq 0}$-adapted. In this case $(\mathcal{F}_n)_{n\geq 0}$ is called the *natural filtration* of $(X_n)_{n\geq 0}$.

Remark 4.1.0.2. In general, if $(\mathcal{F}_n)_{n\geq 0}$ is a filtration, we denote by

$$\mathcal{F}_\infty := \bigvee_{n\geq 0} \mathcal{F}_n = \sigma\left(\bigcup_{n\geq 0} \mathcal{F}_n\right)$$

the *tail σ-algebra.*

Definition 4.1.0.3 (Martingale). Let $(\Omega, \mathcal{F}, (\mathcal{F}_n)_{n\geq 0}, \mathbb{P})$ be a filtered probability space. A stochastic process $(X_n)_{n\geq 0}$ is called a *martingale*, if

(i) $\mathbb{E}[|X_n|] < \infty$ for all $n \geq 0$;
(ii) X_n is $\mathcal{F}_n$-measurable (adapted) for all $n \geq 0$;
(iii) $\mathbb{E}[X_n \mid \mathcal{F}_m] = X_m$ a.s. for all $m \leq n$.

Remark 4.1.0.4. The last point is equivalent to say

$$\mathbb{E}[X_{n+1} \mid \mathcal{F}_n] = X_n \quad \text{a.s.,}$$

which can be obtained by using the tower property (Theorem 3.2.3.11) and induction.

Example 4.1.0.5. Let $(X_n)_{n\geq 0}$ be a sequence independent random variables such that $\mathbb{E}[X_n] = 0$ for all $n \geq 0$ (i.e., $X_n \in L^1(\Omega, \mathcal{F}, (\mathcal{F}_n)_{n\geq 0}, \mathbb{P})$). Moreover, let $\mathcal{F}_n = \sigma(X_1, \ldots, X_n)$ and $S_n = \sum_{i=1}^n X_i$ with $\mathcal{F}_0 = \{\emptyset, \Omega\}$ and $S_0 = 0$. Then $(S_n)_{n\geq 0}$ is an $\mathcal{F}_n$-martingale.

Proof of Example 4.1.0.5. We need to check the assumptions for a martingale:

(i) The first point is clear by the assumption on $X_1, \ldots, X_n$ and linearity of the expectation:

$$\mathbb{E}[|S_n|] \leq \sum_{i=1} \mathbb{E}[|X_i|] < \infty.$$

(ii) It is clear that S_n is $\mathcal{F}_n$-measurable, since it is a function of $X_1, \ldots, X_n$, which are $\mathcal{F}_n$-measurable.

(iii) Note that

$$\begin{aligned}\mathbb{E}[S_{n+1} \mid \mathcal{F}_n] &= \mathbb{E}[\underbrace{X_1 + \cdots + X_n}_{S_n} + X_{n+1} \mid \mathcal{F}_n] \\ &= \underbrace{\mathbb{E}[S_n \mid \mathcal{F}_n]}_{S_n} + \mathbb{E}[X_{n+1} \mid \mathcal{F}_n] \\ &= S_n + \mathbb{E}[X_{n+1} \mid \mathcal{F}_n] = S_n.\end{aligned}$$

We conclude thus that $(S_n)_{n\geq 0}$ is a martingale. □

Example 4.1.0.6. Let $(\Omega, \mathcal{F}, (\mathcal{F}_n)_{n\geq 0}, \mathbb{P})$ be a filtered probability space and let $Y \in L^1(\Omega, \mathcal{F}, (\mathcal{F}_n)_{n\geq 0}, \mathbb{P})$. Define a sequence $(X_n)_{n\geq 0}$ by

$$X_n := \mathbb{E}[Y \mid \mathcal{F}_n].$$

Then $(X_n)_{n\geq 0}$ is an $\mathcal{F}_n$-martingale.

Proof of Example 4.1.0.6. Again, we show the assumptions for a martingale:

(i) Since $|X_n| \leq \mathbb{E}[|Y| \mid \mathcal{F}_n]$, we get

$$\mathbb{E}[|X_n|] \leq \mathbb{E}[|Y|] < \infty.$$

(ii) $\mathbb{E}[Y \mid \mathcal{F}_n]$ is $\mathcal{F}_n$-measurable by definition.
(iii) With the tower property (Theorem 3.2.3.11), we get

$$\mathbb{E}[X_{n+1} \mid \mathcal{F}_n] = \mathbb{E}[\underbrace{\mathbb{E}[Y \mid \mathcal{F}_{n+1}]}_{X_{n+1}} \mid \mathcal{F}_n] = \mathbb{E}[Y \mid \mathcal{F}_n] = X_n.$$

We conclude that $(X_n)_{n\geq 0}$ is a martingale. □

Definition 4.1.0.7 (Regular). Let $(\Omega, \mathcal{F}, (\mathcal{F}_n)_{n\geq 0}, \mathbb{P})$ be a filtered probability space. A martingale $(X_n)_{n\geq 0}$ is said to be *regular*, if there exists a random variable $Y \in L^1(\Omega, \mathcal{F}, (\mathcal{F}_n)_{n\geq 0}, \mathbb{P})$ such that

$$X_n = \mathbb{E}[Y \mid \mathcal{F}_n], \qquad \forall n \geq 0.$$

Proposition 4.1.0.1. *Let $(\Omega, \mathcal{F}, (\mathcal{F}_n)_{n\geq 0}, \mathbb{P})$ be a filtered probability space. Let $(X_n)_{n\geq 0}$ be a martingale. Then the map*

$$n \mapsto \mathbb{E}[X_n]$$

is constant, i.e., we have

$$\mathbb{E}[X_n] = \mathbb{E}[X_0], \qquad \forall n \geq 0.$$

Proof. By Definition 4.1.0.3, we have

$$\mathbb{E}[X_n] = \mathbb{E}[\mathbb{E}[X_n \mid \mathcal{F}_0]] = X_0.$$

□

Definition 4.1.0.8 (Discrete stopping time). Let $(\Omega, \mathcal{F}, (\mathcal{F}_n)_{n\geq 0}, \mathbb{P})$ be a filtered probability space. A random variable $T : \Omega \to \bar{\mathbb{N}} = \mathbb{N} \cup \{\infty\}$ is called a *stopping time* if for every $n \geq 0$ we have

$$\{T \leq n\} \in \mathcal{F}_n.$$

Remark 4.1.0.9. Another, more general definition is used for *continuous* stochastic processes and may be given in terms of a filtration. Let $(I, \leq)$ be an ordered index set (often $I = [0, \infty)$) or a compact subset thereof, thought of as the set of possible *times*), and let $(\Omega, \mathcal{F}, (\mathcal{F}_n)_{n\geq 0}, \mathbb{P})$ be a filtered probability space. Then a random variable $T : \Omega \to I$ is called a *stopping time* if $\{T \leq t\} \in \mathcal{F}_t$ for all $t \in I$. Often, to avoid confusion, we call it a $\mathcal{F}_t$*-stopping time* and explicitly specify the filtration. Speaking concretely, for T to be a stopping time, it should be possible to decide whether or not $\{T \leq t\}$ has occurred on the basis of the knowledge of $\mathcal{F}_t$, i.e., $\{T \leq t\}$ is $\mathcal{F}_t$-measurable.

Proposition 4.1.0.2. *Let $(\Omega, \mathcal{F}, (\mathcal{F}_n)_{n\geq 0}, \mathbb{P})$ be a filtered probability space. Then the following conditions hold:*

(i) *Constant times are stopping times.*
(ii) *The map*

$$T : (\Omega, \mathcal{F}) \to (\bar{\mathbb{N}}, \mathcal{P}(\bar{\mathbb{N}}))$$

is a stopping time if and only if $\{T \leq n\} \in \mathcal{F}_n$ for all $n \geq 0$.
(iii) *If S and T are stopping times, then $S \wedge T$, $S \vee T$ and $S + T$ are also stopping times.*

(iv) *Let* $(T_n)_{n\geq 0}$ *be a sequence of stopping times. Then* $\sup_n T$, $\inf_n T_n$, $\liminf_n T_n$ *and* $\limsup_n T_n$ *are also stopping times.*

(v) *Let* $(X_n)_{n\geq 0}$ *be a sequence of adapted random variables with values in some measure space* $(E, \mathcal{E})$ *and let* $H \in \mathcal{E}$. *Then, with the convention that* $\inf \emptyset = \infty$, *we get that*

$$D_H := \inf\{n \in \mathbb{N} \mid X_n \in H\}$$

is a stopping time.

Proof. (i) This is clear.

(ii) Note that

$$\{T = n\} = \{T \leq n\} \setminus \{T \leq n-1\} \in \mathcal{F}_n,$$

and, conversely, we have

$$\{T \leq n\} = \bigcup_{k=0}^{n} \{T = k\} \in \mathcal{F}_n.$$

(iii) Note that

$$\{S \vee T \leq n\} = \{S \leq n\} \cap \{T \leq n\} \in \mathcal{F}_n,$$
$$\{S \wedge T \leq n\} = \{S \leq n\} \cup \{T \leq n\} \in \mathcal{F}_n.$$

Thus, we have

$$\{S + T = n\} = \bigcup_{k=0}^{n} \underbrace{\{S = k\}}_{\in \mathcal{F}_k \subset \mathcal{F}_n} \cap \underbrace{\{T = n-k\}}_{\in \mathcal{F}_{n-k} \subset \mathcal{F}_n} \in \mathcal{F}_n.$$

(iv) Note first that

$$\left\{\sup_k T_k \leq n\right\} = \bigcap_k \{T_k \leq n\} \in \mathcal{F}_n$$

and

$$\left\{\inf_k T_k \leq n\right\} = \bigcup_k \{T_k \leq n\} \in \mathcal{F}_n.$$

Thus we can rewrite

$$\limsup_k T_k = \inf_k \sup_{m \geq k} T_m$$

and

$$\liminf_k T_k = \sup_k \inf_{m \leq k} T_m$$

and use the relation above.

(v) Note that for all $n \in \mathbb{N}$, we get

$$\{D_H \leq n\} = \bigcup_{k=0}^{n} \underbrace{\{X_k \in H\}}_{\in \mathcal{F}_k \subset \mathcal{F}_n} \in \mathcal{F}_n.$$

□

Remark 4.1.0.10. We say that a stopping time T is *bounded* if there exists a constant $C > 0$ such that for all $\omega \in \Omega$ we have

$$T(\omega) \leq C.$$

Without loss of generality, we can always assume that $C \in \mathbb{N}$. In this case, we shall denote by X_T the random variable given by

$$X_T(\omega) := X_{T(\omega)}(\omega) = \sum_{n=0}^{\infty} X_n(\omega) \mathbb{1}_{\{T(\omega)=n\}}.$$

Note that the sum on the right-hand side is perfectly defined since $T(\omega)$ is bounded.

Theorem 4.1.0.3. *Let $(\Omega, \mathcal{F}, (\mathcal{F}_n)_{n\geq 0}, \mathbb{P})$ be a filtered probability space. Let T be a bounded stopping time and let $(X_n)_{n\geq 0}$ be a martingale. Then we have*

$$\mathbb{E}[X_T] = \mathbb{E}[X_0].$$

Proof. Assume that $T \leq N \in \mathbb{N}$. Then we get

$$\begin{aligned}\mathbb{E}[X_T] &= \mathbb{E}\left[\sum_{n=0}^{\infty} X_n \mathbb{1}_{\{T=n\}}\right] = \mathbb{E}\left[\sum_{n=0}^{N} X_n \mathbb{1}_{\{T=n\}}\right] = \sum_{n=0}^{N} \mathbb{E}[X_n \mathbb{1}_{\{T=n\}}] \\ &= \sum_{n=0}^{N} \mathbb{E}[\mathbb{E}[X_N \mid \mathcal{F}_n] \mathbb{1}_{\{T=n\}}] = \sum_{n=0}^{N} \mathbb{E}[\mathbb{E}[X_n \mathbb{1}_{\{T=n\}} \mid \mathcal{F}_n]] \\ &= \sum_{n=0}^{N} \mathbb{E}[X_n \mathbb{1}_{\{T=n\}}] = \mathbb{E}\left[X_n \sum_{n=0}^{N} \mathbb{1}_{\{T=n\}}\right] = \mathbb{E}[X_n] = \mathbb{E}[X_0].\end{aligned}$$

□

Definition 4.1.0.11 (Stopping time σ-algebra). Let $(\Omega, \mathcal{F}, (\mathcal{F}_n)_{n\geq 0}, \mathbb{P})$ be a filtered probability space. Let T be a stopping time for $(\mathcal{F}_n)_{n\geq 0}$. We define the σ-algebra of events prior of T to be

$$\mathcal{F}_T := \{A \in \mathcal{F} \mid A \cap \{T \leq n\} \in \mathcal{F}_n, \forall n \geq 0\}.$$

Remark 4.1.0.12. It remains to show that $\mathcal{F}_T$ is indeed a σ-algebra.

Proposition 4.1.0.4. *If T is a stopping time, then $\mathcal{F}_T$ is a σ-algebra.*

Proof. It is clear that for a filtered probability space $(\Omega, \mathcal{F}, (\mathcal{F}_n)_{n\geq 0}, \mathbb{P})$, we have $\Omega \in \mathcal{F}_T$. Moreover, if $A \in \mathcal{F}_T$, then

$$A^{\mathsf{C}} \cap \{T \leq n\} = \underbrace{\{T \leq n\}}_{\in \mathcal{F}_n} \setminus \underbrace{(A \cap \{T \leq n\})}_{\in \mathcal{F}_n} \in \mathcal{F}_n,$$

and hence $A^{\mathsf{C}} \in \mathcal{F}_n$. Further, if $(A_i)_{i\geq 0} \in \mathcal{F}_T$, then

$$\bigcup_{i\geq 0} A_i \cap \{T \leq n\} = \bigcup_{i=1}^{\infty} A_i \cap \underbrace{\{T \leq n\}}_{\in \mathcal{F}_n}.$$

Hence, we get $\bigcup_{i\geq 0} A_i \in \mathcal{F}_T$. Therefore, $\mathcal{F}_T$ is a σ-algebra. □

Remark 4.1.0.13. If $T = n_0$ is constant, then $\mathcal{F}_T = \mathcal{F}_{n_0}$.

Exercise 4.1.0.14. Show that

$$\mathcal{F}_T = \left\{ \bigcup_{n \in \bar{\mathbb{N}}} A_n \cap \{T = n\} \,\middle|\, A_\infty \in \mathcal{F}, A_n \in \mathcal{F}_n \right\}.$$

Exercise 4.1.0.15. Show that a random variable L with values in $\bar{\mathbb{N}}$ is a stopping time if and only if $\left(\mathbb{1}_{\{L\leq n\}}\right)_{n\geq 0}$ is $(\mathcal{F}_n)$-adapted and for the case it is a stopping time, we get

$$L = \inf\{n \geq 0 \mid \mathbb{1}_{\{L \leq n\}} = 1\}.$$

Proposition 4.1.0.5. *Let S and T be two stopping times. Then the following conditions hold:*

(i) *If $S \leq T$, then $\mathcal{F}_S \subset \mathcal{F}_T$.*
(ii) *$\mathcal{F}_{S\wedge T} = \mathcal{F}_S \cap \mathcal{F}_T$.*
(iii) *$\{S \leq T\}$, $\{S = T\}$ and $\{S < T\}$ are $\mathcal{F}_S \cap \mathcal{F}_T$-measurable.*

Proof. (i) Note that for $n \in \mathbb{N}$ and $A \in \mathcal{F}_S$ we get

$$A \cap \{T \leq n\} = A \cap \underbrace{\{S \leq n\} \cap \{T \leq n\}}_{\{T \leq n\}} = (A \cap \{S \leq n\}) \cap \underbrace{\{T \leq n\}}_{\in \mathcal{F}_n} \in \mathcal{F}_n.$$

Therefore, we have $A \in \mathcal{F}_T$.

(ii) Since $S \wedge T \leq S$, we get by (i) that $\mathcal{F}_{S \wedge T} \subset \mathcal{F}_S$ and similarly that $\mathcal{F}_{S \wedge T} \subset \mathcal{F}_T$. Let now $A \in \mathcal{F}_S \cap \mathcal{F}_T$. Then

$$A \cap \{S \wedge T \leq n\} = \left(\underbrace{A \cap \{S \leq n\}}_{\in \mathcal{F}_n, (\text{since } A \in \mathcal{F}_S)} \right) \cup \left(\underbrace{A \cap \{T \leq n\}}_{\in \mathcal{F}_n, (\text{since } A \in \mathcal{F}_T)} \right) \in \mathcal{F}_n.$$

Therefore, we have $A \in \mathcal{F}_{S \wedge T}$.

(iii) Note that

$$\{S \leq T\} \cap \{T = n\} = \{S \leq n\} \cap \{T = n\} \in \mathcal{F}_n.$$

Therefore, we have $\{S \leq T\} \in \mathcal{F}_T$. Note also that

$$\{S < T\} \cap \{T = n\} = \{S < n\} \cap \{T = n\} \in \mathcal{F}_n.$$

Therefore, we have $\{S < T\} \in \mathcal{F}_T$. Finally, note that

$$\{S = T\} \cap \{T = n\} = \{S = n\} \cap \{T = n\} \in \mathcal{F}_n.$$

Therefore, we have $\{S = T\} \in \mathcal{F}_T$. Similarly, one can show that these events are also $\mathcal{F}_S$-measurable (we leave this as an exercise). □

Proposition 4.1.0.6. *Let $(\Omega, \mathcal{F}, (\mathcal{F}_n)_{n \geq 0}, \mathbb{P})$ be a filtered probability space. Let $(X_n)_{n \geq 0}$ be a stochastic process, which is adapted, i.e., X_n is $\mathcal{F}_n$-measurable for all $n \geq 0$. Let T be a finite stopping time, i.e., $T < \infty$ almost surely, such that X_T is well-defined. Then X_T is $\mathcal{F}_T$-measurable.*

Proof. Let $\Lambda \in \mathcal{B}(\mathbb{R})$ be a Borel measurable set. We want to show that

$$\{X_T \in \Lambda\} \in \mathcal{F}_T,$$

which means that, for all $n \geq 0$, we have

$$\{X_T \in \Lambda\} \cap \{T \leq n\} \in \mathcal{F}_n.$$

Note that

$$\begin{aligned}\{X_T \in \Lambda\} \cap \{T \leq n\} &= \bigcup_{k=1}^{n} \{X_T \in \Lambda\} \cap \{T \leq k\} \\ &= \bigcup_{k=1}^{n} \underbrace{\{X_k \in \Lambda\}}_{\in \mathcal{F}_k \subset \mathcal{F}_n} \cap \underbrace{\{T = k\}}_{\in \mathcal{F}_k \subset \mathcal{F}_n},\end{aligned}$$

which implies that $\{X_T \in \Lambda\} \cap \{T \leq n\} \in \mathcal{F}_n$ and thus the claim follows. $\square$

Theorem 4.1.0.7. *Let $(\Omega, \mathcal{F}, (\mathcal{F}_n)_{n\geq 0}, \mathbb{P})$ be a filtered probability space. Let $(X_n)_{n\geq 0}$ be a martingale and let S and T be two bounded stopping times such that $S \leq T$ almost surely. Then we have*

$$\mathbb{E}[X_T \mid \mathcal{F}_S] = X_S \quad \text{a.s.}$$

Proof. Since we assume that $T \leq C \in \mathbb{N}$, we note that

$$|X_T| \leq \sum_{i=0}^{C} |X_i| \in L^1(\Omega, \mathcal{F}, (\mathcal{F}_n)_{n\geq 0}, \mathbb{P}).$$

Let now $A \in \mathcal{F}_S$. We need to show that

$$\mathbb{E}[X_T \mathbb{1}_A] = \mathbb{E}[X_S \mathbb{1}_A].$$

Let us define the *random time*

$$R(\omega) := S(\omega)\mathbb{1}_A(\omega) + T(\omega)\mathbb{1}_{A^{\complement}}(\omega).$$

We can observe that R is a stopping time. Indeed, we have

$$\{R \leq n\} = \underbrace{(A \cap \{S \leq n\})}_{\in \mathcal{F}_n} \cup \underbrace{(A^{\complement} \cap \{T \leq n\})}_{\in \mathcal{F}_n}.$$

Consequently, since S, T and R are bounded, we have

$$\mathbb{E}[X_S] = \mathbb{E}[X_T] = \mathbb{E}[X_R] = \mathbb{E}[X_0].$$

Therefore, we get

$$\mathbb{E}[X_R] = \mathbb{E}[X_S \mathbb{1}_A + X_T \mathbb{1}_{A^\complement}]$$

and

$$\mathbb{E}[X_T] = \mathbb{E}[X_T \mathbb{1}_A + X_T \mathbb{1}_{A^\complement}]$$

and thus

$$\mathbb{E}[X_S \mathbb{1}_A] = \mathbb{E}[X_T \mathbb{1}_A].$$

Moreover, since X_S is $\mathcal{F}_S$-measurable, we conclude that

$$\mathbb{E}[X_T \mid \mathcal{F}_S] = X_S \quad \text{a.s.}$$

□

Exercise 4.1.0.16. Let T be a stopping time and $\Lambda \in \mathcal{F}_T$. Define

$$T_\Lambda(\omega) := \begin{cases} T(\omega) & \text{if } \omega \in \Lambda, \\ \infty & \text{if } \omega \notin \Lambda. \end{cases}$$

Prove that T_Λ is a stopping time.

Proposition 4.1.0.8. *Let $(\Omega, \mathcal{F}, (\mathcal{F}_n)_{n\geq 0}, \mathbb{P})$ be a filtered probability space. Let $(X_n)_{n\geq 0}$ be a stochastic process such that for all $n \geq 0$*

$$\mathbb{E}[|X_n|] < \infty$$

and with X_n being $\mathcal{F}_n$-measurable. If for all bounded stopping times T, we have

$$\mathbb{E}[X_T] = \mathbb{E}[X_0].$$

Then $(X_n)_{n\geq 0}$ is a martingale.

Proof. Let $0 \leq m < n < \infty$ and $\Lambda \in \mathcal{F}_m$. Define for all $\omega \in \Omega$

$$T(\omega) := m\mathbb{1}_{\Lambda^\complement}(\omega) + n\mathbb{1}_\Lambda(\omega).$$

Then it is easy to see that T is a stopping time. Therefore, we have

$$\mathbb{E}[X_0] = \mathbb{E}[X_T] = \mathbb{E}[X_m \mathbb{1}_{\Lambda^\complement} + X_n \mathbb{1}_\Lambda] = \mathbb{E}[X_m].$$

Hence, we get

$$\mathbb{E}[X_m \mathbb{1}_\Lambda] = \mathbb{E}[X_n \mathbb{1}_\Lambda],$$

and thus

$$\mathbb{E}[X_n \mid \mathcal{F}_m] = X_m \quad \text{a.s.} \qquad \square$$

4.2. Submartingales and Supermartingales

Definition 4.2.0.1 (Submartingale and supermartingale). Let $(\Omega, \mathcal{F}, (\mathcal{F}_n)_{n\geq 0}, \mathbb{P})$ be a filtered probability space. A stochastic process $(X_n)_{n\geq 0}$ is called a *submartingale* (respectively, *supermartingale*) if

(i) $\mathbb{E}[|X_n|] < \infty$ for all $n \geq 0$,
(ii) $(X_n)_{n\geq 0}$ is $\mathcal{F}_n$-adapted,
(iii) $\mathbb{E}[X_n \mid \mathcal{F}_m] \geq X_m$ a.s. for all $m \leq n$ (respectively, $\mathbb{E}[X_n \mid \mathcal{F}_m] \leq X_m$ a.s. for all $m \leq n$).

Remark 4.2.0.2. A stochastic process $(X_n)_{n\geq 0}$ is a martingale if and only if it is a submartingale and a supermartingale at the same time. A martingale is in particular both a submartingale and a supermartingale. If $(X_n)_{n\geq 0}$ is a submartingale, then the map $n \mapsto \mathbb{E}[X_n]$ is increasing. If $(X_n)_{n\geq 0}$ is a supermartingale, then the map $n \mapsto \mathbb{E}[X_n]$ is decreasing.

Example 4.2.0.3. Let $(\Omega, \mathcal{F}, (\mathcal{F}_n)_{n\geq 0}, \mathbb{P})$ be a filtered probability space. Let $S_n = \sum_{j=1}^n Y_j$, where $(Y_n)_{n\geq 1}$ is a sequence of i.i.d. random variables. Moreover, let $S_0 = 0$, $\mathcal{F}_0 = \{\emptyset, \Omega\}$ and $\mathcal{F}_n = \sigma(Y_1, \dots, Y_n)$. Then we get

$$\mathbb{E}[S_{n+1} \mid \mathcal{F}_n] = S_n + \mathbb{E}[Y_{n+1}].$$

If $\mathbb{E}[Y_{n+1}] > 0$, then $\mathbb{E}[S_{n+1} \mid \mathcal{F}_n] \geq S_n$ and thus $(S_n)_{n\geq 0}$ is a submartingale. On the other hand, if $\mathbb{E}[Y_{n+1}] < 0$, then $\mathbb{E}[S_{n+1} \mid \mathcal{F}_n] \leq S_n$ and thus $(S_n)_{n\geq 0}$ is a supermartingale.

Proposition 4.2.0.1. *Let $(\Omega, \mathcal{F}, (\mathcal{F}_n)_{n\geq 0}, \mathbb{P})$ be a filtered probability space. If $(M_n)_{n\geq 0}$ is a martingale and φ is a convex function such that $\varphi(M_n) \in L^1(\Omega, \mathcal{F}, (\mathcal{F}_n)_{n\geq 0}, \mathbb{P})$ for all $n \geq 0$, then*

$$(\varphi(M_n))_{n\geq 0}$$

is a submartingale.

Proof. The first two conditions for a martingale are clearly satisfied. Now for $m \leq n$, we get

$$\mathbb{E}[M_n \mid \mathcal{F}_m] = M_m \quad \text{a.s.},$$

since $(M_n)_{n\geq 0}$ is assumed to be a martingale. Hence, with Jensen's inequality (Theorem 3.2.3.10), we get

$$\varphi(\mathbb{E}[M_n \mid \mathcal{F}_m]) = \varphi(M_m) \leq \mathbb{E}[\varphi(M_n) \mid \mathcal{F}_m] \quad \text{a.s.}$$ □

Corollary 4.2.0.2. *Let $(\Omega, \mathcal{F}, (\mathcal{F}_n)_{n\geq 0}, \mathbb{P})$ be a filtered probability space. If $(M_n)_{n\geq 0}$ is a martingale, then the following conditions hold:*

(i) *$(|M_n|)_{n\geq 0}$ and $(M_n^+)_{n\geq 0}$ are submartingales;*
(ii) *if for all $n \geq 0$, $\mathbb{E}[M_n^2] < \infty$, then $(M_n^2)_{n\geq 0}$ is a submartingale.*

Theorem 4.2.0.3. *Let $(\Omega, \mathcal{F}, (\mathcal{F}_n)_{n\geq 0}, \mathbb{P})$ be a filtered probability space. Let $(X_n)_{n\geq 0}$ be a submartingale and let T be a stopping time bounded by $C \in \mathbb{N}$. Then*

$$\mathbb{E}[X_T] \leq \mathbb{E}[X_C].$$

Exercise 4.2.0.4. Prove Theorem 4.2.0.3.

Theorem 4.2.0.4 (Doob's decomposition). *Let $(\Omega, \mathcal{F}, (\mathcal{F}_n)_{n\geq 0}, \mathbb{P})$ be a filtered probability space. Let $(X_n)_{n\geq 0}$ be a submartingale. Then there exists a martingale $M = (M_n)_{n\geq 0}$ with $M_0 = 0$ and a sequence $A = (A_n)_{n\geq 0}$, such that $A_{n+1} \geq A_n$ a.s. with $A_0 = 0$ a.s., which is called an* increasing process, *and with A_{n+1} being $\mathcal{F}_n$-measurable, which we will call* predictable, *such that*

$$X_n = X_0 + M_0 + A_n.$$

Moreover, this decomposition is almost surely unique.

Proof. Let us define $A_0 = 0$ and for $n \geq 1$

$$A_n := \sum_{k=1}^{n} \mathbb{E}[X_k - X_{k-1} \mid \mathcal{F}_{k-1}].$$

Since $(X_n)_{n\geq 0}$ is a submartingale, we get

$$\mathbb{E}[X_k - X_{k-1} \mid \mathcal{F}_{k-1}] \geq 0,$$

and hence $A_{n+1} - A_n \geq 0$. Therefore, we get that $(A_n)_{n\geq 0}$ is an increasing process. Moreover, from the definition of the conditional expectation, A_n is $\mathcal{F}_{n-1}$-measurable for $n \geq 1$. Thus A_n is predictable as well. We also note that

$$\mathbb{E}[X_n \mid \mathcal{F}_{n-1}] - X_{n-1} = \mathbb{E}[X_n - X_{n-1} \mid \mathcal{F}_{n-1}] = A_n - A_{n-1}.$$

Hence, we get

$$\underbrace{\mathbb{E}[X_n \mid \mathcal{F}_{n-1}] - A_n}_{\mathbb{E}[X_n - A_n \mid \mathcal{F}_{n-1}]} = X_{n-1} - A_{n-1}.$$

If we set $M_n = X_n - A_n - X_0$, it follows that $M = (M_n)_{n \geq 0}$ is a martingale with $M_0 = 0$. This proves the existence part. For uniqueness, we note that if we have two such decompositions

$$X_n = X_0 + M_n + A_n = X_0 + L_n + C_n,$$

where L_n denotes the martingale part and C_n the increasing process part, it follows that

$$L_n - M_n = A_n - C_n.$$

Now since $A_n - C_n$ is $\mathcal{F}_{n-1}$-measurable, we get that $L_n - M_n$ is also $\mathcal{F}_{n-1}$-measurable. Thus, we get

$$L_n - M_n = \mathbb{E}[L_n - M_n \mid \mathcal{F}_{n-1}] = L_{n-1} - M_{n-1},$$

because of the martingale property. By induction, we have a chain of equalities

$$L_n - M_n = L_{n-1} - M_{n-1} = \cdots = L_0 - M_0 = 0.$$

Therefore, we have $L_n = M_n$ and also $A_n = C_n$. □

Corollary 4.2.0.5. *Let $(\Omega, \mathcal{F}, (\mathcal{F}_n)_{n \geq 0}, \mathbb{P})$ be a filtered probability space and let $X = (X_n)_{n \geq 0}$ be a supermartingale. Then there exists almost surely a unique decomposition*

$$X_n = X_0 + M_n - A_n,$$

where $M = (M_n)_{n \geq 0}$ is a martingale with $M_0 = 0$ and $A = (A_n)_{n \geq 0}$ is a increasing process with $A_0 = 0$.

Proof. Let $Y_n := -X_n$ for all $n \geq 0$. Then the stochastic process obtained by $(Y_n)_{n\geq 0}$ is a submartingale. Theorem 4.2.0.4 tells us that there exists a unique decomposition

$$Y_n = Y_0 + L_n + C_n,$$

where L_n denotes the martingale part and C_n the increasing process part. Hence, we get

$$X_n = X_0 - L_n - C_n$$

and if we take $M_n = -L_n$ and $A_n = C_n$, the claim follows. □

Consider now a *stopped process*. Let $(\Omega, \mathcal{F}, (\mathcal{F}_n)_{n\geq 0}, \mathbb{P})$ be a filtered probability space. Let T be a stopping time and let $(X_n)_{n\geq 0}$ be a stochastic process. We denote by $X^T = (X_n^T)_{n\geq 0}$ the process $(X_{n\wedge T})_{n\geq 0}$.

Proposition 4.2.0.6. *Let $(\Omega, \mathcal{F}, (\mathcal{F}_n)_{n\geq 0}, \mathbb{P})$ be a filtered probability space. Let $(X_n)_{n\geq 0}$ be a martingale (respectively, sub- or supermartingale) and let T be a stopping time. Then $(X_{n\wedge T})_{n\geq 0}$ is also a martingale (respectively, sub- or supermartingale).*

Proof. Note that

$$\{T \geq n+1\} = \{T \leq n\}^{\mathsf{C}} \in \mathcal{F}_n.$$

Hence, we have

$$\begin{aligned}\mathbb{E}[X_{n+1\wedge T} - X_{n\wedge T} \mid \mathcal{F}_n] &= \mathbb{E}[(X_{n+1\wedge T} - X_{n\wedge T})\mathbb{1}_{\{T\geq n+1\}} \mid \mathcal{F}_n] \\ &= \mathbb{1}_{\{T\geq n+1\}}\mathbb{E}[X_{n+1} - X_n \mid \mathcal{F}_n].\end{aligned}$$

If $(X_n)_{n\geq 0}$ is a martingale, we deduce that

$$\mathbb{E}[X_{n+1\wedge T} - X_{n\wedge T} \mid \mathcal{F}_n] = 0.$$

Moreover, $X_{n\wedge T}$ is $\mathcal{F}_n$-measurable. Therefore, we have

$$\mathbb{E}[X_{n+1\wedge T} \mid \mathcal{F}_n] = X_{n\wedge T}.$$

The same holds for sub- and supermartingales. □

Theorem 4.2.0.7. *Let $(\Omega, \mathcal{F}, (\mathcal{F}_n)_{n\geq 0}, \mathbb{P})$ be a filtered probability space. Let $(X_n)_{n\geq 0}$ be a submartingale (respectively, supermartingale) and let S and T be two bounded stopping times, such that $S \leq T$ a.s. Then*

$$\mathbb{E}[X_T \mid \mathcal{F}_S] \geq X_S \quad a.s. \quad (respectively, \quad \mathbb{E}[X_T \mid \mathcal{F}_S] \leq X_S \quad a.s.).$$

Proof. Let us assume that $(X_n)_{n\geq 0}$ is a supermartingale. Let $A \in \mathcal{F}_S$ such that $S \leq T \leq C \in \mathbb{N}$. We already know that $(X_{n\wedge T})_{n\geq 0}$ is a supermartingale. Therefore, we get

$$\begin{aligned}
\mathbb{E}[X_T \mathbb{1}_A] &= \sum_{j=0}^{C} \mathbb{E}[X_{\underbrace{C \wedge T}_{T}} \mathbb{1}_A \mathbb{1}_{\{S=j\}}] = \sum_{j=0}^{C} \mathbb{E}[X_{C\wedge T} \mathbb{1}_{\underbrace{A \cap \{S = j\}}_{\in \mathcal{F}_j}}] \\
&\leq \sum_{j=0}^{C} \mathbb{E}[X_{j\wedge T} \mathbb{1}_{A\cap\{S=j\}}] = \sum_{j=0}^{C} \mathbb{E}[X_j \mathbb{1}_A \mathbb{1}_{\{S=j\}}] \\
&= \mathbb{E}\left[\sum_{j=0}^{C} X_j \mathbb{1}_{\{S=j\}} \mathbb{1}_A\right] = \mathbb{E}[X_S \mathbb{1}_A] = \mathbb{E}[X_T \mid \mathcal{F}_S] \leq X_S. \square
\end{aligned}$$

Corollary 4.2.0.8. *Let $(\Omega, \mathcal{F}, (\mathcal{F}_n)_{n\geq 0}, \mathbb{P})$ be a filtered probability space. Let $(X_n)_{n\geq 0}$ be a submartingale (respectively, supermartingale) and let T be a bounded stopping time. Then*

$$\mathbb{E}[X_T] \geq \mathbb{E}[X_0] \qquad (respectively, \quad \mathbb{E}[X_T] \leq \mathbb{E}[X_0]).$$

Moreover, if $S \leq T$, for S and T two bounded stopping times, we have

$$\mathbb{E}[X_T] \geq \mathbb{E}[X_S] \qquad (respectively, \quad \mathbb{E}[X_T] \leq \mathbb{E}[X_S]).$$

Exercise 4.2.0.5. Let $(\Omega, \mathcal{F}, (\mathcal{F}_n)_{n\geq 0}, \mathbb{P})$ be a filtered probability space. Let $X = (X_n)_{n\geq 0}$ be a supermartingale and let T be a stopping time. Show that then

$$X_T \in L^1(\Omega, \mathcal{F}, (\mathcal{F}_n)_{n\geq 0}, \mathbb{P}),$$

and

$$\mathbb{E}[X_T] \leq \mathbb{E}[X_0]$$

in each case of the following situations:

(i) T is bounded,
(ii) X is bounded and T is finite,
(iii) $\mathbb{E}[T] < \infty$ and for some $k \geq 0$, we have

$$|X_n(\omega) - X_{n-1}(\omega)| \leq k, \quad \forall \omega \in \Omega.$$

4.3. Martingale Inequalities

Let $(\Omega, \mathcal{F}, (\mathcal{F}_n)_{n\geq 0}, \mathbb{P})$ be a filtered probability space. Let $X = (X_n)_{n\geq 0}$ be a stochastic process, such that X_n is $\mathcal{F}_n$-measurable for all $n \geq 0$. Define a stochastic process

$$X_n^* := \sup_{j\leq n} |X_j|.$$

Note that $(X_n^*)_{n\geq 0}$ is increasing and $\mathcal{F}_n$-adapted. Therefore, if $X_n \in L^1(\Omega, \mathcal{F}, (\mathcal{F}_n)_{n\geq 0}, \mathbb{P})$ for all $n \geq 0$, then $(X_n^*)_{n\geq 0}$ is a submartingale.

4.3.1. Maximal inequality and Doob's inequality

Recall Markov's inequality (see Proposition 2.3.1.2) in terms of $(X_n^*)_{n\geq 0}$, which is given by

$$\mathbb{P}(X_n^* \geq \alpha) \leq \frac{\mathbb{E}[X_n^*]}{\alpha},$$

with the obvious bound

$$\mathbb{E}[X_n^*] \leq \sum_{j=1}^{n} \mathbb{E}[|X_j|].$$

We shall see for instance that when $(X_n)_{n\geq 0}$ is a martingale, one can replace $\mathbb{E}[X_n^*]$ by $\mathbb{E}[|X_n|]$.

Proposition 4.3.1.1. *Let $(\Omega, \mathcal{F}, (\mathcal{F}_n)_{n\geq 0}, \mathbb{P})$ be a filtered probability space. Let $(X_n)_{n\geq 0}$ be a submartingale and let $\lambda > 0$ and $k \in \mathbb{N}$.*

Moreover, define

$$A := \left\{ \max_{0\le n\le k} X_n \ge \lambda \right\},$$

$$B := \left\{ \min_{0\le n\le k} X_n \le -\lambda \right\}.$$

Then the following conditions hold:

(i)
$$\lambda \mathbb{P}(A) \le \mathbb{E}[X_k \mathbb{1}_A],$$

(ii)
$$\lambda \mathbb{P}(B) \le \mathbb{E}[X_k \mathbb{1}_{B^\complement}] - \mathbb{E}[X_0].$$

Remark 4.3.1.1. If $(X_n)_{n\ge 0}$ is a martingale, then $(|X_n|)_{n\ge 0}$ is a submartingale. Moreover, using (i), we get

$$\lambda \mathbb{P}(X_k^* \ge \alpha) \le \mathbb{E}[|X_k| \mathbb{1}_A] \le \mathbb{E}[|X_k|]$$

and thus

$$\mathbb{P}(X_k^* \ge \alpha) \le \frac{\mathbb{E}[|X_k|]}{\alpha}.$$

Proof of Proposition 4.3.1.1. (i) Let us introduce

$$T := \inf\{n \in \mathbb{N} \mid X_n \le \lambda\} \wedge k.$$

Then T is a stopping time, which is bounded by k. We thus have

$$\mathbb{E}[X_T] \le \mathbb{E}[X_k].$$

We note that $X_T = X_k$ if $T = k$, which happens for $\omega \in A^\complement$. Hence, we get

$$\mathbb{E}[X_T] = \mathbb{E}[X_T \mathbb{1}_A + X_T \mathbb{1}_{A^\complement}] = \mathbb{E}[X_T \mathbb{1}_A] + \mathbb{E}[X_k \mathbb{1}_{A^\complement}] \le \underbrace{\mathbb{E}[X_k]}_{\mathbb{E}[X_k(\mathbb{1}_A + \mathbb{1}_{A^\complement})]}.$$

Moreover, we note that

$$\mathbb{E}[X_T \mathbb{1}_A] \geq \lambda \mathbb{E}[\mathbb{1}_A] = \lambda \mathbb{P}(A).$$

Therefore, we get

$$\lambda \mathbb{P}(A) \leq \mathbb{E}[X_k \mathbb{1}_A].$$

(ii) Let us define

$$S := \inf\{n \in \mathbb{N} \mid X_n \leq -\lambda\} \wedge k.$$

Again S is a stopping time, which is bounded by k. We hence have

$$\mathbb{E}[X_S] \geq \mathbb{E}[X_0].$$

Thus, we have

$$\mathbb{E}[X_0] \leq \mathbb{E}[X_S \mathbb{1}_B] + \mathbb{E}[X_S \mathbb{1}_{B^\complement}] \leq -\lambda \mathbb{P}[B] + \mathbb{E}[X_k \mathbb{1}_{B^\complement}].$$

Therefore, we get

$$\lambda \mathbb{P}(B) \leq \mathbb{E}[X_k \mathbb{1}_{B^\complement}] - \mathbb{E}[X_0].$$

□

Proposition 4.3.1.2 (Kolmogorov's inequality). *Let $(\Omega, \mathcal{F}, (\mathcal{F}_n)_{n\geq 0}, \mathbb{P})$ be a filtered probability space. Let $(X_n)_{n\geq 0}$ be a martingale such that for all $n \geq 0$ we have $\mathbb{E}[X_n^2] < \infty$. Then*

$$\mathbb{P}\left(\max_{0\leq k\leq n} |X_k| \geq \lambda\right) \leq \frac{\mathbb{E}[X_k^2]}{\lambda^2}.$$

Proof. We use the fact that $(X_n^2)_{n\geq 0}$ is a positive submartingale. Therefore, we get

$$\underbrace{\lambda^2 \mathbb{P}\left(\max_{0\leq k\leq n} |X_k|^2 \geq \lambda^2\right)}_{\mathbb{P}(\max_{0\leq k\leq k} |X_k|\geq \lambda)} \leq \mathbb{E}\left[X_k^2 \mathbb{1}_{\{\max_{0\leq k\leq n} |X_k|^2 \geq \lambda^2\}}\right] \leq \mathbb{E}[X_k^2].$$

□

Theorem 4.3.1.3 (Maximal inequality). *Let $(\Omega, \mathcal{F}, (\mathcal{F}_n)_{n\geq 0}, \mathbb{P})$ be a filtered probability space. Let $(X_n)_{n\geq 0}$ be a submartingale. Then for all $\lambda \geq 0$ and $n \in \mathbb{N}$, we get*

$$\lambda \mathbb{P}\left(\max_{0\leq k\leq n} |X_k| \geq \lambda\right) \leq \mathbb{E}[X_0] + 2\mathbb{E}[|X_n|].$$

Proof. Let A and B be defined as in Proposition 4.3.1.1. Then we have

$$\begin{aligned}\lambda\mathbb{P}\left(\max_{0\leq k\leq n}|X_k|\geq\lambda\right) &= \lambda\mathbb{P}(A\cup B)\leq\mathbb{E}[X_k\mathbb{1}_A]-\mathbb{E}[X_0]+\mathbb{E}[X_k\mathbb{1}_{B^{\complement}}]\\ &\leq\mathbb{E}[|X_0|]+\mathbb{E}[|X_n|]+\mathbb{E}[|X_n|]\\ &=\mathbb{E}[|X_0|]+2\mathbb{E}[|X_n|].\end{aligned}$$

□

Theorem 4.3.1.4 (Doob's inequality). *Let $(\Omega,\mathcal{F},(\mathcal{F}_n)_{n\geq0},\mathbb{P})$ be a filtered probability space. Let $p>1$ and $q>1$ such that $\frac{1}{p}+\frac{1}{q}=1$. Then the following conditions hold:*

(i) *If $(X_n)_{n\geq0}$ is a submartingale, then for all $n\geq0$ we have*

$$\left\|\max_{0\leq k\leq n}X_k^+\right\|_p\leq q\left\|X_n^+\right\|_p,$$

(ii) *If $(X_n)_{n\geq0}$ is a martingale, then for all $n\geq0$ we have*

$$\left\|\max_{0\leq k\leq n}|X_k|\right\|_p\leq q\|X_n\|_p.$$

Remark 4.3.1.2. Recall that if $X\in L^1(\Omega,\mathcal{F},(\mathcal{F}_n)_{n\geq0},\mathbb{P})$, then

$$\|X\|_p=\mathbb{E}[|X|^p]^{1/p}.$$

Moreover, if $p=q=2$ and $(X_n)_{n\geq0}$ is a martingale, then for all $n\geq0$ we have

$$\mathbb{E}\left[\max_{0\leq k\leq n}X_k^2\right]\leq4\mathbb{E}[X_n^2].$$

In general, we have

$$|X_n|^2\leq\max_{0\leq k\leq n}|X_k|^p.$$

Therefore, we get

$$\mathbb{E}[|X_n|^p]\leq\mathbb{E}\left[\max_{0\leq k\leq n}|X_k|^p\right]\leq^{\text{Doob}}q^p\mathbb{E}[|X_n|^p],$$

where we have used Theorem 4.3.1.4 for the inequality. We shall also recall that for $X \in L^p(\Omega, \mathcal{F}, (\mathcal{F}_n)_{n\geq 0}, \mathbb{P})$, we can write

$$\mathbb{E}[|X|^p] = \mathbb{E}\left[\int_0^{|X|} p\lambda^{p-1}\mathrm{d}\lambda\right] = \mathbb{E}\left[\int_0^{\infty} \mathbb{1}_{\{|X|\geq\lambda\}} p\lambda^{p-1}\mathrm{d}\lambda\right]$$
$$= \int_0^{\infty} p\lambda^{p-1}\mathbb{P}(|X|\geq\lambda)\mathrm{d}\lambda$$

by using Fubini's theorem.

Proof of Theorem 4.3.1.4. It is enough to prove (ii). Since $(X_n)_{n\geq 0}$ is a submartingale, we know that $(X_n^+)_{n\geq 0}$ is a submartingale. Hence, we have

$$\lambda\mathbb{P}\left(\max_{0\leq k\leq n} X_k^+ \geq \lambda\right) \geq \mathbb{E}\left[X_n^+ \mathbb{1}_{\{\max_{0\leq k\leq n} X_k^+ \geq\lambda\}}\right].$$

Now define $Y_n := \max_{0\leq k\leq n} X_k^+$. Then for any $k > 0$, we have

$$\mathbb{E}[(Y_n\wedge k)^p] = \int_0^{\infty} p\lambda^{p-1}\mathbb{P}(Y_n\wedge k\geq\lambda)\mathrm{d}\lambda = \int_0^{n} p\lambda^{p-1}\mathbb{P}(Y_n\geq\lambda)\mathrm{d}\lambda$$
$$\leq \int_0^n p\lambda^{p-1}\left(\frac{1}{\lambda}\mathbb{E}[X_n^+\mathbb{1}_{\{Y_n\geq\lambda\}}]\right)\mathrm{d}\lambda$$
$$= \mathbb{E}\left[\int_0^n p\lambda^{p-1}X_n^+\mathbb{1}_{\{Y_n\geq\lambda\}}\mathrm{d}\lambda\right] = \mathbb{E}\left[\int_0^{Y_n\wedge k} p\lambda^{p-2}X_n^+\mathrm{d}\lambda\right]$$
$$= \mathbb{E}\left[\frac{p}{p-1}(Y_n\wedge k)^{p-1}X_n^+\right]$$
$$\leq q\mathbb{E}[(X_n^+)^p]^{1/p}\mathbb{E}[(Y_n\wedge k)^p]^{1/q},$$

where we have used that $q = \frac{p}{p-1}$ and Markov's inequality (Proposition 2.3.1.2). Therefore, we obtain

$$\mathbb{E}[(Y_n\wedge k)^p] \leq q\mathbb{E}[(X_n^+)^p]^{1/p}\mathbb{E}[(Y_n\wedge k)^p]^{1/q}.$$

Since $\mathbb{E}[(Y_n\wedge k)^p] \neq 0$, we can divide by it to get

$$\mathbb{E}[(Y_n\wedge k)^p]^{1-1/q=1/p} \leq q\mathbb{E}[(X_n^+)^p]^{1/p}$$

and thus

$$\|(Y_k\wedge n)\|_p \leq q\|X_n^+\|_p.$$

Now for $k\to\infty$, the *monotone convergence theorem* implies that

$$\|Y_n\|_p \leq q\|X_n^+\|_p. \qquad \square$$

Corollary 4.3.1.5. *Let $(\Omega, \mathcal{F}, (\mathcal{F}_n)_{n\geq 0}, \mathbb{P})$ be a filtered probability space. Let $(X_n)_{n\geq 0}$ be a martingale and $p > 1$, $q > 1$ such that $\frac{1}{p} + \frac{1}{q} = 1$. Then we get*

$$\left\| \sup_{n\geq 0} |X_n| \right\|_p \leq q \sup_{n\geq 0} \|X_n\|_p.$$

Exercise 4.3.1.3. Prove Corollary 4.3.1.5 by using Theorem 4.3.1.4.

4.4. Almost Sure Convergence for Martingales

Let $(\Omega, \mathcal{F}, (\mathcal{F}_n)_{n\geq 0}, \mathbb{P})$ be a filtered probability space. We start with a useful remark. If $(X_n)_{n\geq 0}$ is a submartingale, we get in particular that $X_n \in L^1(\Omega, \mathcal{F}, (\mathcal{F}_n)_{n\geq 0}, \mathbb{P})$ for all $n \geq 0$. Moreover, we know that we can write

$$\mathbb{E}[X_n] = \mathbb{E}[X_n^+] - \mathbb{E}[X_n^-]$$

and hence

$$\mathbb{E}[X_n^-] = \mathbb{E}[X_n^+] - \mathbb{E}[X_n].$$

The submartingale property implies that $\mathbb{E}[X_0] \leq \mathbb{E}[X_n]$ and thus

$$\mathbb{E}[X_n^-] \leq \mathbb{E}[X_n^+] - \mathbb{E}[X_0].$$

Therefore, if $\sup_{n\geq 0} \mathbb{E}[X_n^+] < \infty$, then $\mathbb{E}[X_n^-] \leq \sup_{n\geq 0} \mathbb{E}[X_n^+] - \mathbb{E}[|X_0|] < \infty$. Since $|X_n| = X_n^+ + X_n^-$, we have that $\sup_{n\geq 0} \mathbb{E}[X_n^+] < \infty$ if and only if $\sup_{n\geq 0} \mathbb{E}[|X_n|] < \infty$.

Lemma 4.4.0.1 (Doob's upcrossing inequality). *Let $(\Omega, \mathcal{F}, (\mathcal{F}_n)_{n\geq 0}, \mathbb{P})$ be a filtered probability space. Let $X = (X_n)_{n\geq 0}$ be a supermartingale and $a < b$ two real numbers. Then for all $n \geq 0$ we get*

$$(b-a)\mathbb{E}[N_n([a,b], X)] \leq \mathbb{E}[(X_n - a)^-], \tag{4.4.0.1}$$

where $N_n([a,b], x) = \sup\{k \geq 0 \mid T_k(x) \leq n\}$, i.e., the number of upcrossings of the interval $[a,b]$ by the sequence $x = (x_n)_n$ by time n, *and* $(T_k)_{k\geq 0}$ *is a sequence of stopping times. Moreover, as* $n \to \infty$, *we have*

$$N_n([a,b],x) \uparrow N([a,b],x) = \sup\{k \geq 0 \mid T_k(x) < \infty\},$$

i.e., the total number of upcrossings of the interval $[a,b]$.

Lemma 4.4.0.2. *A sequence of real numbers* $x = (x_n)_n$ *converges in* $\bar{\mathbb{R}} = \mathbb{R} \cup \{\pm\infty\}$ *if and only if* $N([a,b],x) < \infty$ *for all rationals* $a < b$.

Proof. Suppose that x converges. Then if for some $a < b$ we had that $N([a,b],x) = \infty$, that would imply that $\liminf_n x_n \leq a < b \leq \limsup_n x_n$, which is a contradiction. On the other hand, suppose that x does not converge. Then $\liminf_n x_n < \limsup_n x_n$ and so taking $a < b$ rationals between these two numbers gives that $N([a,b],x) = \infty$. □

Proof of Lemma 4.4.0.1. We will omit the dependence on X from T_k and S_k and we will write $N = N_n([a,b],X)$ to simplify notation. By the definition of the times $(T_k)_{k\geq 0}$ and $(S_k)_{k\geq 0}$, it is clear that for all k we have

$$X_{T_k} - X_{S_k} \geq b - a. \tag{4.4.0.2}$$

Moreover, we have

$$\begin{aligned}\sum_{k=1}^{n}(X_{T_k\wedge n} - X_{S_k\wedge n}) &= \sum_{k=1}^{N}(X_{T_k} - X_{S_k}) + \sum_{k=N+1}^{n}(X_n - X_{S_k\wedge n})\mathbb{1}_{\{N<n\}} \\ &= \sum_{k=1}^{N}(X_{T_k} - X_{S_k}) + (X_n - X_{S_{N+1}})\mathbb{1}_{\{S_{N+1}\leq n\}},\end{aligned}$$

since the only term contributing in the second sum appearing in the middle of the last equation chain is $N+1$, by the definition of N. Indeed, if $S_{N+2} \leq n$, then that would imply that $T_{N+1} \geq n$, which would contradict the definition of N. Using induction on $k \geq 0$, it is easy to see that $(T_k)_{k\geq 0}$ and $(S_k)_{k\geq 0}$ are stopping times. Hence for all $n \geq 0$, we have that $S_k \wedge n \leq T_k \wedge n$ are bounded stopping times and thus we get that $\mathbb{E}[X_{S_k\wedge n}] \geq \mathbb{E}[X_{T_k\wedge n}]$, for all $k \geq 0$. Therefore, taking expectations in the equations above and using the inequality (4.4.0.2), we get

$$0 \geq \mathbb{E}\left[\sum_{k=1}^{n}(X_{T_k\wedge n} - X_{S_k\wedge n})\right] \geq (b-a)\mathbb{E}[N] - \mathbb{E}[(X_n - a)^-],$$

since $(X_n - X_{S_{N+1}})\mathbb{1}_{\{S_{N+1}\leq n\}} \geq -(X_n - a)^-$. Rearranging gives the desired inequality. □

Theorem 4.4.0.3 (Almost sure martingale convergence theorem). *Let $(\Omega, \mathcal{F}, (\mathcal{F}_n)_{n\geq 0}, \mathbb{P})$ be a filtered probability space. Let $X = (X_n)_{n\geq 0}$ be a submartingale such that $\sup_{n\geq 0}\mathbb{E}[|X_n|] < \infty$. Then the sequence $(X_n)_{n\geq 0}$ converges almost surely to a random variable $X_\infty \in L^1(\Omega, \mathcal{F}_\infty, (\mathcal{F}_n)_{n\geq 0}, \mathbb{P})$ as $n \to \infty$, where $\mathcal{F}_\infty = \sigma\left(\bigcup_{n\geq 0}\mathcal{F}_n\right)$ is the tail σ-algebra.*

Proof. Let $a < b \in \mathbb{Q}$. By Lemma 4.4.0.1, we get that

$$\mathbb{E}[N_n([a,b],X)] \leq (b-a)^{-1}\mathbb{E}[(X_n - a)^-] \leq (b-a)^{-1}\mathbb{E}[|X_n| + a].$$

Using the *monotone convergence theorem*, since $N_n([a,b],X) \uparrow N([a,b],X)$ as $n \to \infty$, we get that

$$\mathbb{E}[N([a,b],X)] \leq (b-a)^{-1}\left(\sup_n \mathbb{E}[|X_n|] + a\right) < \infty,$$

by the assumption on X being bounded in $L^1(\Omega, \mathcal{F}, (\mathcal{F}_n)_{n\geq 0}, \mathbb{P})$. Therefore, we get that $N([a,b],X) < \infty$ a.s. for every $a < b \in \mathbb{Q}$. Hence,

$$\mathbb{P}\left(\bigcap_{a<b\in\mathbb{Q}}\{N([a,b),X) < \infty\}\right) = 1.$$

Writing $\Omega_0 = \bigcap_{a<b\in\mathbb{Q}}\{N([a,b),X) < \infty\}$, we have that $\mathbb{P}(\Omega_0) = 1$ and by Lemma 4.4.0.2 on Ω_0 we have that X converges to a possible infinite limit X_∞. So we can define

$$X_\infty := \begin{cases} \lim_{n\to\infty} X_n & \text{on } \Omega_0, \\ 0 & \text{on } \Omega \setminus \Omega_0, \end{cases}$$

Then X_∞ is $\mathcal{F}_\infty$-measurable and by *Fatou's lemma* and the assumption on X being in $L^1(\Omega, \mathcal{F}, (\mathcal{F}_n)_{n\geq 0}, \mathbb{P})$, we get

$$\mathbb{E}[|X_\infty|] = \mathbb{E}\left[\liminf_{n\to\infty} |X_n|\right] \leq \liminf_{n\to\infty} \mathbb{E}[|X_n|] < \infty.$$

Hence, we have $X_\infty \in L^1(\Omega, \mathcal{F}_\infty, (\mathcal{F}_n)_{n\geq 0}, \mathbb{P})$. □

Corollary 4.4.0.4. *Let $(\Omega, \mathcal{F}, (\mathcal{F}_n)_{n\geq 0}, \mathbb{P})$ be a filtered probability space. Let $(X_n)_{n\geq 0}$ be a non-negative supermartingale. Then $(X_n)_{n\geq 0}$ converges almost surely to a limit $X_\infty \in L^1(\Omega, \mathcal{F}_\infty, (\mathcal{F}_n)_\infty, \mathbb{P})$ which satisfies*

$$X_n \geq \mathbb{E}[X_\infty \mid \mathcal{F}_n] \quad a.s.$$

Proof. Note that $(-X_n)_{n\geq 0}$ is a submartingale. Thus $(-X_n)^+ = 0$ for all $n \geq 0$, which implies that $\sup_{n\geq 0} \mathbb{E}[-X_n^+] = 0 < \infty$. Hence, we have

$$X_n \xrightarrow[\text{a.s. and } L^1]{n\to\infty} X_\infty \in L^1(\Omega, \mathcal{F}_\infty, (\mathcal{F}_n)_{n\geq 0}, \mathbb{P}).$$

Moreover, for all $m \geq n$ we have

$$X_n \geq \mathbb{E}[X_m \mid \mathcal{F}_n].$$

Using *Fatou's lemma*, we get

$$X_n \geq \liminf_{m\to\infty} \mathbb{E}[X_m \mid \mathcal{F}_n] \geq \mathbb{E}\left[\liminf_{m\to\infty} X_m \mid \mathcal{F}_n\right] = \mathbb{E}[X_\infty \mid \mathcal{F}_n]. \quad □$$

Example 4.4.0.1 (Simple random walk on $\mathbb{Z}$). Let $Y_n := 1 + Z_1 + \cdots + Z_n$, for Z_j i.i.d. random variables with $\mathbb{P}(Z_j = \pm 1) = \frac{1}{2}$, $Y_0 = 1$, $\mathcal{F}_0 = \{\emptyset, \Omega\}$ and $\mathcal{F}_n = \sigma(Z_1, \ldots, Z_n)$. Then we have already seen that $(Y_n)_{n\geq 0}$ is a martingale. Let $T := \inf\{n \geq 0 \mid Y_n = 0\}$. We need to show that $T < \infty$ a.s. Let $X_n := Y_{n\wedge T}$. Then $(X_n)_{n\geq 0}$ is also a martingale. Moreover, $X_n \geq 0$ for all $n \geq 0$ and $(X_n)_{n\geq 0}$ converges a.s. to a random variable $X_\infty \in L^1(\Omega, \mathcal{F}_\infty, (\mathcal{F}_n)_{n\geq 0}, \mathbb{P})$.

Since $X_n = Y_{n \wedge T}$, we get $X_\infty = Y_T$. The convergence of $(X_n)_{n\geq 0}$ implies that $T < \infty$ a.s., indeed, on the set $\{T = \infty\}$ we get $|X_{n+1} - X_n| = 1$. Consequently, on $\{T = \infty\}$, we get that $(X_n)_{n\geq 0}$ is not a Cauchy sequence and therefore cannot converge. This implies that $\mathbb{P}(T = \infty) = 0$ and thus $T < \infty$. Hence, we have

$$\lim_{n\to\infty} X_n = Y_T = 0.$$

We also note that $\mathbb{E}[X_n] = 1 > \mathbb{E}[X_\infty] = 0$ for all $n \geq 0$ and so X_n does not converge to X_∞ in L^1.

4.5. L^p-convergence for Martingales

Theorem 4.5.0.1. *Let $(\Omega, \mathcal{F}, (\mathcal{F}_n)_{n\geq 0}, \mathbb{P})$ be a filtered probability space. Let $(X_n)_{n\geq 0}$ be a martingale. Then $(X_n)_{n\geq 0}$ converges a.s. to a random variable $X_\infty \in L^1(\Omega, \mathcal{F}_\infty, (\mathcal{F}_n)_{n\geq 0}, \mathbb{P})$ if and only if there exists a random variable $Z \in L^1(\Omega, \mathcal{F}, (\mathcal{F}_n)_{n\geq 0}, \mathbb{P})$ such that $X_n = \mathbb{E}[Z \mid \mathcal{F}_n]$ for all $n \geq 0$, where $\mathcal{F}_\infty = \sigma\left(\bigcup_{n\geq 0} \mathcal{F}_n\right)$.*

Remark 4.5.0.1. We shall see that one can always represent X_n as

$$X_n = \mathbb{E}[X_\infty \mid \mathcal{F}_n].$$

Proof of Theorem 4.5.0.1. For the $\Rightarrow$ implication, we note that for all $m \geq n$ and for all $A \in \mathcal{F}_n$ we have

$$\mathbb{E}[X_n \mathbb{1}_A] = \mathbb{E}[X_m \mathbb{1}_A].$$

Therefore, we see that $X_m \xrightarrow[L^1]{m\to\infty} X_\infty$, implies that

$$\lim_{m\to\infty} \mathbb{E}[X_m \mathbb{1}_A] = \mathbb{E}[X_\infty \mathbb{1}_A]$$

and thus

$$X_n = \mathbb{E}[X_\infty \mid \mathcal{F}_n].$$

For the $\Leftarrow$ implication, we note that if $X_n = \mathbb{E}[Z \mid \mathcal{F}_n]$, then

$$|X_n| \leq \mathbb{E}[|Z| \mid \mathcal{F}_n]$$

and thus

$$\mathbb{E}[|X_n|] \leq \mathbb{E}[|Z|].$$

This implies that

$$\sup_{n \geq 0} \mathbb{E}[|X_n|] \leq \mathbb{E}[|Z|] < \infty.$$

Hence, we know now that $X_n \xrightarrow[\text{a.s.}]{n\to\infty} X_\infty$. It remains to show that

$$X_n \xrightarrow[L^1]{n\to\infty} X_\infty.$$

First, we assume that Z is bounded, i.e., for all $\omega \in \Omega$, we have

$$|Z(\omega)| \leq M \in \mathbb{R}_{\geq 0}.$$

Then $|X_n(\omega)| \leq M$ and thus L^1-convergence follows from *Lebesgue's dominated convergence theorem.* For the general case, let $\varepsilon > 0$ and $M \in \mathbb{R}_{\geq 0}$, such that

$$\mathbb{E}[|Z - Z\mathbb{1}_{\{|Z| \leq M\}}|] < \varepsilon.$$

Thus, for all $n \geq 0$, we have

$$\begin{aligned} \mathbb{E}[|X_n - \mathbb{E}[Z\mathbb{1}_{\{|Z|\leq M\}} \mid \mathcal{F}_n]|] &\leq \mathbb{E}[\mathbb{E}[|Z|\mathbb{1}_{\{|Z|>M\}} \mid \mathcal{F}_n]] \\ &= \mathbb{E}[|Z|\mathbb{1}_{\{|Z|>M\}}] < \varepsilon. \end{aligned}$$

Moreover, from the bounded case, it follows that

$$\left(\mathbb{E}[Z\mathbb{1}_{\{|Z|\leq M\}} \mid \mathcal{F}_n]\right)_{n\geq 0}$$

converges in L^1. Hence, there exists $n_0 \in \mathbb{N}$ such that for all $m, n \geq n_0$, we have

$$\mathbb{E}[|\mathbb{E}[Z\mathbb{1}_{|||Z|\leq M\}} \mid \mathcal{F}_m] - \mathbb{E}[Z\mathbb{1}_{\{|Z|\leq M\}} \mid \mathcal{F}_n]|] < \varepsilon.$$

Now a simple application of the *triangular inequality* and the above estimates gives, for all $m, n \geq n_0$

$$\mathbb{E}[|X_m - X_n|] \leq 3\varepsilon.$$

Therefore, we get that $(X_n)_{n\geq 0}$ is a Cauchy sequence in L^1 and hence it converges in L^1. $\square$

Corollary 4.5.0.2. *Let $(\Omega, \mathcal{F}, (\mathcal{F}_n)_{n\geq 0}, \mathbb{P})$ be a filtered probability space. Let $Z \in L^1(\Omega, \mathcal{F}, (\mathcal{F}_n)_{n\geq 0}, \mathbb{P})$. Then the unique martingale $X_n = \mathbb{E}[Z \mid \mathcal{F}_n]$ converges almost surely and in L^1 to $X_\infty = \mathbb{E}[Z \mid \mathcal{F}_\infty]$, where $\mathcal{F}_\infty = \sigma\left(\bigcup_{n\geq 0} \mathcal{F}_n\right)$ is the tail σ-algebra.*

Proof. First, we note that X_∞ is $\mathcal{F}_\infty$-measurable. Now choose $A \in \mathcal{F}_n$. Then we get

$$\lim_{n\to\infty} \mathbb{E}[Z\mathbb{1}_A] = \lim_{n\to\infty} \mathbb{E}[X_n\mathbb{1}_A] = \mathbb{E}[X_\infty\mathbb{1}_A],$$

and hence for all $A \in \bigcup_{n\geq 0} \mathcal{F}_n$, we have

$$\mathbb{E}[Z\mathbb{1}_A] = \mathbb{E}[X_\infty\mathbb{1}_A].$$

The *monotone class theorem* implies that for every $A \in \mathcal{F}_\infty$, we have

$$\mathbb{E}[Z\mathbb{1}_A] = \mathbb{E}[X_\infty\mathbb{1}_A],$$

which on the other hand implies that

$$X_\infty = \mathbb{E}[Z \mid \mathcal{F}_\infty].$$

□

Exercise 4.5.0.2. Prove Kolmogorov's 0-1 law (Theorem 2.7.2.1) by using Corollary 4.5.0.2.

Theorem 4.5.0.3 **(L^p martingale convergence theorem).** *Let $(\Omega, \mathcal{F}, (\mathcal{F}_n)_{n\geq 0}, \mathbb{P})$ be a filtered probability space. Let $(X_n)_{n\geq 0}$ be a martingale. Assume that there exists $p \geq 1$ such that*

$$\sup_{n\geq 0} \mathbb{E}[|X_n|^p] < \infty.$$

Then we have

$$X_n \xrightarrow[\text{a.s. and } L^p]{n\to\infty} X_\infty.$$

Moreover, we have

$$\mathbb{E}[|X_\infty|^p] = \sup_{n\geq 0} \mathbb{E}[|X_n|^p],$$

and

$$\mathbb{E}[(X_\infty^*)^p] \leq \left(\frac{p}{p-1}\right)^p \mathbb{E}[|X_\infty|^p],$$

where

$$X_\infty^* = \sup_{n\geq 0} |X_n|.$$

Remark 4.5.0.3. Let us summarize what we have seen so far.

- If $(X_n)_{n\geq 0}$ is bounded in L^1, we get $X_n \xrightarrow[\text{a.s.}]{n\to\infty} X_\infty \in L^1$.
- $X_n \xrightarrow[\text{a.s. and } L^1]{n\to\infty} X_\infty$ if and only if $X_n = \mathbb{E}[X_\infty \mid \mathcal{F}_n]$.
- If $(X_n)_{n\geq 0}$ is bounded in L^p with $p \geq 1$, then $X_n \xrightarrow[\text{a.s. and } L^p]{n\to\infty} X_\infty$.

Proof of Theorem 4.5.0.3. We first note that since $\sup_{n\geq 0} \mathbb{E}[|X_n|^p] < \infty$, we also have that

$$\sup_{n\geq 0} \mathbb{E}[|X_n|] < \infty.$$

Thus $X_n \xrightarrow[\text{a.s.}]{n\to\infty} X_\infty$. Using Doob's inequality (Theorem 4.3.1.4), we get

$$\mathbb{E}[(X_\infty^*)^p] \leq \left(\frac{p}{p-1}\right)^p \sup_{n\geq 0} \mathbb{E}[|X_n|^p] < \infty,$$

and therefore $X_\infty^* \in L^p(\Omega, \mathcal{F}, (\mathcal{F}_n)_{n\geq 0}, \mathbb{P})$. Moreover, for all $n \geq 0$ we get $|X_n| \leq X_\infty^*$ and

$$|X_n - X_\infty|^p \leq 2^p (X_\infty^*) p.$$

Using *Lebesgue's dominated convergence theorem*, we get

$$\mathbb{E}[|X_n - X_\infty|^p] \xrightarrow{n\to\infty} 0,$$

and thus

$$X_n \xrightarrow[\text{L}^\text{p}]{\text{n}\to\infty} X_\infty.$$

Finally, we note that $(|X_n|^p)_{n\geq 0}$ is a positive submartingale. Hence, we know that

$$(\mathbb{E}[|X_n|^p])_{n\geq 0}$$

is increasing, which implies that

$$\mathbb{E}[|X_\infty|^p] = \lim_{n\to\infty} \mathbb{E}[|X_n|^p] = \sup_{n\geq 0} \mathbb{E}[|X_n|^p]. \qquad \square$$

4.6. Uniform Integrability

Definition 4.6.0.1 (Uniformly integrable). Let $(\Omega, \mathcal{F}, (\mathcal{F}_n)_{n\geq 0}, \mathbb{P})$ be a filtered probability space. A family $(X_i)_{i\in I}$ of random variables in $L^1(\Omega, \mathcal{F}, (\mathcal{F}_n)_{n\geq 0}, \mathbb{P})$, indexed by an arbitrary index set I, is called *uniformly integrable* if

$$\lim_{a\to\infty} \sup_{i\in I} \mathbb{E}[|X_i| \mathbb{1}_{\{|X_i|>a\}}] = 0.$$

Remark 4.6.0.2. A single random variable in $L^1(\Omega, \mathcal{F}, (\mathcal{F}_n)_{n\geq 0}, \mathbb{P})$ is always uniformly integrable (this follows from *Lebesgue's dominated convergence theorem*). If $|I| < \infty$, then using

$$|X_i| \leq \sum_{j\in I} |X_j|,$$

we get

$$\mathbb{E}[|X_i| \mathbb{1}_{\{|X_i|>a\}}] \leq \sum_{j\in I} \mathbb{E}[|X_j| \mathbb{1}_{\{X_j|>a\}}] \xrightarrow{a\to\infty} 0.$$

Let $(X_i)_{i\in I}$ be uniformly integrable For a large enough, we then have

$$\sup_{i\in I} \mathbb{E}[|X_i| \mathbb{1}_{\{|X_i|>a\}}] \leq 1.$$

Hence, we have

$$\sup_{i\in I} \mathbb{E}[|X_i|] = \sup_{i\in I} \mathbb{E}[|X_i|(\mathbb{1}_{\{|X_i|\leq a\}} + \mathbb{1}_{\{|X_i|>a\}})] \leq 1 + a < \infty,$$

which implies that $(X_i)_{i\in I}$ is bounded in $L^1(\Omega, \mathcal{F}, (\mathcal{F}_n)_{n\geq 0}, \mathbb{P})$.

Example 4.6.0.3. Let $Z \in L^1(\Omega, \mathcal{F}, (\mathcal{F}_n)_{n\geq 0}, \mathbb{P})$. Then the family

$$\Theta := \{X \in L^1(\Omega, \mathcal{F}, (\mathcal{F}_n)_{n\geq 0}, \mathbb{P}) \mid |X| \leq Z\}$$

is uniformly integrable. Indeed, we have

$$\sup_{X\in\Theta} \mathbb{E}[|X| \mathbb{1}_{\{|X|>a\}}] \leq \mathbb{E}[Z \mathbb{1}_{\{Z>a\}}] \xrightarrow{a\to\infty} 0.$$

Example 4.6.0.4. Let $\phi : \mathbb{R}_{\geq 0} \to \mathbb{R}_{\geq 0}$ be a measurable map such that

$$\frac{\phi(x)}{x} \xrightarrow{x\to\infty} \infty.$$

Then, for all $C > 0$, the family

$$\Theta_C := \{X \in L^1(\Omega, \mathcal{F}, (\mathcal{F}_n)_{n\geq 0}, \mathbb{P}) \mid \mathbb{E}[\phi(|X|)] \leq C\}$$

is uniformly integrable. Indeed, for a large enough, we have

$$\begin{aligned}\mathbb{E}[|X|\mathbb{1}_{\{|X|>a\}}] &= \mathbb{E}\left[\frac{|X|}{\phi(|X|)}\phi(|X|)\mathbb{1}_{\{|X|>a\}}\right] \leq \sup_{x>a}\left(\frac{x}{\phi(x)}\right)\mathbb{E}[\phi(|X|)] \\ &\leq C\sup_{x>a}\left(\frac{x}{\phi(x)}\right) \xrightarrow{a\to\infty} 0.\end{aligned}$$

Thus, we have

$$\sup_{X\in\Theta}\mathbb{E}[|X|\mathbb{1}_{\{|X|>a\}}] \leq C\sup_{x>a}\left(\frac{x}{\phi(x)}\right).$$

Proposition 4.6.0.1. *Let $(\Omega, \mathcal{F}, (\mathcal{F}_n)_{n\geq 0}, \mathbb{P})$ be a filtered probability space. Let $(X_i)_{i\in I}$ be a family of random variables bounded in $L^1(\Omega, \mathcal{F}, (\mathcal{F}_n)_{n\geq 0}, \mathbb{P})$, i.e., $\sup_{i\in I}\mathbb{E}[|X_i|] < \infty$. Then $(X_i)_{i\in I}$ is uniformly integrable if and only if for all $\varepsilon > 0$ there is a $\delta > 0$ such that for all $A \in \mathcal{F}$, if $\mathbb{P}[A] < \delta$ then $\sup_{i\in I}\mathbb{E}[|X_i|\mathbb{1}_A] < \varepsilon$.*

Proof. For the $\Rightarrow$ implication, let $\varepsilon > 0$. Then there exists $a > 0$ such that

$$\sup_{i\in I}\mathbb{E}[|X_i|\mathbb{1}_{\{|X_i|>a\}}] < \frac{\varepsilon}{2}.$$

Now let $\delta := \frac{\varepsilon}{2a}$ and $A \in \mathcal{F}$ such that $\mathbb{P}(A) < \delta$. Then

$$\begin{aligned}\mathbb{E}[|X_i|\mathbb{1}_A] &= \mathbb{E}[|X_i|\mathbb{1}_A\mathbb{1}_{\{|X_i|>a\}}] + \mathbb{E}[|X_i|\mathbb{1}_A\mathbb{1}_{\{|X_i|\leq a\}}] \\ &\leq \mathbb{E}[|X_i|\mathbb{1}_{\{|X_i|>a\}}] + a\underbrace{\mathbb{E}[\mathbb{1}_A]}_{\mathbb{P}(A)} < \frac{\varepsilon}{2} + a\delta < \frac{\varepsilon}{2} + a\frac{\varepsilon}{2a} = \varepsilon.\end{aligned}$$

For the $\Leftarrow$ implication, let $C := \sup_{i\in I} \mathbb{E}[|X_i|] < \infty$. From Markov's inequality (Proposition 2.3.1.2), we get

$$\mathbb{P}(|X_i| > a) \leq \frac{\mathbb{E}[|X_i|]}{a} \leq \frac{C}{a}.$$

Now let $\delta > 0$ such that (ii) holds. If $\frac{C}{a} < \delta$, then for all $i \in I$

$$\mathbb{E}[|X_i|\mathbb{1}_{\{|X_i|>a\}}] < \varepsilon. \qquad \square$$

Corollary 4.6.0.2. *Let $(\Omega, \mathcal{F}, (\mathcal{F}_n)_{n\geq 0}, \mathbb{P})$ be a filtered probability space. Let X be a bounded random variable, i.e., $X \in L^1(\Omega, \mathcal{F}, (\mathcal{F}_n)_{n\geq 0}, \mathbb{P})$. Then the family*

$$\{\mathbb{E}[X \mid \mathcal{G}] \mid \mathcal{G} \subset \mathcal{F},\ \mathcal{G} \text{ is a } \sigma\text{-algebra}\}$$

is uniformly integrable.

Proof. Let $\varepsilon > 0$. Then there exists a $\delta > 0$ such that for all $A \in \mathcal{F}$, if $\mathbb{P}(A) < \delta$, then $\mathbb{E}[|X|\mathbb{1}_A] < \varepsilon$. Moreover, for all $a > 0$, we have

$$\mathbb{P}(\mathbb{E}[X \mid \mathcal{G}] > a) \leq \frac{\mathbb{E}[|\mathbb{E}[X \mid \mathcal{G}]|]}{a} \leq \frac{\mathbb{E}[|X|]}{a}.$$

Note that for a large enough, i.e., $\frac{\mathbb{E}[|X|]}{a} < \delta$, we have

$$\begin{aligned}\mathbb{E}[|\mathbb{E}[X \mid \mathcal{G}]|\mathbb{1}_{\{|\mathbb{E}[X|\mathcal{G}]|>a\}}] &\leq \mathbb{E}[\mathbb{E}[|X|\mathbb{1}_{\{|\mathbb{E}[X|\mathcal{G}]|>a\}} \mid \mathcal{G}]] \\ &\leq \mathbb{E}[|X|\mathbb{1}_{\{\mathbb{E}[X|\mathcal{G}]|>a\}}] < \varepsilon. \qquad \square\end{aligned}$$

Theorem 4.6.0.3. *Let $(\Omega, \mathcal{F}, (\mathcal{F}_n)_{n\geq 0}, \mathbb{P})$ be a filtered probability space. Let $(X_n)_{n\geq 0}$ be a sequence of random variables in $L^1(\Omega, \mathcal{F}, (\mathcal{F}_n)_{n\geq 0}, \mathbb{P})$, which converges in probability to X_∞. Then $X_n \xrightarrow[\mathrm{L}^1]{\mathrm{n}\to\infty} X_\infty$ if and only if $(X_n)_{n\geq 0}$ is uniformly integrable.*

Proof. For the $\Rightarrow$ implication, we first note that $(X_n)_{n\geq 0}$ is bounded in $L^1(\Omega, \mathcal{F}, (\mathcal{F}_n)_{n\geq 0}, \mathbb{P})$ since it converges in L^1. For $\varepsilon > 0$, there exists $N \in \mathbb{N}$ such that for $n \geq N$ we get

$$\mathbb{E}[|X_N - X_n|] < \frac{\varepsilon}{2}.$$

Next we note that $\{X_0, X_1, \ldots, X_N\}$ is uniformly integrable since it is a finite family of bounded random variables. Therefore, there exists a $\delta > 0$, such that for all $A \in \mathcal{F}$, if $\mathbb{P}(A) < \delta$ then $\mathbb{E}[|X_n|\mathbb{1}_A] < \frac{\varepsilon}{2}$ for all $n \in \{0, 1, \dots, N\}$. Finally, for $n \geq N$, we get

$$\mathbb{E}[|X_n|\mathbb{1}_A] \leq \mathbb{E}[|X_N - X_n|\mathbb{1}_A] + \mathbb{E}[|X_N|\mathbb{1}_A] < \frac{\varepsilon}{2} + \frac{\varepsilon}{2} = \varepsilon.$$

Thus $(X_n)_{n\geq 0}$ is uniformly integrable. For the $\Leftarrow$ implication, we note that if $(X_n)_{n\geq 0}$ is uniformly integrable, then the family $(X_n - X_m)_{(n,m)\in\mathbb{N}^2}$ is also uniformly integrable, since

$$\mathbb{E}[|X_n - X_m|\mathbb{1}_A] \leq \mathbb{E}[|X_n|\mathbb{1}_A] + \mathbb{E}[X_m|\mathbb{1}_A]$$

for all $A \in \mathcal{F}$. Now for $\varepsilon > 0$, there exists some a sufficiently large, such that

$$\mathbb{E}[|X_n - X_m|\mathbb{1}_{\{|X_n - X_m| > a\}}] < \varepsilon.$$

Moreover, we note that

$$\begin{aligned}
\mathbb{E}[|X_m - X_m|] &\leq \mathbb{E}[|X_n - X_m|\mathbb{1}_{\{|X_n - X_m| < \varepsilon\}}] \\
&\quad + \mathbb{E}[|X_n - X_m|\mathbb{1}_{\{|X_n - X_m| \geq \varepsilon\}}] \\
&\leq \varepsilon + \mathbb{E}[|X_n - X_m|\mathbb{1}_{\{|X_n - X_m| \geq \varepsilon\}}\mathbb{1}_{\{|X_n - X_m| \leq a\}}] \\
&\quad + \mathbb{E}[|X_n - X_m|\mathbb{1}_{\{|X_n - X_m| \geq \varepsilon\}}\mathbb{1}_{\{|X_n - X_m| > a\}}] \\
&\leq \varepsilon + a\mathbb{P}(|X_n - X_m| \geq \varepsilon) \\
&\quad + \mathbb{E}[|X_n - X_m|\mathbb{1}_{\{|X_n - X_m| > a\}}]. \qquad (4.6.0.1)
\end{aligned}$$

Then, using that $\lim_{n\to\infty} \mathbb{P}(|X_n - X_m| \geq \varepsilon) = 0$, we can show that the right-hand side of (4.6.0.1) converges to zero for a large enough, which implies that $(X_n)_{n\geq 0}$ is a Cauchy sequence in $L^1(\Omega, \mathcal{F}, (\mathcal{F}_n)_{n\geq 0}, \mathbb{P})$ and hence converges in L^1. □

Remark 4.6.0.5. Combining all our previous results, we get that if $(X_n)_{n\geq 0}$ is a martingale, then the following conditions are equivalent:

(i) $(X_n)_{n\geq 0}$ converges a.s. and in L^1,
(ii) $(X_n)_{n\geq 0}$ is uniformly integrable,
(iii) $(X_n)_{n\geq 0}$ is regular and $X_n = \mathbb{E}[X_\infty \mid \mathcal{F}_n]$ a.s.

4.7. Stopping Theorems

If $S \leq T$ are two bounded stopping times and $(X_n)_{n\geq 0}$ a martingale, then

$$\mathbb{E}[X_T \mid \mathcal{F}_S] = X_S \quad a.s.$$

If $(X_n)_{n\geq 0}$ is an *adapted process*, which converges a.s. to X_∞, we can define X_T for all stopping times (finite or not) by

$$X_T = \sum_{n=0}^{\infty} X_n \mathbb{1}_{\{T=n\}} + X_\infty \mathbb{1}_{\{T=\infty\}}.$$

Theorem 4.7.0.1. *Let $(\Omega, \mathcal{F}, (\mathcal{F}_n)_{n\geq 0}, \mathbb{P})$ be a filtered probability space. Let $(X_n)_{n\geq 0}$ be uniformly integrable martingale. Then, for any stopping time T, we have that*

$$\mathbb{E}[X_\infty \mid \mathcal{F}_T] = X_T \quad a.s.$$

In particular, we have

$$\mathbb{E}[X_T] = \mathbb{E}[X_\infty] = \mathbb{E}[X_n], \quad \forall n \geq 0.$$

Moreover, if S and T are two stopping times, such that $S \leq T$, we get that

$$\mathbb{E}[X_T \mid \mathcal{F}_S] = X_S \quad a.s.$$

Proof. We first want to check that X_T is indeed in $L^1(\Omega, \mathcal{F}, (\mathcal{F}_n)_{n\geq 0}, \mathbb{P})$. Note thus that we have

$$\begin{aligned}
\mathbb{E}[|X_T|] &= \sum_{n=0}^{\infty} \mathbb{E}[|X_n| \mathbb{1}_{\{T=n\}}] + \mathbb{E}[|X_\infty| \mathbb{1}_{\{T=\infty\}}] \\
&\leq \sum_{n=0}^{\infty} \mathbb{E}[\mathbb{E}[|X_\infty| \mid \mathcal{F}_n] \mathbb{1}_{\{T=n\}}] + \mathbb{E}[|X_\infty| \mathbb{1}_{\{T=\infty\}}] \\
&= \sum_{n=0}^{\infty} \mathbb{E}[\mathbb{E}[|X_\infty| \mathbb{1}_{\{T=n\}} \mid \mathcal{F}_n]] + \mathbb{E}[|X_\infty| \mathbb{1}_{\{T=\infty\}}] \\
&= \sum_{n=0}^{\infty} \mathbb{E}[|X_\infty| \mathbb{1}_{\{T=n\}}] + \mathbb{E}[|X_\infty| \mathbb{1}_{\{T=\infty\}}] \\
&= \mathbb{E}[|X_\infty|] < \infty.
\end{aligned}$$

Now let $A \in \mathcal{F}_T$. Then we have

$$\mathbb{E}[X_T \mathbb{1}_A] = \sum_{n \in \mathbb{N} \cup \{\infty\}} \mathbb{E}[X_T \mathbb{1}_{A \cap \{T=n\}}] = \sum_{n \in \mathbb{N} \cup \{\infty\}} \mathbb{E}[X_n \mathbb{1}_{A \cap \{T=n\}}]$$
$$= \sum_{n \in \mathbb{N} \cup \{\infty\}} \mathbb{E}[\mathbb{E}[X_\infty \mid \mathcal{F}_n] \mathbb{1}_{A \cap \{T=n\}}]$$
$$= \sum_{n \in \mathbb{N} \cup \{\infty\}} \mathbb{E}[X_\infty \mathbb{1}_{A \cap \{T=n\}}] = \mathbb{E}[X_\infty \mathbb{1}_A],$$

where we have used that $X_\infty \in L^1(\Omega, \mathcal{F}_\infty, (\mathcal{F}_n)_{n \geq 0}, \mathbb{P})$ and Fubini's theorem for the first and last equation. Now since X_T is $\mathcal{F}_T$-measurable, we get that $\mathbb{E}[X_\infty \mid \mathcal{F}_T] = X_T$ a.s. Finally, for $S \leq T$, we have $\mathcal{F}_S \subset \mathcal{F}_T$ and thus

$$X_S = \mathbb{E}[X_\infty \mid \mathcal{F}_S] = \mathbb{E}[\mathbb{E}[X_\infty \mid \mathcal{F}_T] \mid \mathcal{F}_S] = \mathbb{E}[X_T \mid \mathcal{F}_S]. \qquad \square$$

Remark 4.7.0.1. If $(X_n)_{n \geq 0}$ is a uniformly integrable martingale, then the family

$$\{X_T \mid T \text{ a stopping time}\}$$

is uniformly integrable Indeed, we note that

$$\{X_T \mid T \text{ a stopping time}\} = \{\mathbb{E}[X_\infty \mid \mathcal{F}_T] \mid T \text{ a stopping time}\}$$
$$\subset \mathbb{E}[X_\infty \mid \mathcal{G}] \mid \mathcal{G} \text{ a } \sigma\text{-algebra}, \mathcal{G} \subset \mathcal{F}\},$$

where the superset is uniformly integrable. If $N \in \mathbb{N}$, then $(X_{n \wedge N})_{n \geq 0}$ is uniformly integrable. Indeed, if $Y_n = X_{n \wedge N}$, then $(Y_n)_{n \geq 0}$ is a martingale and

$$Y_n = \mathbb{E}[X_N \mid \mathcal{F}_n] = \mathbb{E}[Y_\infty \mid \mathcal{F}_n].$$

Example 4.7.0.2 (Another random walk). Consider a simple random walk with $X_0 = k \geq 0$. Let $m \geq 0$, $0 \leq k \leq m$ and

$$T = \inf\{n \geq 1 \mid X_n = 0 \text{ or } X_n = m\}$$

with $X_n = k + Y_1 + \cdots + Y_n$, where $(Y_n)_{n \geq 1}$ are i.i.d. and $\mathbb{P}(Y_n = \pm 1) = \frac{1}{2}$. We have already seen that $T < \infty$ a.s. Now let $Z_n = X_{n \wedge T}$. Then we get that $(Z_n)_{n \geq 0}$ is a martingale and

$0 \leq Z_n \leq m$. Therefore, $(Z_n)_{n\geq 0}$ is uniformly integrable and hence Z_n converges a.s. and in L^1 to Z_∞ with

$$\mathbb{E}[Z_\infty] = \mathbb{E}[Z_0] = k.$$

Moreover, we have that

$$\mathbb{E}[Z_\infty] = \mathbb{E}[X_T] = \mathbb{E}[m\mathbb{1}_{\{X_T=m\}}] + \mathbb{E}[0\mathbb{1}_{\{X_T=0\}}] = m\mathbb{P}(X_T = m),$$

which implies that $\mathbb{P}(X_T = m) = \frac{k}{m}$ and thus $\mathbb{P}(X_T = 0) = 1 - \mathbb{P}(X_T = m) = \frac{m-k}{m}$.

Now let us assume that $\mathbb{P}(Y_n = 1) = p$ and $\mathbb{P}(Y_n = -1) = 1-p =: q$ for $p \in (0,1)$ and $p \neq \frac{1}{2}$. Moreover, consider

$$Z_n = \left(\frac{q}{p}\right)^{X_n}.$$

Then $(Z_n)_{n\geq 0}$ is a martingale. Indeed, by definition, Z_n is adapted and $Z_n \in L^1(\Omega, \mathcal{F}, (\mathcal{F}_n)_{n\geq 0}, \mathbb{P})$ because

$$k - n \leq X_n \leq k + n,$$

and thus Z_n is bounded. Furthermore, we have that

$$\begin{aligned}
\mathbb{E}[Z_{n+1} \mid \mathcal{F}_n] &= \mathbb{E}\left[\left(\frac{q}{p}\right)^{X_n}\left(\frac{q}{p}\right)^{Y_n}\middle|\mathcal{F}_n\right] = Z_n\mathbb{E}\left[\left(\frac{q}{p}\right)^{Y_n}\middle|\mathcal{F}_n\right] \\
&= Z_n\mathbb{E}\left[\left(\frac{q}{p}\right)^{Y_n}\right] = Z_n(q+p) = Z_n,
\end{aligned}$$

and therefore $(Z_n)_{n\geq 0}$ is a martingale. Now $(Z_{n\wedge T})_{n\geq 0}$ is bounded and hence uniformly integrable, which implies that $(Z_{n\wedge T})_{n\geq 0}$ converges a.s. and in L^1. We also have that

$$\mathbb{E}[Z_T] = \mathbb{E}[Z_0] = \left(\frac{q}{p}\right)^k.$$

On the other hand, we have

$$\mathbb{E}[Z_T] = \left(\frac{q}{p}\right)^m \mathbb{P}(X_T = m) + (1 - \mathbb{P}(X_T = m)).$$

Hence, we get

$$\mathbb{P}(X_T = m) = \frac{\left(\frac{q}{p}\right)^k - 1}{\left(\frac{q}{p}\right)^m - 1}.$$

Theorem 4.7.0.2. *Let $(\Omega, \mathcal{F}, (\mathcal{F}_n)_{n\geq 0}, \mathbb{P})$ be a filtered probability space. Let $(X_n)_{n\geq 0}$ be a supermartingale. Assume that one of the following two conditions is satisfied:*

(i) $X_n \geq 0$ *for all* $n \geq 0$;
(ii) $(X_n)_{n\geq 0}$ *is uniformly integrable.*

Then, for every stopping time T (finite or not), we get that $X_T \in L^1(\Omega, \mathcal{F}, (\mathcal{F}_n)_{n\geq 0}, \mathbb{P})$. Moreover, if S and T are two stopping times, such that $S \leq T$, then in case

(i) $\mathbb{1}_{\{S<\infty\}} X_S \geq \mathbb{E}[X_T \mathbb{1}_{\{T<\infty\}} \mid \mathcal{F}_S]$ *a.s.*;
(ii) $X_S \geq \mathbb{E}[X_T \mid \mathcal{F}_S]$ *a.s.*

Proof. We first deal with the case (i). We have already seen that if T is a bounded stopping time, we have

$$\mathbb{E}[X_T] \leq \mathbb{E}[X_0].$$

Using *Fatou's lemma*, we get

$$\mathbb{E}\left[\liminf_{n\to\infty} X_{n\wedge T}\right] \leq \liminf_{n\to\infty} \mathbb{E}[X_{n\wedge T}] \leq \mathbb{E}[X_0],$$

which implies that $X_T \in L^1(\Omega, \mathcal{F}, (\mathcal{F}_n)_{n\geq 0}, \mathbb{P})$. Now let $S \leq T$ be two stopping times. First assume that $S \leq T \leq N \in \mathbb{N}$. Then we know that

$$\mathbb{E}[X_S] \geq \mathbb{E}[X_T].$$

Now let $A \in \mathcal{F}_S$ and consider the stopping times

$$S^A(\omega) := \begin{cases} S(\omega), & \omega \in A, \\ 0, & \omega \notin A, \end{cases}$$

$$T^A(\omega) := \begin{cases} T(\omega), & \omega \in A, \\ 0, & \omega \notin A. \end{cases}$$

Then we can observe that

$$S^A \leq T^A \leq N,$$

thus $\mathbb{E}[X_{S^A}] \geq \mathbb{E}[X_{T^A}]$ and therefore $\mathbb{E}[X_S \mathbb{1}_A] \geq \mathbb{E}[X_T \mathbb{1}_A]$ for all $A \in \mathcal{F}_S$.

Let us now go back to the general case $S \leq T$ and let $B \in \mathcal{F}_S$. We want to apply the above to $S \wedge k$, $T \wedge k$ and $A := B \cap \{S \leq k\} \in \mathcal{F}_S$. Then we get

$$\mathbb{E}[X_{S\wedge k} \mathbb{1}_{B\cap\{S\leq k\}}] \geq \mathbb{E}[X_{T\wedge k} \mathbb{1}_{B\cap\{S\leq k\}}] \geq \mathbb{E}[X_{T\wedge k} \mathbb{1}_{B\cap\{T\leq k\}}].$$

Hence, we get

$$\mathbb{E}[X_{S\wedge k} \mathbb{1}_{B\cap\{S\leq k\}}] \geq \mathbb{E}[X_{T\wedge k} \mathbb{1}_{B\cap\{T\leq k\}}]$$

and thus

$$\mathbb{E}[X_S \mathbb{1}_B \mathbb{1}_{\{S\leq k\}}] \geq \mathbb{E}[X_T \mathbb{1}_B \mathbb{1}_{\{T\leq k\}}].$$

Using *Lebesgue's dominated convergence theorem*, we get

$$\mathbb{E}[X_S \mathbb{1}_B \mathbb{1}_{\{S<\infty\}}] \geq \mathbb{E}[X_T \mathbb{1}_B \mathbb{1}_{\{T<\infty\}}].$$

Now let $\tilde{X}_S = \mathbb{1}_{\{S<\infty\}} X_S$ and $\tilde{X}_T = \mathbb{1}_{\{T<\infty\}} X_T$. Then, for any $B \in \mathcal{F}_S$, we get

$$\mathbb{E}[\tilde{X}_S \mathbb{1}_B] \geq \mathbb{E}[\tilde{X}_T \mathbb{1}_B] = \mathbb{E}[\mathbb{1}_B \mathbb{E}[\tilde{X}_T \mid \mathcal{F}_S]].$$

Since the last equality is true for all $B \in \mathcal{F}_S$, we can conclude that

$$\tilde{X}_S \geq \mathbb{E}[\tilde{X}_T \mid \mathcal{F}_S].$$

Now let us prove (ii). We know from previous results that in this case $X_n \xrightarrow[\text{a.s. and } L^1]{n\to\infty} X_\infty$. Moreover, we have $X_n \geq \mathbb{E}[X_m \mid \mathcal{F}_n]$ for all $m \geq n$. The L^1-convergence as $n \to \infty$ gives

$$X_n \geq \mathbb{E}[X_\infty \mid \mathcal{F}_n].$$

In fact, the martingale $Z_n := \mathbb{E}[X_\infty \mid \mathcal{F}_n]$ converges a.s. to X_∞. Set $Y_n := X_n - Z_n$. Then $(Y_n)_{n\geq 0}$ is a positive supermartingale and hence it converges a.s. to $X_\infty - Z_\infty = 0$. We can now apply case (i) to deduce that $X_T = Y_T + Z_T \in L^1(\Omega, \mathcal{F}, (\mathcal{F}_n)_{n\geq 0}, \mathbb{P})$ and $Y_S \geq \mathbb{E}[Y_T \mid \mathcal{F}_S]$.[1] Theorem 4.7.0.1 applied to $(Z_n)_{n\geq 0}$ gives

$$Z_S = \mathbb{E}[Z_T \mid \mathcal{F}_S].$$

Hence, we get that

$$Y_S + Z_S \geq \mathbb{E}[Z_T + Y_T \mid \mathcal{F}_S],$$

and thus

$$X_S \geq \mathbb{E}[X_T \mid \mathcal{F}_S]. \qquad \square$$

4.8. Applications of Martingale Limit Theorems

4.8.1. Backward martingales and the law of large numbers

Definition 4.8.1.1 (Backward filtration). Let $(\Omega, \mathcal{F}, \mathbb{P})$ be a probability space. A *backward filtration* is a family $(\mathcal{F}_n)_{n\in -\mathbb{N}}$ of σ-algebras indexed by the negative integers, which we will denote by $(\mathcal{F}_n)_{n\leq 0}$, such that for all $n \leq m \leq 0$ we have

$$\mathcal{F}_n \subset \mathcal{F}_m.$$

Remark 4.8.1.2. We will then define

$$\mathcal{F}_{-\infty} := \bigcap_{n\leq 0} \mathcal{F}_n.$$

It is clear that $\mathcal{F}_{-\infty}$ is also a σ-algebra included in $\mathcal{F}$. A stochastic process $(X_n)_{n\leq 0}$, indexed by the negative integers, is called a *backwards martingale* (respectively, *backwards sub- or supermartingale*) if for all $n \leq 0$, X_n is $\mathcal{F}_n$-measurable, $\mathbb{E}[|X_n|] < \infty$ and for all $n \leq m$ we have

$$\mathbb{E}[X_m \mid \mathcal{F}_n] = X_n$$
$$\text{(respectively, } \mathbb{E}[X_m \mid \mathcal{F}_n] < X_n \text{ or } \mathbb{E}[X_m \mid \mathcal{F}_n] \geq X_n).$$

---

[1]We use the fact that $Y_S \mathbb{1}_{\{S=\infty\}} = 0$ and $Y_T \mathbb{1}_{\{T=\infty\}} = 0$, since $Y_\infty = 0$.

Theorem 4.8.1.1 (Backward convergence theorem). *Let $(\Omega, \mathcal{F}, (\mathcal{F}_n)_{n\leq 0}, \mathbb{P})$ be a backward filtered probability space. Let $(X_n)_{n\leq 0}$ be a backward supermartingale. Assume that*

$$\sup_{n\leq 0} \mathbb{E}[|X_n|] < \infty. \tag{4.8.1.1}$$

Then $(X_n)_{n\leq 0}$ is uniformly integrable and converges almost surely and in L^1 to $X_{-\infty}$ as $n \to -\infty$. Moreover, for all $n \leq 0$, we have

$$\mathbb{E}[X_n \mid \mathcal{F}_{-\infty}] \leq X_{-\infty} \quad a.s.$$

Proof. First we show almost sure convergence. Let therefore $k \geq 1$ be a fixed integer. For $n \in \{1, \ldots, k\}$, let $Y_n^k = X_{n-k}$ and $\mathcal{G}_n^k = \mathcal{F}_{n-k}$. For $n > k$, we take $Y_n^k = X_0$ and $\mathcal{G}_n^k = \mathcal{F}_0$. Then $(Y_n^k)_{n\geq 0}$ is a supermartingale with respect to $(\mathcal{G}_n^k)_{n\geq 0}$. We now apply Doob's upcrossing inequality (Lemma 4.4.0.1) to the submartingale $(-Y_n^k)_{n\geq 0}$ to obtain that for $a < b$ we get

$$\begin{aligned}(b-a)\mathbb{E}[N_k([a,b], -Y^k)] &\leq \mathbb{E}[(-Y_n^k - a)^+] \\ &= \mathbb{E}[(-X_0 - a)^+] \leq |a| + \mathbb{E}[|X_0|].\end{aligned}$$

We note that when $k \uparrow \infty$, $N_k([a,b], -Y^k)$ increases and hence

$$\begin{aligned}N([a,b], -X) := \quad & \sup\{k \in \mathbb{N} \mid \exists m_1 < n_1 < \cdots < m_k < n_k \leq 0; \\ & -X_{m_1} \leq a, -X_{n_1} \geq b, \ldots, -X_{m_k} \leq a, -X_{n_k} \geq b\}.\end{aligned}$$

Using the *monotone convergence theorem*, we get

$$(b-a)\mathbb{E}[N([a,b], -X)] \leq |a| + \mathbb{E}[|X_0|] < \infty.$$

One can easily show that $(X_n)_{n\leq 0}$ converges almost surely to X_∞ as $n \to -\infty$ and *Fatou's lemma* implies then that $\mathbb{E}[|X_{-\infty}|] < \infty$. We want to show that $(X_n)_{n\leq 0}$ is uniformly integrable. Thus, let $\varepsilon > 0$. The sequence $(\mathbb{E}[X_n])_{n\geq 0}$ is increasing and bounded; we can take $k \leq 0$ small enough to get for $n \leq k$,

$$\mathbb{E}[X_n] \leq \mathbb{E}[X_k] + \frac{\varepsilon}{2}.$$

Moreover, the finite family $\{X_k, X_{k+1}, \ldots, X_{-1}, X_0\}$ is uniformly integrable and one can then choose $\alpha > 0$ large enough such that for all $k \leq n \leq 0$ we get

$$\mathbb{E}[|X_n|\mathbb{1}_{\{|X_n|>\alpha\}}] < \varepsilon.$$

We can also choose $\delta > 0$ sufficiently small such that for all $A \in \mathcal{F}$, $\mathbb{P}(A) < \delta$ implies that $\mathbb{E}[|X_n \mid \mathbb{1}_A] < \frac{\varepsilon}{2}$. Now if $n < k$, we get

$$\begin{aligned}\mathbb{E}[|X_n|\mathbb{1}_{\{|X_n|>\alpha\}}] &= \mathbb{E}[-X_n\mathbb{1}_{\{X_n<-\alpha\}}] + \mathbb{E}[X_n\mathbb{1}_{\{X_n>\alpha\}}] \\ &= -\mathbb{E}[X_n\mathbb{1}_{\{X_n<-\alpha\}}] + \mathbb{E}[X_n] - \mathbb{E}[X_n\mathbb{1}_{\{X_n\leq\alpha\}}] \\ &\leq -\mathbb{E}[\mathbb{E}[X_k \mid \mathcal{F}_n]\mathbb{1}_{\{X_n<-\alpha\}}] + \mathbb{E}[X_k] \\ &\quad + \frac{\varepsilon}{2} - \mathbb{E}[\mathbb{E}[X_k \mid \mathcal{F}_n]\mathbb{1}_{\{X_{\leq}\alpha\}}] \\ &= -\mathbb{E}[X_k\mathbb{1}_{\{X_n<-\alpha\}}] + \mathbb{E}[X_k] + \frac{\varepsilon}{2} - \mathbb{E}[X_k\mathbb{1}_{\{X_n\leq\alpha\}}] \\ &= -\mathbb{E}[X_n\mathbb{1}_{\{X_n<-\alpha\}}] + \mathbb{E}[X_n\mathbb{1}_{\{X_n>\alpha\}}] \\ &\quad + \frac{\varepsilon}{2} \leq \mathbb{E}[|X_k|\mathbb{1}_{\{|X_n|>\alpha\}}] + \frac{\varepsilon}{2}.\end{aligned}$$

Next, we observe that

$$\mathbb{P}(|X_n| > \alpha) \leq \frac{1}{\alpha}\mathbb{E}[X_n] \leq \frac{C}{\alpha},$$

where $C := \sup_{n\leq 0}\mathbb{E}[|X_n|] < \infty$. Choose α such that $\frac{C}{\alpha} < \delta$. Consequently, we get

$$\mathbb{E}[|X_k|\mathbb{1}_{\{|X_n|>\alpha\}}] < \frac{\varepsilon}{2}.$$

Hence, for all $n < k$, $\mathbb{E}[|X_n|\mathbb{1}_{\{|X_n|>\alpha\}}] < \varepsilon$. This inequality is also true for $k \leq n \leq 0$ and thus we have that $(X_n)_{n\leq 0}$ is uniformly integrable. In order to conclude, we note that uniformly integrable and almost sure convergence implies L^1-convergence. Then, for $m \leq n$ and $A \in \mathcal{F}_{-\infty} \subset \mathcal{F}_m$, we have

$$\mathbb{E}[X_n\mathbb{1}_A] \leq \mathbb{E}[X_m\mathbb{1}_A] \xrightarrow{m\to-\infty} \mathbb{E}[X_{-\infty}\mathbb{1}_A].$$

Therefore, $\mathbb{E}[\mathbb{E}[X_n \mid \mathcal{F}_{-\infty}]\mathbb{1}_A] \leq \mathbb{E}[X_{-\infty}\mathbb{1}_A]$ and hence

$$\mathbb{E}[X_n \mid \mathcal{F}_{-\infty}] \leq X_{-\infty}. \qquad \square$$

Remark 4.8.1.3. Note that (4.8.1.1) is always satisfied for backward martingales. Indeed, for all $n \leq 0$, we get

$$\mathbb{E}[X_0 \mid \mathcal{F}_n] = X_n,$$

which implies that $\mathbb{E}[|X_n|] \leq \mathbb{E}[|X_0|]$ and thus

$$\sup_{n\leq 0} \mathbb{E}[|X_n|] < \infty.$$

Backward martingales are therefore always uniformly integrable.

Corollary 4.8.1.2. *Let $(\Omega, \mathcal{F}, \mathbb{P})$ be a probability space. Let Z be a random variable in $L^1(\Omega, \mathcal{F}, \mathbb{P})$ and let $(\mathcal{G}_n)_{n\geq 0}$ be a decreasing family of σ-algebras. Then we have*

$$\mathbb{E}[Z \mid \mathcal{G}_n] \xrightarrow[\text{a.s. and } L^1]{n\to\infty} \mathbb{E}[Z \mid \mathcal{G}_\infty],$$

where

$$\mathcal{G}_\infty := \bigcap_{n\geq 0} \mathcal{G}_n.$$

Proof. For $n \geq 0$ define $X_{-n} := \mathbb{E}[Z \mid \mathcal{F}_n]$, where $\mathcal{F}_{-n} = \mathcal{G}_n$. Then $(X_n)_{n\leq 0}$ is a backward martingale with respect to $(\mathcal{F}_n)_{n\leq 0}$. Hence, Theorem 4.8.1.1 implies that $(X_n)_{n\leq 0}$ converges almost surely and in L^1 for $n \to \infty$. Moreover,[2] we get

$$\begin{aligned} X_\infty &= \mathbb{E}[X_0 \mid \mathcal{F}_{-\infty}] = \mathbb{E}[\mathbb{E}[Z \mid \mathcal{F}_0] \mid \mathcal{F}_{-\infty}] \\ &= \mathbb{E}[Z \mid \mathcal{F}_{-\infty}] = \mathbb{E}[Z \mid \mathcal{G}_\infty]. \end{aligned}$$

□

Lemma 4.8.1.3. *Let $(\Omega, \mathcal{F}, \mathbb{P})$ be a probability space. Let $Z \in L^1(\Omega, \mathcal{F}, \mathbb{P})$ and $\mathcal{H}_1$ and $\mathcal{H}_2$ two σ-algebras included in $\mathcal{F}$. Assume that $\mathcal{H}_2$ is independent of $\sigma(Z) \vee \mathcal{H}_1$. Then we get that*

$$\mathbb{E}[Z \mid \mathcal{H}_1 \vee \mathcal{H}_2] = \mathbb{E}[Z \mid \mathcal{H}_1].$$

---

[2]This follows from the last part of Theorem 4.8.1.1.

Proof. Let $A \in \mathcal{H}_1 \vee \mathcal{H}_2$ such that $A = B \cap C$, where $B \in \mathcal{H}_1$ and $C \in \mathcal{H}_2$. Then we get that

$$\begin{aligned}\mathbb{E}[Z\mathbb{1}_A] &= \mathbb{E}[Z\mathbb{1}_B\mathbb{1}_C] = \mathbb{E}[Z\mathbb{1}_B]\mathbb{E}[\mathbb{1}_C] = \mathbb{E}[\mathbb{1}_A]\mathbb{E}[\mathbb{E}[Z \mid \mathcal{H}_1]\mathbb{1}_B] \\ &= \mathbb{E}[\mathbb{E}[Z \mid \mathcal{H}_1]\mathbb{1}_B\mathbb{1}_C] = \mathbb{E}[\mathbb{E}[Z \mid \mathcal{H}_1] \mid \mathbb{1}_A].\end{aligned}$$

Note also that

$$\sigma(W) := \{B \cap C \mid B \in \mathcal{H}_1, C \in \mathcal{H}_2\}) = \mathcal{H}_1 \vee \mathcal{H}_2$$

and W is stable under finite intersections. Thus the *monotone class theorem* implies that for all $A \in \mathcal{H}_1 \vee \mathcal{H}_2$ we have

$$\mathbb{E}[Z\mathbb{1}_A] = \mathbb{E}[\mathbb{E}[Z \mid \mathcal{H}_1]\mathbb{1}_A]. \qquad \square$$

We want to give an alternative proof of the strong law of large numbers (Theorem 2.7.2.3). Recall therefore again the statement:

Theorem 4.8.1.4 (Strong law of large numbers (Theorem 2.7.2.3)). *Let $(\Omega, \mathcal{F}, \mathbb{P})$ be a probability space. Let $(\xi_n)_{n\geq 1}$ be a sequence of i.i.d. random variables such that for all $n \geq 1$ we have $\mathbb{E}[|\xi_n|] < \infty$. Moreover, let $S_0 = 0$ and $S_n = \sum_{j=1}^n \xi_j$. Then*

$$\frac{S_n}{n} \xrightarrow[\text{a.s. and L}^1]{n\to\infty} \mathbb{E}[\xi_1].$$

Alternative proof of Theorem 2.7.2.3 using backward martingales. At first, we want to show that $\mathbb{E}[\xi_1 \mid S_n] = \frac{S_n}{n}$. Indeed, we know that there is a measurable map g such that

$$\mathbb{E}[\xi_1 \mid S_n] = g(S_n).$$

Moreover, we know that, for $k \in \{1, \dots, n\}$, we get that (ξ_1, S_n) and (ξ_k, S_n) have the same law. Now for all bounded and Borel measurable maps h we have

$$\mathbb{E}[\xi_k h(S_n)] = \mathbb{E}[\xi_1 h(S_n)] = \mathbb{E}[g(S_n)h(S_n)].$$

Thus, we get that $\mathbb{E}[\xi_k \mid S_n] = g(S_n)$. Moreover, we have that

$$\sum_{j=1}^n \mathbb{E}[\xi_j \mid S_n] = \mathbb{E}\left[\sum_{j=1}^n \xi_j \,\middle|\, S_n\right] = S_n,$$

but on the other hand, we have that

$$\sum_{j=1}^{n} \mathbb{E}[\xi_j \mid S_n] = ng(S_n).$$

Hence, we have $g(S_n) = \frac{S_n}{n}$. Now take $\mathcal{H}_1 = \sigma(S_n)$ and $\mathcal{H}_2 = \sigma(S_n, \xi_{n+1}, \xi_{n+2}, \ldots)$. Thus, by Lemma 4.8.1.3, we get that

$$\mathbb{E}[\xi_1 \mid S_n, \xi_{n+1}, \xi_{n+2}, \ldots] = \mathbb{E}[\xi_1 \mid S_n].$$

Now define $\mathcal{G} := \sigma(S_n, \xi_{n+1}, \xi_{n+2}, \ldots)$. Then we have $\mathcal{G}_{n+1} \subset \mathcal{G}_n$ because $S_{n+1} = S_n + \xi_{n+1}$. Hence, it follows that $\mathbb{E}[\xi_1 \mid \mathcal{G}_n] = \mathbb{E}[\xi_1 \mid S_n] = \frac{S_n}{n}$ converges almost surely and in L^1 to some random variable, but Kolmogorov's 0–1 law (Theorem 2.7.2.1) implies that this limit is almost surely constant. In particular, $\mathbb{E}\left[\frac{S_n}{n}\right] = \mathbb{E}[\xi_1]$ converges in L^1 to this limit, which is thus $\mathbb{E}[\xi_1]$. □

Exercise 4.8.1.4 (Hewitt–Savage 0–1 law). Let $(\xi_n)_{n\geq 1}$ be i.i.d. random variables with values in some measurable space $(E, \mathcal{E})$. The map $\omega \mapsto (\xi_1(\omega), \xi_2(\omega), \ldots)$ defines a random variable without values in $E^{\mathbb{N}^\times}$. A measurable map F defined on $E^{\mathbb{N}^\times}$ is said to be symmetric if

$$F(x_1, x_2, \ldots) = F(x_{\pi(1)}, x_{\pi(2)}, \ldots)$$

for all permutations π of $\mathbb{N}^\times$ with finite support.

Prove that if F is a symmetric function on $E^{\mathbb{N}^\times}$, then $F(\xi_1, \xi_2, \ldots)$ is almost surely constant.

Hint: Consider $\mathcal{F}_n = \sigma(\xi_1, x_2, \ldots, \xi_n)$, $\mathcal{G}_n = \sigma(\xi_{n+1}, \xi_{n+2}, \ldots)$, $Y = F(\xi_1, \xi_2, \ldots)$, $X = \mathbb{E}[Y \mid \mathcal{F}_n]$ and $Z_n = \mathbb{E}[Y \mid \mathcal{G}_n]$.

4.8.2. Martingales bounded in L^2 and random series

Let $(\Omega, \mathcal{F}, (\mathcal{F}_n)_{n\geq 0}, \mathbb{P})$ be a filtered probability space. Let $(M_n)_{n\geq 0}$ be a martingale in $L^2(\Omega, \mathcal{F}, (\mathcal{F}_n)_{n\geq 0}, \mathbb{P})$, i.e., $\mathbb{E}[M_n^2] < \infty$ for all $n \geq 0$. We say that $(M_n)_{n\geq 0}$ is bounded in $L^2(\Omega, \mathcal{F}, (\mathcal{F}_n)_{n\geq 0}, \mathbb{P})$ if $\sup_{n\geq 0} \mathbb{E}[M_n^2] < \infty$. For $n \leq \nu$, we have that $\mathbb{E}[M_\nu \mid \mathcal{F}_n] = M_n$ implies that $(M_\nu - M_n)$ is orthogonal to $L^2(\Omega, \mathcal{F}, (\mathcal{F}_n)_{n\geq 0}, \mathbb{P})$. Hence, for all $s \leq t \leq n \leq \nu$, $(M_\nu - M_n)$ is orthogonal to $(M_t - M_s)$. In particular, we have

$$\langle M_\nu - M_n, M_t - M_s\rangle = 0 \Longleftrightarrow \mathbb{E}[(M_\nu - M_n)(M_t - M_s)] = 0.$$

Now define $M_n := M_0 + \sum_{k=1}^n (M_k - M_{k-1})$. Then M_n is a sum of orthogonal terms and therefore

$$\mathbb{E}[M_n^2] = \mathbb{E}[M_0^2] + \sum_{k=1}^n \mathbb{E}[(M_k - M_{k-1})^2].$$

Theorem 4.8.2.1. *Let $(M_n)_{n\geq 0}$ be a martingale in $L^2(\Omega, \mathcal{F}, (\mathcal{F}_n)_{n\geq 0}, \mathbb{P})$. Then $(M_n)_{n\geq 0}$ is bounded in $L^2(\Omega, \mathcal{F}, (\mathcal{F}_n)_{n\geq 0}, \mathbb{P})$ if and only if $\sum_{k\geq 1} \mathbb{E}[(M_k - M_{k-1})^2] < \infty$ and in this case*

$$M_n \xrightarrow[\text{a.s. and } L^1]{n\to\infty} M_\infty.$$

Theorem 4.8.2.2. *Suppose that $(X_n)_{n\geq 1}$ is a sequence of independent random variables such that for all $k \geq 1$, $\mathbb{E}[X_k] = 0$ and $\sigma_k^2 = \mathbb{V}(X_k) < \infty$. Then the following conditions hold:*

(i) *$\sum_{k\geq 1} \sigma_k^2 < \infty$ implies that $\sum_{k\geq 1} X_k$ converges almost surely.*
(ii) *If there is a $C > 0$ such that for all $\omega \in \Omega$ and $k \geq 1$, $|X_k(\omega)| \leq C$, then almost sure convergence of $\sum_{k\geq 1} X_k$ implies that $\sum_{k\geq 1} \sigma_k^2 < \infty$.*

Proof. Consider $\mathcal{F}_n := \sigma(X_1, \dots, x_n)$ with $F_0 := \{\emptyset, \Omega\}$, $S_n = \sum_{j=1}^n X_j$ with $S_0 = 0$ and $A_n = \sum_{k=1}^n \sigma_k^2$ with $A_0 = 0$. Moreover, set $M_n = S_n^2 - A_n$. Then $(S_n)_{n\geq 0}$ is a martingale and

$$\mathbb{E}[(S_n - S_{n-1})^2] = \mathbb{E}[X_n^2] = \sigma_n^2.$$

Thus, the property $\sum_{n\geq 1} \sigma_n^2 < \infty$ implies $\sum_{n\geq 1} \mathbb{E}[(S_n - S_{n-1})^2] < \infty$ and hence $(S_n)_{n\geq 0}$ is bounded in $L^2(\Omega, \mathcal{F}, (\mathcal{F}_n)_{n\geq 0}, \mathbb{P})$, which means that S_n converges almost surely. Next we show that $(M_n)_{n\geq 0}$ is a martingale. Note that we have

$$\mathbb{E}[(S_n - S_{n-1})^2 \mid \mathcal{F}_{n-1}] = \mathbb{E}[X_n^2 \mid \mathcal{F}_{n-1}] = \mathbb{E}[X_n^2] = \sigma_n^2.$$

Hence, we get that

$$\sigma_n^2 = \mathbb{E}[(S_n - S_{n-1})^2 \mid \mathcal{F}_{n-1}] = \mathbb{E}[S_n^2 - 2S_{n-1}S_n + S_{n-1}^2 \mid F_{n-1}]$$
$$= \mathbb{E}[S_n^2 \mid \mathcal{F}_{n-1}] - 2S_{n-1}^2 + S_{n-1}^2 = \mathbb{E}[S_n^2 \mid \mathcal{F}_{n-1}] - S_{n-1}^2,$$

which implies that $(M_n)_{n\geq 0}$ is a martingale. Let $T := \inf\{n \in \mathbb{N} \mid |S_n| > \alpha\}$ for some constant α. Then T is a stopping time and thus

$(M_{n\wedge T})_{n\geq 1}$ is a martingale and therefore

$$\mathbb{E}[M_{n\wedge T}] = \mathbb{E}[S_{n\wedge T}^2] - \mathbb{E}[A_{n\wedge T}] = 0.$$

Hence, we get that $\mathbb{E}[S_{n\wedge T}^2] = \mathbb{E}[A_{n\wedge T}]$ and if T is finite, $|S_T - S_{T-1}| < |X_T| \leq C$ for some constant C, thus $|S_{n\wedge T}| \leq C + \alpha$ and hence $\mathbb{E}[A_{n\wedge T}] \leq (C+\alpha)^2$ for all $n \geq 0$. Now since A_n is increasing, we get that $\mathbb{E}[A_{n\wedge T}] \leq (C+\alpha)^2 < \infty$. Since $\sum_{n\geq 0} X_n$ converges almost surely, $\sum_{k=1}^n X_k$ is bounded and therefore exists $\alpha > 0$ such that $\mathbb{P}(T = \alpha) > 0$. Choosing[3] α correctly yields $\sum_{k\geq 1} \sigma_k^2 < \infty$. □

Example 4.8.2.1. Let $(a_n)_{n\geq 1}$ be a sequence of real numbers and let $(\xi_n)_{n\geq 1}$ be i.i.d. random variables with $\mathbb{P}(\xi_n = \pm 1) = \frac{1}{2}$. Then we get that $\sum_{n\geq 1} a_n\xi_n$ converges almost surely if and only if $\sum_{n\geq 1} a_n^2 < \infty$. Indeed, we get that $|a_n\xi_n| = |a_n| \xrightarrow{n\to\infty} 0$ and therefore there exists a constant $C > 0$ such that for all $n \geq 1$, $|a_n| \leq C$. Now for a random variable X, recall that we can consider its characteristic function $\Phi_X(t) = \mathbb{E}\left[e^{itX}\right]$. Moreover, we know that

$$e^{itx} = \sum_{n\geq 0} \frac{i^n t^n x^n}{n!}, \quad e^{itX} = \sum_{n\geq 0} \frac{i^n t^n X^n}{n!}.$$

Further, define $R_n(x) := e^{ix} - \sum_{k=0}^n \frac{i^k x^k}{k!}$. Therefore, we get that

$$|R_n(x)| \leq \min\left(2\frac{|x|^n}{n!}, \frac{|x|^{n+1}}{(n+1)!}\right).$$

Indeed, $|R_0(x)| = |e^{ix} - 1| = \left|\int_0^x ie^{iy}dy\right| \leq \min(2, |x|)$. Moreover, we have $|R_n(x)| = \left|\int_0^x iR_{n-1}(y)dy\right|$. Hence, the claim follows by a simple induction on n. If X is such that $\mathbb{E}[X] = 0$, $\mathbb{E}[X^2] = \sigma^2 < \infty$ and $e^{itX} - (1 + itX - \frac{t^2X^2}{2}) = R_2(tX)$, we get that

$$\mathbb{E}\left[e^{itX}\right] = 1 - \frac{\sigma^2 t^2}{2} + \mathbb{E}[R_2(tX)]$$

and thus $\mathbb{E}[R_2(tX)] \leq t^2\mathbb{E}[|X|^2 \wedge tX^3]$. Using *Lebesgue's dominated convergence theorem*, it follows that $\Phi(t) = 1 - \frac{t^2\sigma^2}{2} + O(t^2)$ as $t \to 0$.

---

[3]Note that $\mathbb{E}\left[\sum_{k\geq 1} \sigma_k^2 \mathbb{1}_{\{T=\infty\}}\right] = \sum_{k\geq 1} \sigma_k^2 \mathbb{P}(T = \infty)$.

Lemma 4.8.2.3. *Let $(\Omega, \mathcal{F}, (\mathcal{F}_n)_{n\geq 0}, \mathbb{P})$ be a filtered probability space. Let $(X_n)_{n\geq 0}$ be a sequence of independent random variables bounded by $k > 0$. Then, if $\sum_{n\geq 0} X_n$ converges almost surely, $\sum_{n\geq 0} \mathbb{E}[X_n]$ and $\sum_{n\geq 0} \mathbb{V}(X_n)$ both converge.*

Proof. If Z is a random variable such that $|Z| < k$, $\mathbb{E}[Z] = 0$ and $\sigma^2 = \mathbb{V}(Z) < \infty$, then for $|t| \leq \frac{1}{k}$ we get that

$$
\begin{aligned}
|\Phi_Z(t)| &\leq 1 - \frac{t^2\sigma^2}{2} + \frac{t^3 k\mathbb{E}[Z^2]}{6} \leq 1 - \frac{t^2\sigma^2}{2} + \frac{t^2\sigma^2}{6} \\
&= 1 - \frac{t^2\sigma^2}{3} \leq \exp\left(-\frac{t^2\sigma^2}{3}\right).
\end{aligned}
$$

Let $Z_n := X_n - \mathbb{E}[X_n]$. Then $|\Phi_{Z_n}(t)| = |\Phi_{X_n}(t)|$ and $|Z_n| \leq 2k$. If $\sum_{n\geq 0} X_n = \infty$, we get that

$$
\prod_{n\geq 0} |\Phi_{X_n}(t)| = \prod_{n\geq 0} |\Phi_{Z_n}(t)| \leq \exp\left(-\frac{1}{3}t^2 \sum_{n\geq 0} \mathbb{V}(X_n)\right) = 0.
$$

This is a contradiction, since $|\Phi_{\sum_{n\geq 0} X_n}(t)| \xrightarrow{n\to\infty} |\Phi(t)|$ with Φ continuous and $\Phi(0) = 1$. Hence, we have $\sum_{n\geq 0} \mathbb{V}(X_n) = \sum_{n\geq 0} \mathbb{V}(Z_n) < \infty$. Since $\mathbb{E}[Z_n] = 0$ and $\sum_{n\geq 0} \mathbb{V}(Z_n) < \infty$, we have $\sum_{n\geq 0} Z_n$ converges almost surely, but $\sum_{n\geq 0} -Z_n = \sum_{n\geq 0} X_n - \sum_{n\geq 0} \mathbb{E}[X_n]$ and thus since $\sum_{n\geq 0} X_n$ converges almost surely, it follows that $\sum_{n\geq 0} \mathbb{E}[X_n]$ converges. □

Theorem 4.8.2.4 (Kolmogorov's three series theorem). *Let $(\Omega, \mathcal{F}, \mathbb{P})$ be a probability space. Let $(X_n)_{n\geq 0}$ be a sequence of independent random variables. Then we get that $\sum_{n\geq 0} X_n$ converges almost surely if and only if for some $k > 0$ (then also for every $k > 0$) the following properties hold:*

(i) $\sum_{n\geq 0} \mathbb{P}(|X_n| > k) < \infty$;

(ii) $\sum_{n\geq 0} \mathbb{E}[X_n^{(k)}]$ *converges, where* $X_n^{(k)} = X_n \mathbb{1}_{\{|X_n|\leq k\}}$;

(iii) $\sum_{n\geq 0} \mathbb{V}(X_n^{(k)}) < \infty$.

Proof. Suppose that for some $k > 0$, (i), (ii) and (iii) hold. Then we get that

$$\sum_{n\geq 0} \mathbb{P}\left[X_n \neq X_n^{(k)}\right] = \sum_{n\geq 0} \mathbb{P}[|X_n| > k] < \infty.$$

It follows from the Borel–Cantelli lemma (Lemma 2.5.4.1) that

$$\mathbb{P}(X_n = X_n^{(k)} \text{for all but finitely many } n) = 1.$$

Hence, we only need to show that $\sum_{n\geq 0} X_n^k$ converges almost surely. Because of (ii), it is enough to show that $\sum_{n\geq 0} Y_n^{(k)}$ converges, where $Y_n^{(k)} = X_n^{(k)} - \mathbb{E}[X_n^{(k)}]$. The convergence of $\sum_{n\geq 0} Y_n^{(k)}$ follows then from (iii). Conversely, assume that $\sum_{n\geq 0} X_n$ converges almost surely and that $k \in (0, \infty)$. Since $X \xrightarrow{n\to\infty} 0$ almost surely, we have that $|X_n| > k$ for only finitely many n. Therefore, the Borel–Cantelli lemma implies (i). Since $X_n = X_n^{(k)}$ for all but finitely many n, $\sum_{n\geq 0} X_n^{(k)}$ converges and it follows from Lemma 4.8.2.3 that (ii) and (iii) have to hold. □

Lemma 4.8.2.5 (Cesàro). *Suppose $(b_n)_{n\geq 1}$ is a sequence of strictly positive real numbers with $b_n \uparrow \infty$ as $n \to \infty$. Let $(v_n)_{n\geq 1}$ be a sequence of real numbers such that $v_n \xrightarrow{n\to\infty} v$. Then we get that*

$$\frac{1}{b_n}\sum_{k=1}^{n}(b_k - b_{k-1})v_k \xrightarrow{n\to\infty} v, \quad b_0 = 0.$$

Proof. Note that

$$\begin{aligned}\left|\frac{1}{b_n}\sum_{k=1}^{n}(b_k - b_{k-1})v_k - v\right| &= \left|\frac{1}{b_n}\sum_{k=1}^{n}(b_k - b_{k-1})(v_k - v)\right| \\ &\leq \frac{1}{b_n}\sum_{k=1}^{N}(b_k - b_{k-1})|v_k - v| \\ &\quad + \frac{1}{b_n}\sum_{k=N+1}^{n}(b_k - b_{k-1})|v_k - v|.\end{aligned}$$

Now we only have to choose N such that $n \geq N$ and $|v_k - v| < \varepsilon$ for any $\varepsilon > 0$. □

Lemma 4.8.2.6 (Kronecker). *Let $(b_n)_{n\geq 1}$ be a sequence of real numbers, strictly positive with $b_n \uparrow \infty$ as $n \to \infty$. Let $(x_n)_{n\geq 1}$ be a sequence of real numbers. Then if $\sum_{n\geq 1} \frac{x_n}{b_n}$ converges, we get that*

$$\frac{x_1 + \cdots + x_n}{b_n} \xrightarrow{n\to\infty} 0.$$

Proof. Let $v_n = \sum_{k=1}^{n} \frac{x_n}{b_n}$ and $v = \lim_{n\to\infty} v_n$. Then we get that $v_n - v_{n-1} = \frac{x_n}{b_n}$. Moreover, we note that

$$\sum_{k=1}^{n} x_k = \sum_{k=1}^{n} b_k(v_k - v_{k-1}) = b_n v_n - \sum_{k=1}^{n} (b_k - b_{k-1}) v_k,$$

which implies that

$$\frac{x_1 + \cdots + x_n}{b_n} = v_n - \frac{1}{b_n} \sum_{k=1}^{n} (b_k - b_{k-1}) v_k \xrightarrow{n\to\infty} v - v = 0. \qquad \square$$

Proposition 4.8.2.7. *Let $(\Omega, \mathcal{F}, \mathbb{P})$ be a probability space. Let $(w_n)_{n\geq 1}$ be a sequence of random variables such that $\mathbb{E}[w_n] = 0$ for all $n \geq 1$ and $\sum_{n\geq 1} \frac{\mathbb{V}(w_n)}{n^2} < \infty$. Then we get that*

$$\frac{1}{n} \sum_{n\geq 1} w_n \xrightarrow[\text{a.s.}]{n\to\infty} 0.$$

Exercise 4.8.2.2. Prove[4] Proposition 4.8.2.7.

Theorem 4.8.2.8. *Let $(\Omega, \mathcal{F}, \mathbb{P})$ be a probability space. Let $(X_n)_{n\geq 1}$ be independent and non-negative random variables such that $\mathbb{E}[X_n] = 1$ for all $n \geq 1$. Define $M_0 = 1$ and for $n \in \mathbb{N}$, let*

$$M_n = \prod_{j=1}^{n} X_j.$$

Then we get that $(M_n)_{n\geq 1}$ is a non-negative martingale, so that $M_\infty := \lim_{n\to\infty} M_n$ exists almost surely. Then the following conditions are equivalent:

[4]From Lemma 4.8.2.6 it is enough to prove that $\sum_{k\geq 1} \frac{w_k}{k}$ converges almost surely.

(i) $\mathbb{E}[M_\infty] = 1$;
(ii) $M_n \xrightarrow[L^1]{n\to\infty} M_\infty$;
(iii) $(M_n)_{n\geq 1}$ *is uniformly integrable;*
(iv) $\prod_n a_n > 0$, *where* $0 < a_n = \mathbb{E}[X_n^{1/2}] \leq 1$;
(v) $\sum_n (1 - a_n) < \infty$.

Moreover, if one of the following (hence all of them) statements hold, then

$$\mathbb{P}(M_\infty = 0) = 1.$$

Exercise 4.8.2.3. Prove Theorem 4.8.2.8.

4.8.3. A martingale central limit theorem

Theorem 4.8.3.1 (Martingale CLT). *Let* $(\Omega, \mathcal{F}, (\mathcal{F}_n)_{n\geq 0}, \mathbb{P})$ *be a filtered probability space. Let* $(X_n)_{n\geq 0}$ *be a sequence of real-valued random variables such that for all* $n \geq 1$ *the following properties hold:*

(i) $\mathbb{E}[X_n \mid \mathcal{F}_{n-1}] = 0$,
(ii) $\mathbb{E}[X_n^2 \mid \mathcal{F}_{n-1}] = 1$,
(iii) $\mathbb{E}[|X_n^3| \mid \mathcal{F}_{n-1}] \leq k < \infty$.

Moreover, let $S_n = \sum_{j=1}^n X_j$. *Then we get that*

$$\frac{S_n}{\sqrt{n}} \xrightarrow[law]{n\to\infty} \mathcal{N}(0,1).$$

Proof. Define $\Phi_{n,j}(u) := \mathbb{E}[\mathrm{e}^{iu\frac{X_j}{\sqrt{n}}} \mid \mathcal{F}_{j-1}]$. Using Taylor expansion yields

$$\exp\left(iu\frac{X_j}{\sqrt{n}}\right) = 1 + iu\frac{X_j}{\sqrt{n}} - \frac{u^2}{2n}X_j^2 - \frac{iu^3}{6n^{3/2}}\bar{X}_j^3,$$

where $\bar{X}_j$ is a random number between 0 and X_j. Therefore, we get

$$\Phi_{n,j}(u) = 1 + iu\frac{1}{\sqrt{n}}\mathbb{E}[X_j \mid \mathcal{F}_{j-1}] - \frac{u^2}{2n}\mathbb{E}[X_j^2 \mid \mathcal{F}_{j-1}] - \frac{iu^3}{6n^{3/2}}\mathbb{E}[\bar{X}_j^3 \mid \mathcal{F}_{j-1}]$$

and thus

$$\Phi_{n,j}(u) - 1 + \frac{u^2}{2n} = -\frac{iu^3}{6n^{3/2}}\mathbb{E}[\bar{X}_j^2 \mid \mathcal{F}_{j-1}].$$

Hence, we get

$$\mathbb{E}[e^{iu\frac{S_p}{\sqrt{n}}}] = \mathbb{E}[e^{iu\frac{S_{p-1}}{\sqrt{n}}} e^{iu\frac{X_p}{\sqrt{n}}}] = \mathbb{E}[e^{iu\frac{S_{p-1}}{\sqrt{n}}} \mathbb{E}[e^{iu\frac{X_p}{\sqrt{n}}} \mid \mathcal{F}_{p-1}]]$$
$$= \mathbb{E}[e^{iu\frac{S_{p-1}}{\sqrt{n}}} \Phi_{n,p}(u)].$$

Consequently, we get

$$\mathbb{E}[e^{iu\frac{S_p}{\sqrt{n}}}] = \mathbb{E}\left[e^{iu\frac{S_{p-1}}{\sqrt{n}}}\left(1 + \frac{u^2}{2n} - \frac{iu^3}{6n^{3/2}}\bar{X}_p^3\right)\right].$$

Thus, we get that

$$\mathbb{E}\left[e^{iu\frac{S_p}{\sqrt{n}}} - \left(1 - \frac{u^2}{2n}\right)e^{iu\frac{S_{p-1}}{\sqrt{n}}}\right] = \mathbb{E}\left[e^{iu\frac{S_{p-1}}{\sqrt{n}}}\frac{iu^3}{6n^{3/2}}\bar{X}_p^3\right],$$

which implies that

$$\left|\mathbb{E}\left[e^{iu\frac{S_p}{\sqrt{n}}} - \left(1 - \frac{u^2}{2n}\right)e^{iu\frac{S_{p-1}}{\sqrt{n}}}\right]\right| \leq \frac{K|u|^3}{6n^{3/2}}. \tag{4.8.3.1}$$

Let us fix $n \in \mathbb{N}$. For n large enough, we have $0 \leq 1 - \frac{u^2}{2n} \leq 1$. Multiplying both sides of (4.8.3.1) by $(1 - \frac{u^2}{2n})^{n-p}$, we get

$$\left|\left(1 - \frac{u^2}{2n}\right)^{n-p}\mathbb{E}[e^{iu\frac{S_p}{\sqrt{n}}}] - \left(1 - \frac{u^2}{2n}\right)^{n-p+1}\mathbb{E}\left[e^{iu\frac{S_{p-1}}{\sqrt{n}}}\right]\right| \leq \frac{K|u|^3}{6n^{3/2}}. \tag{4.8.3.2}$$

By taking K sufficiently large, we can assume that (4.8.3.2) holds for all n. Now we note that

$$\mathbb{E}[e^{iu\frac{S_n}{\sqrt{n}}}] - \left(1 - \frac{u^2}{2n}\right)^n = \sum_{p=1}^{n}\left\{\left(1 - \frac{u^2}{2n}\right)^{n-p}\right.$$
$$\left.\times\,\mathbb{E}[e^{iu\frac{S_p}{\sqrt{n}}}] - \left(1 - \frac{u^2}{2n}\right)^{n-p+1}\mathbb{E}[e^{iu\frac{S_{p-1}}{\sqrt{n}}}]\right\}.$$

Therefore, we get

$$\left|\mathbb{E}[e^{iu\frac{S_n}{\sqrt{n}}}] - \left(1 - \frac{u^2}{2n}\right)^n\right| \leq n\frac{K|u|^3}{n^{3/2}} = \frac{K|u|^3}{6\sqrt{n}},$$

which implies that

$$\lim_{n\to\infty}\mathbb{E}[e^{iu\frac{S_n}{\sqrt{n}}}] = e^{-\frac{u^2}{2}}.$$

□

Chapter 5

Markov Chains

5.1. Definition and First Properties

In this chapter, E will be a finite or a countable set, endowed with the σ-algebra given by the power set $\mathcal{P}(E)$. We define a *stochastic matrix* on E to be a family $(Q(x,y))_{x,y\in E}$ of real numbers satisfying:

(i) $0 \le Q(x,y) \le 1$ for all $x, y \in E$;
(ii) $\sum_{y\in E} Q(x,y) = 1$ for all $x \in E$.

If $x \in E$ and $A \subset E$, we define

$$\nu(x, A) := \sum_{y\in A} \nu(x,y),$$

and observe that ν is a transition kernel probability from E to E. Conversely, if we start with such a transition kernel, the formula

$$Q(x,y) := \nu(x, \{y\})$$

defines a stochastic matrix on E. For $n \ge 1$, we can define $Q_n = Q^n$. Indeed, $Q_1 = Q$ and by induction

$$Q_{n+1}(x,y) = \sum_{z\in E} Q_n(x,z)Q(z,y).$$

One can also check that Q_n is a stochastic matrix on E. For $n = 0$ we take $Q_0(x,y) = \mathbb{1}_{\{x=y\}}$. For a measurable map $f : E \to \mathbb{R}_{\ge 0}$ we write Qf as the function defined by

$$Qf(x) := \sum_{y\in E} Q(x,y)f(y).$$

Definition 5.1.0.1 (Markov chain). Let $(\Omega, \mathcal{F}, \mathbb{P})$ be a probability space. Let Q be a stochastic matrix on E and let $(X_n)_{n\geq 0}$ be a stochastic process with values in E. We say that $(X_n)_{n\geq 0}$ is a *Markov chain* with *transition matrix* Q if for all $n \geq 0$, the conditional distribution of X_{n+1} given $(X_0, X_1, \ldots, X_n)$ is $Q(X_n, y)$, or equivalently if for all $x_0, x_1, \ldots, x_n, y \in E$, we get that

$$\begin{aligned}\mathbb{P}(X_{n+1} = y \mid X_0 = x_0, X_1 = x_1, \ldots, X_{n-1} = x_{n-1}, X_n = x_n) \\ = Q(x_n, y)\end{aligned}$$

with $\mathbb{P}(X_0 = x_0, \ldots, X_n = x_n) > 0$.

Remark 5.1.0.2. In general, the conditional distribution of X_{n+1} given $X_0, X_1, \ldots, X_n$ depends on all the variables $X_0, X_1, \ldots, X_n$. The fact that this conditional distribution only depends on X_n is called the *Markov property.*

Remark 5.1.0.3. $Q(x, \cdot)$, which is the distribution of X_{n+1} given $X_n = x_n$ does not depend on n. This is called the *homogeneity* of the Markov chain.

Proposition 5.1.0.1. *Let $(\Omega, \mathcal{F}, \mathbb{P})$ be a probability space. A stochastic process $(X_n)_{n\geq 0}$ with values in E is a Markov chain with transition kernel Q if and only if for all $n \geq 0$ and for all $x_0, x_1, \ldots, x_n \in E$ we have*

$$\mathbb{P}(X_0 = x_0, X_1 = x_1, \ldots, X_n = x_n) = \mathbb{P}(X_0 = x_0) \prod_{j=1}^{n} Q(x_{j-1}, x_j). \tag{5.1.0.1}$$

In particular, if $\mathbb{P}(X_0 = x_0) > 0$, then

$$\mathbb{P}(X_n = x_n \mid X_0 = x_0) = Q_n(x_0, x_n).$$

Proof. If $(X_n)_{n\geq 0}$ is a Markov chain with transition matrix Q, we have

$$\begin{aligned}\mathbb{P}(X_0 &= x_0, \ldots, X_n = x_n, X_{n+1} = x_{n+1}) \\ &= \mathbb{P}(X_0 = x_0, \ldots, X_n = x_n) \\ &\quad \times \mathbb{P}(X_{n+1} = x_{n+1} \mid X_0 = x_0, \ldots, X_n = x_n) \\ &= \mathbb{P}(X_1 = x_1, \ldots, X_n = x_n) \\ &= Q(x_{n+1}, x_n).\end{aligned}$$

Thus, we can conclude by induction. Conversely, if (5.1.0.1) is satisfied, then we get that

$$\begin{aligned}\mathbb{P}(X_{n+1} = y \mid X_0 = x_0, \dots, X_n = x_n) \\ = \frac{\mathbb{P}(X_0 = x_0) \prod_{j=0}^{n-1} Q(x_j, x_{j+1}) Q(x_n, y)}{\mathbb{P}(X_0 = x_0) \prod_{j=0}^{n-1} Q(x_j, x_{j+1})} = Q(x_n, y).\end{aligned}$$

For conclusion, we note that

$$Q_n(x_0, x) = \sum_{x_1, \dots, x_{n+1} \in E} \prod_{j=1}^{n} Q(x_{j-1} x_j).$$

□

Remark 5.1.0.4. Proposition 5.1.0.1 shows that for a Markov chain, $(X_0, X_1, \dots, X_n)$ is completely determined by the initial distribution (that of X_0) and the transition matrix Q. For now, we want to denote by $\mathbb{P}(A \mid Z)$ the conditional expectation $\mathbb{E}[\mathbb{1}_A \mid Z]$.

Proposition 5.1.0.2. *Let $(\Omega, \mathcal{F}, \mathbb{P})$ be a probability space. Let $(X_n)_{n \geq 0}$ be a Markov chain with transition matrix Q. Then the following conditions hold:*

(i) *For all $n \geq 0$ and for all measurable maps $f : E \to \mathbb{R}_{\geq 0}$, we have*

$$\mathbb{E}[f(X_{n+1}) \mid X_0, X_1, \dots, X_n] = \mathbb{E}[f(X_{n+1}) \mid X_n] = Qf(X_n).$$

More generally, for all $\{i_1, \dots, i_n\} \subset \{0, 1, \dots, n-1\}$, we have

$$\mathbb{E}[f(X_{n+1}) \mid X_{i_1}, X_{i_2}, \dots, X_{i_n}, X_n] = Qf(X_n).$$

(ii) *For all $n \geq 0, p \geq 1$ and for all $y_1, \dots, y_p \in E$, we have*

$$\begin{aligned}\mathbb{P}(X_{n+1} = y_1, \dots, X_{n+p} = y_p \mid X_0, \dots, X_n) \\ = Q(X_n, y_1) \prod_{j=1}^{p-1} Q(y_j, y_{j+1})\end{aligned}$$

and thus

$$\mathbb{P}(X_{n+p} = y_p \mid X_n) = Q_p(X_n, y_p).$$

If we take $Y_p = X_{n+p}$, *then* $(Y_p)_{p\geq 0}$ *is also a Markov chain with transition matrix* Q.

Proof. For (i), we note that

$$\mathbb{E}[f(X_{n++}) \mid X_0, X_1, \ldots, X_n] = \sum_{y\in E} Q(X_n, y)f(y) = Qf(X_n).$$

Now if $\{i_1, \ldots, i_n\} \subset \{0, 1, \ldots, n-1\}$, then we get that[1]

$$\begin{aligned}
&\mathbb{E}[f(X_{n+1}) \mid X_{i_1}, \ldots, X_{i_n}, X_n] \\
&\quad = \mathbb{E}[\mathbb{E}[f(X_{n+1}) \mid X_0, X_1, \ldots, X_n] \mid X_{i_{1,\ldots,X_{i_n}}}, X_n] \\
&\quad = \mathbb{E}[Qf(X_n) \mid X_{i_1}, \ldots, X_{i_n}, X_n] = Qf(X_n).
\end{aligned}$$

For (ii), note that it follows immediately from (5.1.0.1) that

$$\begin{aligned}
&\mathbb{P}(X_{n+1} = y_1, \ldots, X_{n-p} = y_p \mid X_0 = x_0, \ldots, X_n = x_n) \\
&\quad = Q(x_n, y_1) \prod_{j=1}^{p-1} Q(y_j, y_{j+1}).
\end{aligned}$$

The formula for $\mathbb{P}(X_{n-p} = y_p \mid X_n)$ follows then also for $y_1, \ldots, y_{p-1} \in E$. Finally, we note that

$$\mathbb{P}(Y_0 =, Y_1 = y_1, \ldots, Y_p = y_p) = \mathbb{P}[X_n = y_n] \prod_{j=0}^{p} Q(y_{j-1}, y_j)$$

and hence we can apply Proposition 5.1.0.1. □

Example 5.1.0.5. Let $(\Omega, \mathcal{F}, \mathbb{P})$ be a probability space. Let $(X_n)_{n\geq 0}$ be a sequence of independent random variables with values in E, with the same distribution μ. Then $(X_n)_{n\geq 0}$ is a Markov chain with transition matrix

$$Q(x, y) := \mu(y), \quad \forall x, y \in E.$$

---

[1] Recall that if $X \in (E, \mathcal{E})$, $Y \in (F, \mathcal{F})$, the conditional distribution of Y given X is defined through any transition kernel $\nu : E \to F$ for which we have that for all measurable maps $h : (F; \mathcal{F}) \to \mathbb{R}_{\geq 0}$, $\mathbb{E}[h(y) \mid X] = \int_{\mathbb{R}} h(y)\nu(X, \mathrm{d}y)$.

Example 5.1.0.6 (Simple random walk on $\mathbb{Z}^d$). Let $(\Omega, \mathcal{F}, \mathbb{P})$ be a probability space. Let $\eta, \xi_1, \ldots, \xi_n, \ldots$ be independent random variables in $\mathbb{Z}^d$. We assume that $\xi_1, \ldots \xi_n, \ldots$ have the same distribution μ. For $n \geq 0$ we define

$$X_n := \eta + \xi_1 + \cdots + \xi_n.$$

Then we get that $(X_n)_{n\geq 0}$ is a Markov chain with transition matrix

$$Q(x, y) = \mu(y - x), \quad \forall x, y \in E.$$

Indeed, we have

$$\begin{aligned}
&\mathbb{P}(X_{n+1} = y \mid X_0 = x_0, \ldots, X_n = x_n) \\
&\quad = \mathbb{P}(\xi_{n+1} = y - x_n \mid X_0 = x_0, \ldots, X_n = x_n) \\
&\quad = \mathbb{P}(\xi_{n+1} = y - x_n) = \mu(y - x_n).
\end{aligned}$$

If $(e_1, \ldots, e_d)$ is the canonical basis of $\mathbb{R}^d$, and if $\mu(e_i) = \mu(-e_i) = \frac{1}{2d}$ for all $i \in \{1, \ldots, d\}$, then the Markov chain is called a *simple random walk on* $\mathbb{Z}^d$.

Example 5.1.0.7 (Simple random walk on a graph). Let $\mathcal{P}_2(E)$ be the subsets of E with two elements and let $A \subset \mathcal{P}_2(E)$. For $x \in E$, we define

$$A_x := \{y \in E \mid \{x, y\} \in A\}.$$

Assume that $A_x \neq \emptyset$ for all $x \in E$. Then we define further a transition matrix on E by setting for $x, y \in E$

$$Q(x, y) := \begin{cases} \dfrac{1}{|A_x|} & \text{if } \{x, y\} \in A, \\ 0 & \text{otherwise.} \end{cases}$$

A Markov chain with transition matrix Q is called a *simple random walk on the graph* (E, A).

5.2. The Canonical Markov Chain

We start with the existence of a Markov chain associated with a given transition matrix.

Proposition 5.2.0.1. *Let $(\Omega, \mathcal{F}, \mathbb{P})$ be a probability space. Let Q be a stochastic matrix on E. There exists a probability space $(\Omega', \mathcal{F}', \mathbb{P}')$ on which there exists, for all $x \in E$, a stochastic process $(X_n^x)_{n\geq 0}$ which is a Markov chain with transition matrix Q, starting from $X_0^x = x$.*

Proof. We can take $\Omega = [0, 1)$ together with the Borel σ-algebra and the Lebesgue measure. For $\omega \in [0, 1)$, we consider its series expansion

$$\omega = \sum_{n=0}^{\infty} \mathcal{E}_n(\omega) 2^{-n-1},$$

with $\mathcal{E}_n(\omega) \in \{0, 1\}$. We can further take the sequence $(\mathcal{E}_n)_{n\geq 0}$ of i.i.d. random variables with $\mathbb{P}(\mathcal{E}_n = 1) = \mathbb{P}(\mathcal{E}_n = 0) = \frac{1}{2}$. If $\varphi : \mathbb{N} \times \mathbb{N} \to \mathbb{N}$ is a bijection, then the random variables $\eta_{i,j} := \mathcal{E}_{\varphi(i,j)}$ for $i, j \in \mathbb{N}$ are also i.i.d. random variables. Now define

$$U_i := \sum_{j=0}^{\infty} \eta_{i,j} 2^{-j-1}.$$

Therefore, we get that $U_0, U_1, U_2, \ldots$ are i.i.d. random variables. Define $E := \{y_1, y_2, \ldots, y_k, \ldots\}$ and fix $x \in E$. We set $X_0^x = x$ and $X_1^x = y_k$ whenever

$$\sum_{i\leq j<k} Q(x, y_j) < U_1 \leq \sum_{1\leq j\leq k} Q(x, y_j),$$

such that $\mathbb{P}(X_1^x = y) = Q(x, y)$ for all $y \in E$. We can now proceed by induction $X_{n+1}^x = y_k$ if

$$\sum_{1\leq j<k} Q(X_n^x, y_j) < U_{n+1} \leq \sum_{1\leq j\leq k} Q(X_n^x, y_j).$$

Using the independence of the U_i's, we can check that for all $k \geq 1$, we have

$$\mathbb{P}(X^x_{n+1} = y_k \mid X^x_0 = x, X^x_1 = x_1, \ldots, X^x_n = x_n)$$
$$= \mathbb{P}\left(\sum_{1 \leq j < k} Q(x_n, y_j) < U_{n+1} \right.$$
$$\left. \leq \sum_{1 \leq j \leq k} Q(x_n, y_j) \middle| X^x_0 = x, \ldots, X^x_n = x_n \right)$$
$$= Q(x_n, y_k).$$

□

In the sequel, we shall take $\Omega = E^{\mathbb{N}}$. Then an element $\omega \in \Omega$ is a sequence $\omega = (\omega_0, \omega_1, \ldots)$ of elements in E. Define the coordinate map

$$X_n(\omega) := \omega_n.$$

We take $\mathcal{F}$ to be the smallest σ-algebra on Ω which make all X_n's measurable. This σ-algebra is generated by

$$C := \{\omega \in \Omega \mid \omega_0 = x_0, \omega_1 = x_1, \ldots, \omega_n = x_n\}$$

for $n \in \mathbb{N}$ and $x_0, \ldots, x_n \in E$.

Lemma 5.2.0.2. *Let $(G, \mathcal{G})$ be a measurable space and let $\psi : G \to \Omega$ be a map. Then ψ is measurable if and only if for all $n \geq 0$, $X_n \circ \psi$ is measurable.*

Proof. We only need to show that if $X_n \circ \psi$ is measurable for all $n \geq 0$, then ψ is measurable as well. Consider the σ-algebra $\mathcal{A}$ on Ω given by

$$\mathcal{A} := \{A \in \mathcal{F} \mid \psi^{-1}(A) \in \mathcal{G}\},$$

which contains all the sets of the form $X_n^{-1}(y)$ for $y \in E$. Hence, all the X_n's are measurable with respect to $\mathcal{A}$ and the claim follows. □

Theorem 5.2.0.3. *Let $(\Omega, \mathcal{F}, \mathbb{P})$ be a probability space. Let Q be a stochastic matrix on E. For all $x \in E$, there exists a unique* *probability measure* $\mathbb{P}_x$, *on* $\Omega = E^{\mathbb{N}}$, *such that the sequence of coordinate functions* $(X_n)_{n\geq 0}$ *is a Markov chain with transition matrix* Q *and* $\mathbb{P}_x(X_0 = x) = 1$.

Proof. By Proposition 5.2.0.1, there exists $(\Omega', \mathcal{F}', \mathbb{P}')$ and $(X_n^x)_{n\geq 0}$ which is a Markov chain with transition matrix Q and such that $X_0^x = x$. We then define $\mathbb{P}_x$ as the image of $\mathbb{P}'$ by the map

$$\Omega' \to \Omega,$$
$$\omega' \mapsto (X_n^x(\omega'))_{n\geq 0}.$$

This map is measurable because of Lemma 4.4.0.7. We have $\mathbb{P}_x(X_0 = x) = \mathbb{P}'(X_0^x = x) = 1$ and for $x_0, x_1, \ldots, x_n \in E$ we have

$$\begin{aligned}
&\mathbb{P}_x(X_0 = x_0, X_1 = x_1, \ldots, X_n = x_n) \\
&= \mathbb{P}'(X_0^x = x_0, X_1^x = x_1, \ldots, X_n^x = x_n) \\
&= \mathbb{P}'(X_0^x = x_0)\prod_{j=1}^{n} Q(x_{j-1}, x_j) = \mathbb{P}_x(X_0 = x_0)\prod_{j=1} Q(x_{j-1}, x_j).
\end{aligned}$$

Therefore, it follows that the sequence of the coordinate functions is a Markov chain. For uniqueness, we note that if $\mathbb{P}'_x$ is another such probability measure, $\mathbb{P}_x$ and $\mathbb{P}'_x$ coincide on cylinders. It then follows from an application of the *monotone class theorem* that $\mathbb{P}_x = \mathbb{P}'_x$. □

Remark 5.2.0.1. It follows for all $n \geq 0$ and for all $x, y \in E$, that we have

$$\mathbb{P}_x(X_n = y) = Q_n(x, y).$$

Remark 5.2.0.2. If μ is a probability measure on E, we can define

$$\mathbb{P}_\mu := \sum_{x\in E} \mu(x)\mathbb{P}_x,$$

which defines a new probability measure on Ω. By writing an explicit formula for $\mathbb{P}_\mu(X_0 = x_0, \ldots, X_n = x_n)$, one immediately checks that under $\mathbb{P}_\mu$ we get that $(X_n)_{n\geq 0}$ is a Markov chain with transition matrix Q, where X_0 has the distribution μ.

Remark 5.2.0.3. Let $(X'_n)_{n\geq 0}$ be a Markov chain with transition matrix Q and initial distribution μ. Then for all measurable subsets $B \subset \Omega = E^{\mathbb{N}}$, we have

$$\mathbb{P}((X'_n)_{n\geq 0} \in B) = \mathbb{P}_\mu(B).$$

This is only true when B is a cylinder as in the proof above. This shows that all results proven for the canonical Markov chains are true for a general Markov chain with the same transition matrix. One of the advantages of using a canonical Markov chain is that one can use translation operators. For all $k \in \mathbb{N}$, we define

$$\begin{aligned} \Theta_k : \Omega &\to \Omega, \\ (\omega_n)_{n\geq 0} &\mapsto \Theta_k((\omega_n)_{n\geq 0}) := (\omega_{k+n})_{n\geq 0}. \end{aligned}$$

Lemma 5.2.0.2 shows that these maps are measurable. We define $\mathcal{F}_n := \sigma(X_0, \dots, X_n)$ and we use the notation $\mathbb{E}_x$ when we integrate with respect to $\mathbb{P}_x$.

Theorem 5.2.0.4 (Simple Markov property). *Let $(\Omega, \mathcal{F}, \mathbb{P})$ be a probability space. Let F and G be two measurable, positive maps on Ω and let $n \geq 0$. Assume that F is $\mathcal{F}_n$-measurable. Then for all $x \in E$, we get*

$$\mathbb{E}_x[FG \circ \Theta_n] = \mathbb{E}_x[F\mathbb{E}_{X_n}[G]].$$

Equivalently, we have

$$\mathbb{E}_x[G \circ \Theta_n \mid \mathcal{F}_n] = \mathbb{E}_{X_n}[G].$$

We say that the conditional distribution of $\Theta_n(\omega)$ given $(X_0, X_1, \dots, X_n)$ is $\mathbb{P}_{X_n}$.

Remark 5.2.0.4. Note that Theorem 5.2.0.4 is also true if one replace $\mathbb{E}_x$ with $\mathbb{E}_\mu$.

Proof of Theorem 5.2.0.4. It is enough to prove the first statement. For this we can restrict ourselves to the case where $F =$

$\mathbb{1}_{\{X_0=x_0,X_1=x_1,\ldots,X_n=x_n\}}$ for $x_0, x_1, \ldots, x_n \in E$. Let us first consider the case where $G = \mathbb{1}_{\{X_0=y_0,\ldots,X_p=y_p\}}$. In this case, if $y \in E$, then

$$\mathbb{E}_y[G] = \mathbb{1}_{\{Y_0=y\}} \prod_{j=1}^{p} Q(y_{j-1}, y_j)$$

and

$$\begin{aligned}
&\mathbb{E}_x[FG \circ \Theta_n] \\
&\quad = \mathbb{P}_x(X_0 = x_0, \ldots, X_n = x_n, X_n = y_0, X_{n+1} = y_1, \ldots, X_{n+p} = y_p) \\
&\quad = \mathbb{1}_{\{X_0=x\}} \prod_{j=1}^{n} Q(x_{j-1}, x_j) \mathbb{1}_{\{y_0=y_n\}} \prod_{j=1}^{p} Q(y_{j-1}, y_j).
\end{aligned}$$

Now it follows from the *monotone class theorem* that the above holds for $G = \mathbb{1}_A$, for $A \in \mathcal{F}_n$ and thus we can conclude. □

Remark 5.2.0.5. We would like to make sense of Theorem 5.2.0.4 for n replaced by a stopping time T. Indeed let us assume that we want to look at the problem of knowing whether a Markov chain starting from x visits back x. In other words, let us define

$$N_x := \sum_{n=0}^{\infty} \mathbb{1}_{\{X_n=x\}},$$

and ask the question, whether we have that $\mathbb{P}_x(N_x = \infty) = 1$. It is in fact enough to check that it comes back at least once at x. If $H_x := \inf\{n \geq 1 \mid X_n = x\}$, with the convention that $\inf \emptyset = \infty$, we have $\mathbb{P}_x(N_x = \infty) = 1$ if and only if $\mathbb{P}_x(H_x < \infty) = 1$, which is trivial. If $\mathbb{P}_x(N_x < \infty) = 1$, and if Theorem 5.2.0.4 is true for stopping times, then $\theta_{H_x}(\omega) = \left(\omega_{H_x(\omega)=n}\right)_{n\geq 0}$ has the law $\mathbb{P}_x$. But then, since $N_x(\omega) = 1 + N_x(\Theta_{H_x}(\omega))$, we see that N_x has the same distribution as $1 + N_x$ under $\mathbb{P}_x$. This is only possible if $N_x = \infty$ almost surely.

Theorem 5.2.0.5 (Strong Markov property). *Let* $(\Omega, \mathcal{F}, (\mathcal{F}_n)_{n\geq 0}, \mathbb{P})$ *be a filtered probability space. Let* T *be a stopping time for the filtration* $(\mathcal{F}_n)_{n\geq 0}$. *Let* F *and* G *be two measurable and positive* *functions on Ω. Assume further that F is $\mathcal{F}_T$-measurable. Then, for all $x \in E$, we get*

$$\mathbb{E}_x[\mathbb{1}_{\{T<\infty\}} FG \circ \Theta_T] = \mathbb{E}_x[\mathbb{1}_{\{T<\infty\}} F \mathbb{E}_x[G]],$$

or equivalently

$$\mathbb{E}_x[\mathbb{1}_{\{T<\infty\}} G \circ \Theta_T \mid \mathcal{F}_T] = \mathbb{1}_{\{T<\infty\}} \mathbb{E}_{X_T}[G].$$

Proof. Note that, for $n \geq 0$, we have

$$\mathbb{E}_x[\mathbb{1}_{\{T=n\}} FG \circ \Theta_T] = \mathbb{E}_x[\mathbb{1}_{\{T=n\}} FG \circ \Theta_n] = \mathbb{E}_n[\mathbb{1}_{\{T=n\}} \mathbb{E}_{X_n}[G]].$$

The last equality follows from a previous theorem after obtaining that $\mathbb{1}_{\{T=n\}} F$ is $\mathcal{F}_n$-measurable. We then only need to sum over n. □

Corollary 5.2.0.6. *Let $(\Omega, \mathcal{F}, (\mathcal{F}_n)_{n\geq 0}, \mathbb{P})$ be a filtered probability space. Let T be a stopping time such that $\mathbb{P}_x(T < \infty) = 1$. Let us assume that there exists $y \in E$ such that $\mathbb{P}_x(X_T = y) = 1$. Then, under $\mathbb{P}_x$, $\Theta_T(\omega)$ is independent of $\mathcal{F}_T$ and has the law $\mathbb{P}_y$.*

Proof. We only have to note that

$$\mathbb{E}_x[FG \circ \Theta_T(\omega)] = \mathbb{E}_x[F \mathbb{E}_{X_T}[G]] = \mathbb{E}_x[F \mathbb{E}_y[G]] = \mathbb{E}_x[F] \mathbb{E}_y[G]. \quad \square$$

5.3. Classification of States

From now on we will only use the *canonical* Markov chain. Recall that, for $x \in E$, we have

$$H_x := \inf\{n \geq 1 \mid X_n = x\}, \tag{5.3.0.1}$$

$$N_x := \sum_{n\geq 0} \mathbb{1}_{\{X_n = x\}}. \tag{5.3.0.2}$$

Proposition 5.3.0.1 (Recurrence and transience). *Let $x \in E$. Then we have two situations, which can occur:*

(i) *$\mathbb{P}_x(H_x < \infty) = 1$ and $N_x = \infty$ $\mathbb{P}_x$-a.s. In this case x is said to be* recurrent.

(ii) $\mathbb{P}_x(H_x < \infty) < 1$ *and* $N_x < \infty$ $\mathbb{P}_x$*-a.s. More precisely, we have*

$$\mathbb{E}_x[N_x] = \frac{1}{\mathbb{P}_x(H_x < \infty)} < \infty.$$

In this case x is said to be transient.

Proof. For $k \geq 1$, it follows from the strong Markov property (Theorem 5.2.0.5) that

$$\begin{aligned}\mathbb{P}_x(N_x \geq k+1) &= \mathbb{E}_x[\mathbb{1}_{\{H_x<\infty\}}\mathbb{1}_{\{N_x \geq k\}} \circ \Theta_{H_x}] \\ &= \mathbb{E}_x[\mathbb{1}_{\{H_x<\infty\}}\mathbb{E}_x[\mathbb{1}_{\{N_x \geq k\}}]] \\ &= \mathbb{P}_x(H_x < \infty)\mathbb{P}_x(N_x \geq k).\end{aligned}$$

Now since $\mathbb{P}_x(N_x \geq 1) = 1$, it follows by induction that

$$\mathbb{P}_x(N_x \geq k) = \mathbb{P}_x(H_x < \infty)^{k-1}.$$

If $\mathbb{P}_x(H_x < \infty) = 1$, then $\mathbb{P}_x(N_x = \infty) = 1$. If $\mathbb{P}_x(H_x < \infty) < 1$, then

$$\begin{aligned}\mathbb{E}_x[N_x] &= \sum_{k\geq 1} \mathbb{P}_x(N_x \geq k) = \sum_{k \geq 1} \mathbb{P}_x(H_x < \infty)^{k-1} \\ &= \frac{1}{1 - \mathbb{P}_x(H_x < \infty)} = \frac{1}{\mathbb{P}_x(H_x < \infty)} < \infty.\end{aligned}$$

□

Definition 5.3.0.1 (Potential kernel). The *potential kernel* of the chain is given by the function

$$\begin{aligned}U : E \times E &\to [0, \infty], \\ (x, y) &\mapsto \mathbb{E}_x[N_y].\end{aligned}$$

Proposition 5.3.0.2. *The following properties hold:*

(i) *For all $x, y \in E$, we have*

$$U(x, y) = \sum_{n \geq 0} Q_n(x, y).$$

(ii) $U(x, x) = \infty$ *if and only if x is recurrent.*

(iii) *For all $x, y \in E$, it follows that if $x \neq y$ then we get*

$$U(x,y) = \mathbb{P}_x(H_y < \infty)U(y,y).$$

Proof. (i) Note that we have

$$U(x,y) = \mathbb{E}_x\left[\sum_{n\geq 0} \mathbb{1}_{\{N_x=y\}}\right] = \sum_{n\geq 0} \mathbb{P}_x(X_n = y) = \sum_{n\geq 0} Q_n(x,y).$$

(ii) We leave this point as an exercise.
(iii) This follows from the strong Markov property (Theorem 5.2.0.5). In particular, we get

$$\begin{aligned}\mathbb{E}_x[N_y] &= \mathbb{E}_x[\mathbb{1}_{\{H_y<\infty\}} N_y \circ \Theta_{H_y}] = \mathbb{E}_x[\mathbb{1}_{\{H_y<\infty\}}\mathbb{E}_y[N_y]] \\ &= \mathbb{P}_x(H_y < \infty)U(y,y).\end{aligned}$$

□

Example 5.3.0.2. Let us consider the Markov chain, starting at 0, on $\mathbb{Z}^d$ with transition matrix given by

$$Q((x_1,\ldots,x_d),(y_1,\ldots,y_d)) := \frac{1}{2^d}\prod_{i=1}^{d} \mathbb{1}_{\{|y_i-x_i|=1\}}.$$

This Markov chain has the same distribution as $(Y_n^1,\ldots,Y_n^d)_{n\geq 0}$, where $Y^1,\ldots,Y^d$ are i.i.d. random variables of a simple random walk on $\mathbb{Z}$, starting at 0. Hence, we get

$$Q_n(0,0) = \mathbb{P}[Y_n^1 = 0,\ldots,Y_n^d = 0] = \mathbb{P}(Y_n^1 = 0)^d.$$

Moreover, we get $\mathbb{P}(Y_n^1 = 0) = 0$ if n is odd and for $n = 2k$ we get

$$\mathbb{P}(Y_{2k}^1 = 0) = 2^{-2k}\binom{2k}{k}.$$

Therefore, we get that

$$U(0,0) = \sum_{k\geq 0} Q_{2k}(0,0) = \sum_{k\geq 0}\left(2^{-2k}\binom{2k}{k}\right)^d.$$

Now using Stirling's formula (Theorem 1.1.1.2), we get

$$2^{-2k}\binom{2k}{k} \approx_{k\to\infty} \frac{\left(\frac{2k}{\mathrm{e}}\right)^{2k}\sqrt{4\pi k}}{2^{2k}\left(\left(\frac{k}{\mathrm{e}}\right)^k\sqrt{2\pi k}\right)^2} \approx_{k\to\infty} \sqrt{\frac{1}{\pi k}}.$$

Consequently, 0 is recurrent if $d = 1$ and transient if $d \geq 2$.

Now let us denote the set of all recurrent points by R. Then, we can obtain the following lemma.

Lemma 5.3.0.3. *Let $x \in R$ and let $y \in E$ such that $U(x, y) > 0$. Then $y \in R$ and*

$$\mathbb{P}_y(H_x < \infty) = 1.$$

In particular, we get that $U(y, x) > 0$.

Proof. Let us first show that $\mathbb{P}_y(H_x < \infty) = 1$. Note that we have

$$\begin{aligned} 0 = \mathbb{P}_x(N_x < \infty) &\geq \mathbb{P}_x(H_y < \infty, H_x \circ \Theta_{H_y} = \infty) \\ &= \mathbb{E}_x[\mathbb{1}_{\{H_y<\infty\}}\mathbb{1}_{\{H_x=\infty\}} \circ \Theta_{H_y}] \\ &= \mathbb{E}_x[\mathbb{1}_{\{H_y<\infty\}}\mathbb{P}_y(H_x = \infty)] = \mathbb{P}_x(H_y < \infty)\mathbb{P}_y(H_x = \infty). \end{aligned}$$

Thus $U(x, y) > 0$ implies that $\mathbb{P}_x(H_y < \infty) > 0$ and hence $\mathbb{P}_y(H_x = \infty) = 0$. Therefore, we can find $n_1, n_2 \geq 1$ such that

$$Q_{n_1}(x, y) \geq 0, \qquad Q_{n_2}(y, x) > 0.$$

Then, for all $p \geq 0$, we get that

$$Q_{n_1+p+n_2}(y, y) \geq Q_{n_2}(y, x)Q_p(x, x)Q_{n_1}(x, y),$$

and thus we have

$$\begin{aligned} U(y, y) &\geq \sum_{p=0}^{\infty} Q_{n_1+p+n_2}(y, y) \\ &\geq Q_{n_2}(y, x)\left(\sum_{p=0}^{\infty} Q_p(x, x)\right) Q_{n_1}(x, y) = \infty, \end{aligned}$$

since $x \in R$. □

Remark 5.3.0.3. We can observe that Lemma 5.3.0.3 has the following important consequence: If $x \in R$ and $y \in E \setminus R$, then $U(x, y) = 0$, which means that one cannot go from a recurrent point to a transient point.

Theorem 5.3.0.4 (Classification of states). *Let R be the set of all recurrent points and let N_x be defined as in (5.3.0.2). Then, there exists a partition of R, denoted $(R_i)_{i\in I}$ for some index set I, i.e.,*

$$R = \bigcup_{i\in I} R_i,$$

such that the following properties hold:

(i) *If $x \in R$ and $i \in I$ such that $x \in R_i$, we get that*
   - $N_y = \infty$, $\mathbb{P}_x$*-a.s. for all* $y \in R_i$,
   - $N_y = 0$, $\mathbb{P}_x$*-a.s. for all* $y \in E \setminus R_i$.

(ii) *If $x \in E \setminus R$ and $T := \inf\{n \geq 0 \mid X_n \in R\}$, we get that*
   - $T = \infty$ *and* $N_y < \infty$, $\mathbb{P}_x$*-a.s. for all* $y \in E$,
   - $T < \infty$ *and then there exists $j \in I$ such that for all $n \geq T$ we get $X_n \in R_j$,* $\mathbb{P}_x$*-a.s.*

Proof. For $x, y \in R$, we define a relation $x \sim y$ if $U(x, y) > 0$. It follows from Lemma 5.3.0.3 that this is an equivalence relation[2] on R. Let $i \in I$ and $x \in R_i$. Then we have that $U(x, y) = 0$ for all $y \in E \setminus R_i$ and hence $N_y = 0$, $\mathbb{P}_x$-a.s. for all $y \in E \setminus R_i$. If $y \in R_i$, we get from Lemma 5.3.0.3 that $\mathbb{P}_x(H_y < \infty) = 1$ and it follows from the strong Markov property (Theorem 5.2.0.5) that

$$\begin{aligned}
\mathbb{P}_x(N_y = \infty) &= \mathbb{E}_x[\mathbb{1}_{\{H_y<\infty\}}\mathbb{1}_{\{N_y=\infty\}} \circ \Theta_{H_y}] \\
&= \mathbb{P}_x(H_y < \infty)\mathbb{P}_y(N_y = \infty) = 1.
\end{aligned}$$

If $x \in E \setminus R$ and $T = \infty$, it easily follows from the strong Markov property that $N_y < \infty$ for all $y \in E \setminus R$. If $T < \infty$, let $j \in I$ such that $X_T \in R_j$. Then apply the strong Markov property with T and the first part of the theorem, which then implies that $X_n \in R_j$ for all $n \geq T$. □

Definition 5.3.0.4 (Irreducibility). A chain is called *irreducible* if $U(x, y) > 0$ for all $x, y \in E$.

---

[2]For transitivity observe that if $Q_n(x, y) > 0$ and $Q_m(y, z) > 0$, then $Q_{n+m}(x, z)$.

Corollary 5.3.0.5. *If the chain is irreducible, then the following properties hold:*

(i) *either all states are recurrent and there exists a recurrent chain and for all $x \in E$ we get*

$$\mathbb{P}_x(N_y = \infty, \forall y \in E) = 1,$$

(ii) *or all states are transient and then for all $x \in E$ we get*

$$\mathbb{P}_x(N_y < \infty, \forall y \in E) = 1.$$

Moreover, if $|E| < \infty$, then only the first case can occur.

Proof. If there exists a recurrent state, Lemma 5.3.0.3 shows that all states are recurrent with $U(x, y) > 0$ for all $x, y \in E$ and there is only one recurrent chain. If $|E| < \infty$ and if all states are transient, then we get $\mathbb{P}_x$-a.s. that

$$\sum_{y \in E} N_y < \infty,$$

but we also know

$$\sum_{y \in E} N_y = \sum_{y \in E} \sum_{n \geq 0} \mathbb{1}_{\{X_n = y\}} = \sum_{n \geq 0} \sum_{y \in E} \mathbb{1}_{\{X_n = y\}} = \infty. \qquad \square$$

Example 5.3.0.5. Let us first recall that the results obtained for the canonical Markov chain hold for any arbitrary Markov chain. Now let $(Y_n)_{n \geq 0}$ be a Markov chain with transition matrix Q. For instance, if $Y_0 = y$ and $N_x^Y = \sum_{n \geq 0} \mathbb{1}_{\{Y-n=x\}}$, we have for that $k \in \bar{\mathbb{N}}$

$$\mathbb{P}(N_x^Y = k) = \mathbb{P}_y(N_x = k),$$

since the left-hand side is given by $\mathbb{P}((Y_n)_{n \geq 0} \in B)$ with $B := \{\omega \in E^{\mathbb{N}} \mid N_x(\omega) = k\}$.

Theorem 5.3.0.6. *Let $(\xi_j)_{j \geq 1}$ be i.i.d. random variables having the law μ with $\xi_j \in \mathbb{Z}$ for all $j \geq 1$ and assume that $\mathbb{E}[|\xi_1|] < \infty$ and $m = \mathbb{E}[\xi_1]$. Moreover, set $Y_n = Y_0 + \sum_{j=1}^n \xi_j$. Then the following properties hold:*

(i) *If $m \neq 0$, then all states are transient.*

(ii) *If $m = 0$, then all states are recurrent. Moreover, the chain is irreducible if and only if the subgroup given by $\{y \in \mathbb{Z} \mid \mu(y) > 0\}$ is in $\mathbb{Z}$.*

Proof. (i) If $m \neq 0$, we know from the strong law of large numbers that $(Y_n) \xrightarrow[\text{a.s.}]{n\to\infty} \infty$ and hence all states are transient.

(ii) Assume $m = 0$ and 0 is transient. Thus $U(0,0) < \infty$. We can assume that $Y_0 = 0$. For $x \in \mathbb{Z}$ we get

$$U(0,x) \leq U(x,x) = U(0,0).$$

Therefore, we get that for all $n \geq 1$

$$\sum_{|x|\leq n} U(0,x) \leq (2n+1)U(0,0) \leq Cn,$$

where $C = 3U(0,0)$. Moreover, by Theorem 5.3.0.4, there exists some N sufficiently large such that for all $n \geq N$ we get

$$\mathbb{P}(|Y_n| \leq \varepsilon n) > Y_2,$$

with $\varepsilon = \frac{1}{6C}$, or equivalently

$$\sum_{|x|\leq \varepsilon n} Q_n(0,x) \geq \frac{1}{2}.$$

If $p \geq n \geq N$, we get

$$\sum_{|x|\leq \varepsilon n} Q_p(0,x) \geq \sum_{|x|\leq \varepsilon p} Q_p(0,x) > \frac{1}{2},$$

and by summing over p, we get

$$\sum_{|x|\leq \varepsilon n} U(0,n) \geq \sum_{p=N}^{n} \sum_{|x|\leq \varepsilon p} Q_p(0,x) > \frac{n-N}{\varepsilon}.$$

But if $\varepsilon n \geq 1$, we have

$$\sum_{|x|\leq \varepsilon n} U(0,x) \leq C\varepsilon n = \frac{n}{6},$$

which leads to a contradiction for n sufficiently large. For the last statement, write $G = \langle x \in \mathbb{Z} \mid \mu(x) > 0 \rangle$. Then, since $Y_0 = 0$, we get

$$\mathbb{P}(Y_n \in G \mid \forall n \in \mathbb{N}) = 1.$$

If $G \neq \mathbb{Z}$, then obviously $(Y_n)_{n \geq 0}$ is irreducible. If $G = \mathbb{Z}$, then write

$$H := \{x \in \mathbb{Z} \mid U(0, x) > 0\}.$$

It follows that H is a subgroup of $\mathbb{Z}$. Indeed, for $x, y \in H$, we get that

$$Q_{n+p}(0, x + y) \geq Q_n(0, x)Q_p(x, x + y) = Q_n(0, x)Q_p(0, y),$$

and thus we get that $x + y \in H$. For $x \in H$ and with 0 being recurrent, we get that $U(0, x) > 0$ implies that $U(x, x) > 0$ and therefore $U(x, 0) = U(0, -x)$, which implies that $-x \in H$. Now since $H \supset \{x \in \mathbb{Z} \mid \mu(x) > 0\}$, we get that $H = \mathbb{Z}$ and the claim follows. □

Example 5.3.0.6. Let $\mu = \frac{1}{2}\delta_{-2} + \frac{1}{2}\delta_2$. Then all states are recurrent but there are only two recurrence classes. Here δ_x denotes the Dirac measure at the point x.

5.3.1. Random variable on a graph

Assume that $|E| < \infty$. Moreover, let $A \subset \mathcal{P}(E)$ and for $x \in E$ define

$$A_x := \{y \in E \mid \{x, y\} \in A\} \neq \emptyset.$$

A graph $\mathcal{G}$ is said to be *connected* if every pair of vertices in the graph is connected.

Proposition 5.3.1.1. *The simple random walk on a finite graph is recurrent and irreducible.*

Exercise 5.3.1.1. Prove[3] Proposition 5.3.1.1.

[3]Note that connected also implies irreducible and apply Corollary 5.3.0.5.

5.4. Invariant Measures

Definition 5.4.0.1 (Invariant measure). Let μ be a positive measure on E such that $\mu(x) < \infty$ for all $x \in E$ and $\mu \not\equiv 0$. We say that μ is *invariant* for the transition matrix Q if for all $y \in E$ we get

$$\mu(y) = \sum_{x \in E} \mu(x)Q(x,y).$$

In terms of matrices, this means

$$\mu Q = \mu.$$

Moreover, since for all $n \in \mathbb{N}$ we have $Q_n = Q^n$, we get that $\mu Q_n = \mu$.

5.4.1. Interpretation

Assume that $\mu(E) < \infty$, which is always the case when E is finite. Without loss of generality, we can assume that $\mu(E) = 1$. Then for all measurable maps $f : E \to \mathbb{R}_{\geq 0}$, we get that

$$\begin{aligned}\mathbb{E}_\mu[f(X_1)] &= \sum_{x \in E} \mu(x) \sum_{y \in E} Q(x,y)f(y) = \sum_{y \in E} f(y) \sum_{x \in E} \mu(x)Q(x,y) \\ &= \sum_{y \in E} \mu(y)f(y).\end{aligned}$$

Hence, with respect to $\mathbb{P}_\mu$, we get that $X_1 \overset{\text{law}}{=} X_0 \sim \mu$. Using the fact that $\mu Q_n = \mu$, we show that with respect to $\mathbb{P}_\mu$, we get that $X_n \sim \mu$. For all measurable maps $F : \Omega \to \mathbb{R}_{\geq 0}$ we have

$$\mathbb{E}_\mu[F \circ \Theta_1] = \mathbb{E}_\mu[\mathbb{E}_{X_1}[F]] = \sum_{x \in E} \mu(x)\mathbb{E}_x[F] = \mathbb{E}_\mu[F],$$

which implies that with respect to $\mathbb{P}_\mu$, we get that $(X_{1+n})_{n \geq 0}$ has the same law[4] as $(X_n)_{n \geq 0}$.

Example 5.4.1.1. For the random variable on $\mathbb{Z}^d$, we get $Q(x,y) = (y,x)$. It is then easy to see that the counting measure on $\mathbb{Z}^d$ is invariant.

[4]Note that the same holds for $(X_{k+n})_{n \geq 0}$ for all $k \geq 0$.

Definition 5.4.1.2 (Reversible measure). Let μ be positive measure on E such that $\mu(x) < \infty$ for all $x \in E$. μ is said to be *reversible* if for all $x, y \in E$ we get that

$$\mu(x)Q(x,y) = \mu(y)Q(y,x).$$

Proposition 5.4.1.1. *Any reversible measure is invariant.*

Proof. If μ is reversible, we get that

$$\sum_{x \in E} \mu(x)Q(x,y) = \sum_{x \in E} \mu(y)Q(y,x) = \mu(y). \qquad \square$$

Remark 5.4.1.3. There are invariant measures which are not reversible, for example the counting measure is not reversible if $\mathcal{G}$ is not symmetric, i.e., $\mathcal{G}(x) = \mathcal{G}(-x)$.

Example 5.4.1.4 (Random walk on a graph). The measure $\mu(x) := |A_x|$ is reversible. Indeed, if $\{x, y\} \in A$, we get

$$\mu(x)Q(x,y) = |A_x|\frac{1}{|A_x|} = 1 = \mu(y)Q(y,x).$$

Example 5.4.1.5 (Ehrenfest's model). The Markov chain on $\{1, \ldots, k\}$ with transition matrix

$$\begin{cases} Q(j, j+1) = \frac{k-j}{k}, & 0 \leq j \leq k-1, \\ Q(j, j-1) = \frac{j}{k}, & 1 \leq j \leq k \end{cases}$$

is called *Ehrenfest's model.* A measure μ is reversible if and only if for $0 \leq j \leq k-1$ we have

$$\mu(j)\frac{k-j}{k} = \mu(j+1)\frac{j+1}{k}.$$

One can easily check that $\mu(j) = \binom{k}{j}$ is actually a solution.

Theorem 5.4.1.2. *Let $x \in E$ be recurrent. The formula*

$$\mu(y) = \mathbb{E}_x\left[\sum_{k=0}^{H_x - 1} \mathbb{1}_{\{X_k = y\}}\right]$$

defines an invariant measure μ on E. Moreover, $\mu(y) > 0$ if and only if y is in the same recurrence class as x.

Proof. Note first that if y is not in the same recurrence class as x, then

$$\mathbb{E}_x[N_y] = U(x, y) = 0,$$

which implies that $\mu(y) = 0$. Moreover, for $y \in E$, we get that

$$\begin{aligned}
\mu(y) &= \mathbb{E}_x\left[\sum_{k=1}^{H_x} \mathbb{1}_{\{X_k=y\}}\right] = \sum_{z\in E} \mathbb{E}_x\left[\sum_{k=1}^{H_x} \mathbb{1}_{\{X_{k-1}=z, X_k=y\}}\right] \\
&= \sum_{z\in E}\sum_{k\geq 1} \mathbb{E}_x\left[\mathbb{1}_{\{k\leq H_x, X_{k-1}=z\}}\mathbb{1}_{\{X_k=y\}}\right] \\
&= \sum_{z\in E}\sum_{k\geq 1} \mathbb{E}_x\left[\mathbb{1}_{\{k\leq H_x, X_{k-1}=z\}}\right] \\
&= \sum_{z\in E} \mathbb{E}_x\left[\sum_{k=1}^{H_x} \mathbb{1}_{\{X_{k-1}=z\}}\right] Q(z, y) = \sum_{z\in E} \mu(z) Q(z, y).
\end{aligned}$$

We have shown that $\mu Q = \mu$. Thus, it follows that $\mu Q_n = \mu$ for all $n \geq 0$. In particular, we get

$$\mu(x) = 1 = \sum_{z\in E} \mu(z) Q_n(z, x).$$

Let y be in the same recurrence class as x. Then, there exists some $n \geq 0$ such that $Q_n(y, x) > 0$, which implies that

$$\mu(y) < \infty.$$

We can also find $m \geq 0$ such that $Q_m(x, y) > 0$ and

$$\mu(y) = \sum_{z\in E} \mu(z) Q_m(z, y) \geq Q_m(x, x) > 0. \qquad \square$$

Remark 5.4.1.6. If there are several recurrence classes R_i with $i \in I$ for some index set I and if we set

$$\mu_i(y) = \mathbb{E}_{x_i}\left[\sum_{k=0}^{H_{x_i}-1} \mathbb{1}_{\{X_k=y\}}\right],$$

we can obtain an invariant measure with disjoint supports.

Theorem 5.4.1.3. *Let us assume that the Markov chain is irreducible and recurrent. Then, the invariant measure is unique up to a multiplicative constant.*

Proof. Let μ be an invariant measure. We can show by induction that for $p \geq 0$ and for all $x, y \in E$, we get

$$\mu(y) \geq \mu(x)\mathbb{E}_x\left[\sum_{k=0}^{p\wedge(H_x-1)} \mathbb{1}_{\{X_k=y\}}\right]. \tag{5.4.1.1}$$

First, if $x = y$, the statement is obvious. Therefore, let us suppose that $x \neq y$. If $p = 0$, Equation (5.4.1.1) is immediate. Let us assume (5.4.1.1) holds for any p. Then we have

$$\begin{aligned}
\mu(y) &= \sum_{z\in E}\mu(z)Q(z,y) \geq \mu(x)\sum_{z\in E}\mathbb{E}_x\left[\sum_{k=0}^{p\wedge(H_x-1)} \mathbb{1}_{\{X_k=z\}}\right]Q(z,y) \\
&= \mu(x)\sum_{z\in E}\sum_{k=0}^{p}\mathbb{E}_x[\mathbb{1}_{\{X_k=z,k\leq H_x-1\}}]Q(z,y) \\
&= \mu(x)\sum_{z\in E}\sum_{k=0}^{p}\mathbb{E}_x[\mathbb{1}_{\{X_k=z,k\leq H_x-1\}}]\mathbb{1}_{\{X_{k-1}=y\}} \\
&= \mu(x)\mathbb{E}_x\left[\sum_{k=0}^{p\wedge(H_x-1)} \mathbb{1}_{\{X_{k-1}=y\}}\right] = \mu(x)\mathbb{E}_x\left[\sum_{k=1}^{(p+1)\wedge(H_x)} \mathbb{1}_{\{X_k=y\}}\right].
\end{aligned}$$

This establishes the result for $p+1$. Now, if we let $p \to \infty$ in (5.4.1.1), we get

$$\mu(y) \geq \mu(x)\mathbb{E}_x\left[\sum_{k=0}^{H_x-1} \mathbb{1}_{\{X_k=y\}}\right].$$

Let us fix $x \in E$. The measure $\nu_x(y) = \mathbb{E}_x\left[\sum_{k=0}^{H_x-1}\mathbb{1}_{\{X_k=y\}}\right]$ is invariant and we have $\mu(y) \geq \mu(x)\nu_x(y)$. Hence, for all $n \geq 1$, we get

$$\begin{aligned}
\mu(x) &= \sum_{z\in E}\mu(z)Q_n(z,x) \geq \sum_{x\in E}\mu(x)\nu_x(z)Q_n(z,x) \\
&= \mu(x)\nu_x(x) = \mu(x).
\end{aligned}$$

Therefore, we get that $\mu(z) = \mu(x) = \nu_x(z)$ for all z such that $Q_n(z, x) > 0$. Since the chain is irreducible, there exists some $n \geq 0$ such that $Q_n(z, x) > 0$. Finally, this implies that

$$\mu(z) = \mu(x)\nu_x(z). \qquad \square$$

Corollary 5.4.1.4. *Consider an irreducible and recurrent chain. Then, at least one of the following properties hold:*

(i) *There exists an invariant probability measure μ and for all $x \in E$ we have*

$$\mathbb{E}_x[H_x] = \frac{1}{\mu(x)}.$$

(ii) *All invariant measures have infinite total mass and for all $x \in E$ we get*

$$\mathbb{E}_x[H_x] = \infty.$$

Remark 5.4.1.7. In the first case of Corollary 5.4.1.4, the basis is said to be *positive recurrent* and in the second case of Corollary 5.4.1.4 it is said to be *negative recurrent.* Moreover, if E is finite, then only the first case can occur.

Proof of Corollary 5.4.1.4. We know that in this situation all invariant measures are proportional. Hence, they all have finite mass or infinite mass. For case (i), let μ be the invariant probability measure and let $x \in E$. Moreover, define

$$\nu_x(y) := \mathbb{E}_x\left[\sum_{k=0}^{H_x-1} \mathbb{1}_{\{X_k=y\}}\right].$$

Then, for some $C > 0$, we get $\mu = C\nu$. We can determine C by using

$$1 = \mu(E) = C\nu_x(E),$$

which implies that $C = \frac{1}{\nu_x(E)}$ and thus

$$\mu(x) = \frac{\nu_x(x)}{\nu_x(E)} = \frac{1}{\nu_x(E)}.$$

On the other hand, we have

$$\begin{aligned}\nu_x(E) &= \sum_{y\in E} \mathbb{E}_x\left[\sum_{k=0}^{H_x-1} \mathbb{1}_{\{X_k=y\}}\right] \\ &= \mathbb{E}_x\left[\sum_{k=0}^{H_x-1}\left(\sum_{y\in E} \mathbb{1}_{\{X_k=y\}}\right)\right] = \mathbb{E}_x[H_x].\end{aligned}$$

For the case (ii), it is easy to see that ν_x is infinite and thus

$$\mathbb{E}_x[H_x] = \nu_x(E) = \infty. \qquad \square$$

Part III
Appendices

Appendix A

Basics of Measure Theory

A.1. Measurable Spaces

In this appendix, we want to recall the most important notions of measure theory and its fundamentals in order to understand the theory of Lebesgue integration and the crucial theorems arising within. At first, we want to handle the abstract setting of a measure space. The following definitions should lead to a proper understanding for the theory of probability.

Definition A.1.0.1 (σ-algebra and measurable sets). Let E be any set. A *σ-algebra* $\mathcal{A}$ on E is a collection of subsets of E, which satisfies the following conditions:

(i) The ground space has to be in $\mathcal{A}$, i.e., $E \in \mathcal{A}$,
(ii) If $A \in \mathcal{A}$, then $A^{\mathsf{C}} \in \mathcal{A}$, where A^{C} denotes the complement of A,
(iii) If $(A_n)_{n\in\mathbb{N}} \subset \mathcal{A}$ is a collection of elements in $\mathcal{A}$ then $\bigcup_{n\in\mathbb{N}} A_n \in \mathcal{A}$.

Definition A.1.0.2 (Measurable set). The elements of a σ-algebra are called *measurable sets*.

Definition A.1.0.3 (Measurable space). A set E endowed with a σ-algebra $\mathcal{A}$ is called a *measurable space*.

Remark A.1.0.4. We should note that Definition A.1.0.1 leads to the following statements:

(i) Every σ-algebra $\mathcal{A}$ is a subset of the power set $\mathcal{P}(E)$ of E, i.e., $\mathcal{A} \subseteq \mathcal{P}(E)$.
(ii) The empty set has to be in $\mathcal{A}$, i.e., $\emptyset \in \mathcal{A}$,
(iii) If $(A_n)_{n\in\mathbb{N}} \subset \mathcal{A}$ is a collection of elements of $\mathcal{A}$, then $\bigcap_{n\in\mathbb{N}} A_n \in \mathcal{A}$, i.e.,

$$\bigcap_{n\in\mathbb{N}} A_n = \left(\bigcup_{n\in\mathbb{N}} A_n^{\mathsf{C}}\right)^{\mathsf{C}}.$$

Example A.1.0.5 (Smallest σ-algebra). Let E be a set. The collection $\mathcal{A} := \{\emptyset, E\}$ is called the *trivial* or the *smallest* σ-algebra on E.

Example A.1.0.6 (Largest σ-algebra). Let E be a set. The collection $\mathcal{A} := \mathcal{P}(E)$ is called the *largest* σ-algebra[1] on E.

Example A.1.0.7 (Countable subsets). Let E be a set. The collection

$$\mathcal{A} := \{A \subset E \mid A \text{ is countable or } A^{\mathsf{C}} \text{ is countable}\}$$

is a σ-algebra on E.

Exercise A.1.0.8. Show that the collections as in Examples A.1.0.5, A.1.0.6 and A.1.0.7 are indeed σ-algebras.

Remark A.1.0.9. Let $\mathcal{A}$ be a σ-algebra and, for all $n \in \mathbb{N}$, let $A_n \in \mathcal{A}$. Then, the following properties hold:

(i) If A_n is a countable set for all $n \in \mathbb{N}$, then $\bigcup_{n\in\mathbb{N}} A_n$ is also a countable set and we know that

$$\bigcup_{n\in\mathbb{N}} A_n \in \mathcal{A}.$$

(ii) If there is an $n_0 \in \mathbb{N}$ such that A_{n_0} is an uncountable set, it follows that $A_{n_0}^{\mathsf{C}}$ is a countable set, i.e.,

$$\left(\bigcup_{n\in\mathbb{N}} A_n\right)^{\mathsf{C}} = \bigcap_{n\in\mathbb{N}} A_n^{\mathsf{C}} \subset A_{n_0}^{\mathsf{C}},$$

which implies that $\left(\bigcup_{n\in\mathbb{N}} A_n\right)^{\mathsf{C}}$ is countable.

---

[1]This is convenient for finite and countable measurable spaces.

One can construct many more interesting σ-algebras by observing that any arbitrary intersection of σ-algebras is again a σ-algebra. Namely, let $(\mathcal{A}_i)_{i\in I}$ be a family of σ-algebras and I an arbitrary index set. Then the set

$$\mathcal{A} := \bigcap_{i\in I} \mathcal{A}_i$$

is also a σ-algebra.

Definition A.1.0.10 (Generated σ-algebra). Let E be a set and let $\mathcal{C}$ be a subset of $\mathcal{P}(E)$. Then there exists a smallest σ-algebra, denoted by $\sigma(\mathcal{C})$, which contains $\mathcal{C}$. This σ-algebra may be defined as

$$\sigma(\mathcal{C}) = \bigcap_{\substack{\mathcal{C}\subset\mathcal{A} \\ \mathcal{A}\ \sigma\text{-algebra}}} \mathcal{A}.$$

Remark A.1.0.11. We can observe that if $\mathcal{C}$ is a σ-algebra itself, then clearly $\sigma(\mathcal{C}) = \mathcal{C}$. Moreover, for two subsets $\mathcal{C} \subset \mathcal{P}(E)$ and $\mathcal{C}' \subset \mathcal{P}(E)$ with $\mathcal{C} \subset \mathcal{C}'$, we get that $\sigma(\mathcal{C}) \subset \sigma(\mathcal{C}')$.

Example A.1.0.12. Let E be a set and let $A \subset E$ be a subset. Moreover, let $\mathcal{C} := A \subset \mathcal{P}(E)$. Then, we get that

$$\sigma(\mathcal{C}) = \{\emptyset, A, A^{\mathsf{C}}, E\}.$$

More generally, let $E = \bigcup_{i\in I} E_i$, where I is a finite or countable index set and $E_i \cap E_j = \emptyset$ for $i \neq j$. Then we say that $(E_i)_{i\in I}$ is a *partition* of E and in fact, the collection

$$\mathcal{A} := \left\{ \bigcup_{j\in J} E_j \,\middle|\, J \subset I \right\}$$

has the structure of a σ-algebra. Now let $\mathcal{C} = \{\{x\} \mid x \in E\}$. Then, we get that

$$\sigma(\mathcal{C}) = \{A \subset E \,|\, A \text{ is countable or } A^{\mathsf{C}} \text{ is countable}\}.$$

A.2. Topological Spaces

The notion of a σ-algebra is related to one of the most general constructions, that of a *topology*. To deal with Euclidean spaces and for the description of a natural notion of σ-algebra, we need to take a closer look at the *topological* point of view. Therefore, we want to recall some point-set aspects of topology.

Definition A.2.0.1 (Topological space). Let X be a set. A *topology* on X is a family $\mathcal{O}$ of subsets of X, which statisfies the following conditions:

(i) The ground set is in $\mathcal{O}$, i.e., $X \in \mathcal{O}$.
(ii) The empty set is in $\mathcal{O}$, i.e., $\emptyset \in \mathcal{O}$.
(iii) (Finite intersection) For $O_1, \dots, O_n \in \mathcal{O}$ with $n \geq 1$, we get that $\bigcap_{i=1}^n O_i \in \mathcal{O}$.
(iv) (Arbitrary union) For $O_i \in \mathcal{O}$ with $i \in I$, where I is any index set, we get that $\bigcup_{i\in I} O_i \in \mathcal{O}$.

Remark A.2.0.2. The elements of $\mathcal{O}$ are called *open sets* and the complements are called *closed sets.* Moreover, we call the pair $(X, \mathcal{O})$ a topological space.

Definition A.2.0.3 (Hausdorff). Let $(X, \mathcal{O})$ be a topological space. The topology $\mathcal{O}$ on X is said to be *Hausdorff* if and only if for all $x, y \in X$ with $x \neq y$ there is exists $O_x \in \mathcal{O}$ with $x \in O_x$ and there exists a $O_y \in \mathcal{O}$ with $y \in O_y$, i.e., $O_x \cap O_y = \emptyset$.

Definition A.2.0.4 (Metric space). Let X be a set. A *metric* on X is a map $\mathsf{d} : X \times X \to \mathbb{R}_{\geq 0}$, which, for all $x, y, z \in X$, satisfies the following conditions:

(i) (zero distance) $\mathsf{d}(x, y) = 0$ if and only if $x = y$.
(ii) (symmetry) $\mathsf{d}(x, y) = \mathsf{d}(y, x)$.
(iii) (triangle inequality) $\mathsf{d}(x, y) \leq \mathsf{d}(x, z) + \mathsf{d}(z, y)$.

A set X endowed with a metric, written (X, d), is called a *metric space.* If we have a norm $\| \cdot \|$ on X, and we then consider X as a normed space $(X, \| \cdot \|)$, we can define a metric by $\mathsf{d}(x, y) := \|x - y\|$. If (X, d) is a metric space, the topology on X is associated with d

and is, for some arbitrary index set I, given by

$$\mathcal{O}_X^{\mathsf{d}} := \left\{ \bigcup_{i\in I} B_{r_i}(x_i) \,\middle|\, x_i \in X,\ r_i \in \mathbb{R}_{\geq 0} \right\},$$

$$B_{r_i}(x_i) := \{y \in X \mid \mathsf{d}(x_i, y) < r_i\}.$$

Definition A.2.0.5 (Basis and separability). A topological space $(X, \mathcal{O})$ is said to have a *countable basis* of open sets $\{w_n\}_{n\in\mathbb{N}}$ if for every open set $O \in \mathcal{O}$, there exists a countable index set $I \subset \mathbb{N}$, such that

$$O = \bigcup_{n\in I} w_n.$$

Moreover, a metric space (X, d) is said to be *separable*, if it contains a sequence $(x_n)_{n\in\mathbb{N}}$ which is dense in X, that is, for all $x \in X$ there exists a subsequence $(x_{n_k})_{k\in\mathbb{N}}$ of $(x_n)_{n\in\mathbb{N}}$, such that $\mathsf{d}(x, x_{n_k}) \xrightarrow{k\to\infty} 0$.

Proposition A.2.0.1. *A metric space is separable if and only if it has a countable basis of open sets.*

Proof. We first prove the direction $\Rightarrow$. We can observe that

$$\mathscr{B} := \{B_r(x_n)\}_{r\in\mathbb{Q}_{\geq 0}, n\in\mathbb{N}}$$

is a basis of open sets, and thus we can write every open set as $O = \bigcup_{\substack{B\in\mathscr{B}\\ B\subset O}} B$. Now let $x \in O$. Then there exists an $\varepsilon > 0$, such that $B_\varepsilon(x) \subset O$ and there is an n_0 with $\mathsf{d}(x_{n_0}, x) < \frac{\varepsilon}{4}$ and thus $x \in B_{\frac{\varepsilon}{2}}(x_{n_0}) \subset O$. Now we prove the direction $\Leftarrow$. Consider the collection $\{w_n\}_{n\in\mathbb{N}}$ of open sets. Then we can choose an $x_n \in w_n$ and check that the sequence $(x_n)_{n\in\mathbb{N}}$ is a dense subset, which gives the claim. □

Definition A.2.0.6 (Product topology). Let $(X, \mathcal{O}_X)$ and $(Y, \mathcal{O}_Y)$ be two topological spaces. The *product topology* for the product space $X \times Y$ is defined, with an arbitrary index set I, by the family of open sets

$$\mathcal{O}_{X\times Y} := \left\{ \bigcup_{i\in I} O_i^X \times O_i^Y \,\middle|\, O_i^X \in \mathcal{O}_X, O_i^Y \in \mathcal{O}_Y\ \forall i \in I \right\}.$$

Definition A.2.0.7 (Continuity). Let $(X, \mathcal{O}_X)$ and $(Y, \mathcal{O}_Y)$ be two topological spaces. A map $f : (X, \mathcal{O}_X) \to (Y, \mathcal{O}_Y)$ is *continuous* if and only if for all $O^Y \in \mathcal{O}_Y$, the image of O^Y under f^{-1} is open, i.e.,

$$f^{-1}(O^Y) = \{x \in X \mid f(x) \in O^Y\} \in \mathcal{O}_X.$$

Definition A.2.0.8 (Canonical projection). Let X and Y be two sets. Then we can define the *canonical projections* to be the surjective maps

$$\begin{aligned} \pi_X : X \times Y &\to X, \\ (x, y) &\mapsto x, \\ \pi_Y : X \times Y &\to Y, \\ (x, y) &\mapsto y. \end{aligned}$$

Remark A.2.0.9. A useful observation is that the product topology is defined in such a way that the canonical projections are continuous, i.e.,

$$\pi_X^{-1}(O^X) = O^X \times Y \in \mathcal{O}_{X \times Y}, \qquad \pi_Y^{-1}(O^Y) = X \times O^Y \in \mathcal{O}_{X \times Y}.$$

Remark A.2.0.10. We can also define a metric on the product of two metric spaces (X, d_X) and (Y, d_Y) given by

$$D_p((x, y), (x', y')) = (\mathsf{d}_X^p(x, x') + \mathsf{d}_Y^d(y, y'))^{\frac{1}{p}}, \quad p \geq 1.$$

Proposition A.2.0.2. *Let $(X, \mathcal{O}_X)$ and $(\mathcal{O}_Y)$ be two topological spaces. If $(X, \mathcal{O}_X)$ and $(Y, \mathcal{O}_Y)$ have a countable basis of open sets, then $(X \times Y, \mathcal{O}_{X \times Y})$ also has a countable basis of open sets. Moreover, Let (X, d_X) and (Y, d_Y) be two metric spaces. If (X, d_X) and (Y, d_Y) are separable, then $(X \times Y, D_p)$ is also separable.*

Proof. First, let $\mathcal{U}_X = \{U_n\}_{n \in \mathbb{N}}$ be a basis of open sets on X and $\mathcal{V}_Y = \{V_n\}_{n \in \mathbb{N}}$ a basis of open sets on Y. Then $\{U_n \times V_m\}_{(n,m) \in \mathbb{N}^2}$ is a basis of open sets for $X \times Y$, which proves the first claim. We leave the second claim as an exercise for the reader. □

A.3. Borel Sets

Topologically, the Borel sets in a topological space define the σ-algebra generated by the open sets. One can build up the Borel sets from the open sets by iterating the operations of complementation and taking countable unions.

Definition A.3.0.1 (Borel σ-algebra). Let $(E, \mathcal{O})$ be a topological space. Then $\sigma(\mathcal{O})$ is called the Borel σ-algebra of E and is denoted by $\mathcal{B}(E)$. Moreover, the elements of $\mathcal{B}(E)$ are called Borel sets.

Remark A.3.0.2. Observe that if $E = \mathbb{R}$, then $\mathcal{B}(\mathbb{R}) \neq \mathcal{P}(\mathbb{R})$. That means that there exist subsets which are not Borel measurable.

Proposition A.3.0.1. *Let $(E, \mathcal{O})$ be a topological space with a countable basis of open sets $\{w_n\}_{n\in\mathbb{N}}$. Then we get*

$$\mathcal{B}(E) = \sigma(\{w_n\}_{n\in\mathbb{N}}).$$

Proof. Since $\{w_n\}_{n\in\mathbb{N}} \subset \mathcal{O}$, we get that $\sigma(\{w_n\}_{n\in\mathbb{N}}) \subset \sigma(\mathcal{O}) = \mathcal{B}(E)$. Moreover, since every open set O can be written as $O = \bigcup_{j\in J\subset\mathbb{N}} w_j$, we deduce that for all $O \in \mathcal{O}$ we get $O \in \sigma(\{w_n\}_{n\in\mathbb{N}})$ and thus $\sigma(\mathcal{O}) \subset \sigma(\{w_n\}_{n\in\mathbb{N}})$. $\square$

Remark A.3.0.3. An important observation is also that the σ-algebra generated by open sets equals the σ-algebra generated by closed sets of the form, that is, if we denote by $\mathcal{O}^{\mathsf{C}} := \{F \subset E \mid F \text{ is closed with respect to the topology } \mathcal{O}\}$, we get that

$$\mathcal{B}(E) = \sigma(\{F\}_{F\in\mathcal{O}^{\mathsf{C}}}).$$

Proof. We first show the direction $\Rightarrow$. For $O \in \mathcal{O}$ define $F := O^{\mathsf{C}}$, which is closed, i.e., $F \in \mathcal{O}^{\mathsf{C}}$. The fact that F is closed implies that $F \in \sigma(\{F\}_{F\in\mathcal{O}^{\mathsf{C}}})$ and thus $F^{\mathsf{C}} = O \in \sigma(\{F\}_{F\in\mathcal{O}^{\mathsf{C}}})$, because of the properties of a σ-algebra. Hence, we get $\mathcal{O} \subset \sigma(\{F\}_{F\in\mathcal{O}^{\mathsf{C}}})$ and therefore $\sigma(\mathcal{O}) \subset \sigma(\{F\}_{F\in\mathcal{O}^{\mathsf{C}}})$. The other direction is similar, so we leave it as an exercise. $\square$

Remark A.3.0.4. Consider the case $E = \mathbb{R}$. Then we get

$$\begin{aligned}\mathcal{B}(\mathbb{R}) &= \sigma(\{[a, \infty)\}_{a\in\mathbb{Q}}) = \sigma(\{(a, \infty)_{a\in\mathbb{Q}})\}) \\ &= \sigma(\{(-\infty, a]\}_{a\in\mathbb{Q}})) = \sigma(\{(-\infty, a)_{a\in\mathbb{Q}})\}).\end{aligned}$$

Proof. Recall that $\mathbb{Q}$ is a dense subset of $\mathbb{R}$. Therefore, it follows that

$$\begin{aligned}\{(\alpha,\beta)\mid\alpha,\beta\in\mathbb{Q},\alpha<\beta\}&=\{(\rho-r,\rho+r)\mid\rho\in\mathbb{Q},r\in\mathbb{Q}_{\geq 0}\}\\&=\{B_r(\rho)\mid\rho\in\mathbb{Q},r\in\mathbb{Q}_{\geq 0}\}\end{aligned}$$

is a countable basis of open sets in $\mathbb{R}$ and thus

$$\mathcal{B}(\mathbb{R})=\sigma(\{(\alpha,\beta)\}_{\substack{\alpha,\beta\in\mathbb{Q}\\ \alpha>\beta}}).$$

Moreover, it is important to observe that $(\alpha,\beta)=(\alpha,\infty)\cap[\beta,\infty)^{\mathsf{C}}$ with

$$(\alpha,\infty)=\bigcup_{n\in\mathbb{N}}\left[\frac{\alpha}{n},\infty\right).$$

Therefore, we get that $(\alpha,\infty)\in\sigma(\{(\alpha,\infty)\}_{\alpha\in\mathbb{Q}})$ and thus $(\alpha,\infty)\cap[\beta,\infty)^{\mathsf{C}}\in\sigma(\{[\alpha,\infty)\}_{\alpha\in\mathbb{Q}})$. It follows from the definition of the Borel σ-algebra that

$$\mathcal{B}(\mathbb{R})=\sigma(\{(\alpha,\beta)\}_{\alpha,\beta\in\mathbb{R}})\subset\sigma(\{[\alpha,\infty\}_{\alpha\in\mathbb{R}})\subset\sigma(\{F\}_{F\in\mathcal{O}_{\mathbb{R}}^{\mathsf{C}}})=\mathcal{B}(\mathbb{R}),$$

which finally implies that $\sigma(\{[\alpha,\infty)\}_{\alpha\in\mathbb{R}})=\mathcal{B}(\mathbb{R})$. □

A.4. Positive Measures

Definition A.4.0.1 (Positive measure). Let $(E,\mathcal{A})$ be a measurable space. A *positive measure* μ on $(E,\mathcal{A})$ is a map $\mu:\mathcal{A}\to[0,\infty]\subset\bar{\mathbb{R}}$, which satisfies the following conditions:

(i) (measure of the empty set is zero) $\mu(\emptyset)=0$.
(ii) (σ-additivity) For all sequences $(A_n)_{n\in\mathbb{N}}\in\mathcal{A}$ of disjoint measurable sets, that is $A_i\cap A_j=\emptyset$ for $i\neq j$, we have

$$\mu\left(\bigcup_{n\in\mathbb{N}}A_n\right)=\sum_{n\in\mathbb{N}}\mu(A_n).$$

Remark A.4.0.2. We call a triple $(E,\mathcal{A},\mu)$, i.e., a measurable space endowed with a specific measure, a *measure space*.

Remark A.4.0.3. A nice observation of $\bar{\mathbb{R}}_{\geq 0}$ is that all sums are convergent, that is, for any sequence $(x_n)_{n\in\mathbb{N}} \subset [0,\infty]$, we get $\sum_{n\in\mathbb{N}} x_n \in [0,\infty]$. We can formulate an equivalent definition of this sum as

$$\sum_{n\in\mathbb{N}} x_n := \sup_{\substack{I\subset\mathbb{N}\\ I \text{ finite}}} \sum_{n\in I} x_n.$$

More general, for any sequence $(x_\alpha)_{\alpha\in A}$ of real numbers $x_\alpha \in [0,\infty]$ with an arbitrary index set A, either countable or uncountable, we can define

$$\sum_{\alpha\in A} x_\alpha := \sup_{\substack{I\subset A\\ A \text{ finite}}} \sum_{\alpha\in I} x_\alpha.$$

Moreover, for any two index sets A and B and for any bijection $\phi : B \xrightarrow{\sim} A$, we get

$$\sum_{\alpha\in A} x_\alpha = \sum_{\beta\in B} x_{\phi(\beta)}.$$

Proposition A.4.0.1. *Let $(E,\mathcal{A},\mu)$ be a measure space. Then the following conditions hold:*

(i) *Let $A, B \in \mathcal{A}$ be two measurable sets such that $A \subset B$. Then $\mu(A) \leq \mu(B)$. Moreover, if $\mu(A) < \infty$, then $\mu(B\setminus A) = \mu(B) - \mu(A)$.*

(ii) *(Inclusion–exclusion) Let $A, B \in \mathcal{A}$ be two measurable sets. Then we get*

$$\mu(A) + \mu(B) = \mu(A\cup B) + \mu(A\cap B).$$

(iii) *Let $(A_n)_{n\in\mathbb{N}} \subset \mathcal{A}$ be an increasing sequence of measurable sets. Then we get*

$$\mu\left(\bigcup_{n\in\mathbb{N}} A_n\right) = \lim_{n\to\infty} \uparrow \mu(A_n).$$

(iv) *Let $(B_n)_{n\in\mathbb{N}} \subset \mathcal{A}$ be a decreasing sequence of measurable sets. Moreover, let $\mu(B_0) < \infty$. Then we get*

$$\mu\left(\bigcap_{n\in\mathbb{N}} B_n\right) = \lim_{n\to\infty} \downarrow \mu(B_n).$$

(v) *(σ-subadditivity) Let $(A_n)_{n\in\mathbb{N}} \subset \mathcal{A}$ be a sequence of measurable sets. Then we get*

$$\mu\left(\bigcup_{n\in\mathbb{N}} A_n\right) \leq \sum_{n\in\mathbb{N}} \mu(A_n).$$

Proof. For (i), observe that $B = (B \setminus A) \sqcup A$ and thus $\mu(B) = \mu(B\setminus A)+\mu(A)$. Moreover, if $\mu(A) < \infty$, then $\mu(B\setminus A) = \mu(B)-\mu(A)$. For (ii), observe that $A \cup B = (A \setminus B) \sqcup (B \setminus A) \sqcup (A \cap B)$ and thus $\mu(A \cup B) = \mu(A \setminus B) + \mu(B \setminus A) + \mu(A \cap B)$. Assume $\mu(A \cap B) = \infty$, then we get (ii), since $\mu(A \cup B), \mu(A), \mu(B) \geq \mu(A \cap B) = \infty$. On the other hand, assume $\mu(A \cap B) < \infty$, then $\mu(A \setminus B) = \mu(A \cap B^{\mathsf{C}})$ and $\mu(B \setminus A) = \mu(B \cap A^{\mathsf{C}})$, which implies that $\mu(A \cup B) = \mu(A) + \mu(B) - 2\mu(A \cap B) + \mu(A \cap B)$. Rearranging things, we get the claim. For (iii), let $C_0 = A_0$ and $C_n = A_n \setminus A_{n-1}$ for all $n \geq 1$. Then we get

$$\bigcup_{n\in\mathbb{N}} C_n = \bigcup_{n\in\mathbb{N}} A_n,$$

where the C_n's are disjoint and moreover, $\mu(C_n) = \mu(A_n) - \mu(A_{n-1})$. Therefore, we get

$$\mu\left(\bigcup_{n\in\mathbb{N}} A_n\right) = \mu\left(\bigcup_{n\in\mathbb{N}} C_n\right) = \sum_{n\in\mathbb{N}} \mu(C_n) = \lim_{N\to\infty} \sum_{n=0}^{N} \mu(C_n)$$
$$= \lim_{N\to\infty} \uparrow \mu(A_N).$$

For (iv), let $A_n = B_0 \setminus B_1$, which implies that $A_n \subset A_{n+1}$ for all $n \in \mathbb{N}$. Thus, we get

$$\mu(A_n) = \mu(B_0) - \mu\left(\bigcup_{n\in\mathbb{N}} B_n\right) = \mu\left(B_0 \setminus \bigcup_{n\in\mathbb{N}} B_n\right)$$
$$= \mu\left(\bigcup_{n\in\mathbb{N}} A_n\right) = \lim_{n\to\infty} \uparrow \mu(A_n) = \lim_{n\to\infty} (\mu(B_0) - \mu(B_n)).$$

Now, since $\mu(B_0) < \infty$, we get that

$$\mu\left(\bigcap_{n\in\mathbb{N}} B_n\right) = \lim_{n\to\infty} \downarrow \mu(B_n),$$

where we have used the fact that

$$\bigcup_{n\in\mathbb{N}} A_n = \bigcup_{n\in\mathbb{N}} (B_0 \setminus B_n) = \bigcup_{n\in\mathbb{N}} (B_0 \cap B_n^{\mathsf{C}}) = B_0 \cap \left(\bigcup_{n\in\mathbb{N}} B_n^{\mathsf{C}}\right)$$

$$= B_0 \cap \left(\bigcap_{n\in\mathbb{N}} B_n\right)^{\mathsf{C}} = B_0 \setminus \bigcap_{n\in\mathbb{N}} B_n,$$

and that $\bigcap_{n\in\mathbb{N}} \subseteq B_0$. For (v), let $C_0 = A_0$ and for $n \geq 1$ let

$$C_n = A_n \setminus \bigcup_{k=0}^{n-1} A_k,$$

where the C_n's are disjoint and moreover, $\bigcup_{n\in\mathbb{N}} A_n = \bigcup_{n\in\mathbb{N}} C_n \subset A_n$. Therefore, we get

$$\mu\left(\bigcup_{n\in\mathbb{N}} A_n\right) = \mu\left(\bigcup_{n\in\mathbb{N}} C_n\right) = \sum_{n\in\mathbb{N}} \mu(C_n) \leq \sum_{n\in\mathbb{N}} \mu(A_n).$$

This can also been proven by induction, which we leave as an exercise. □

Example A.4.0.4 (Dirac measure). Let $(E, \mathcal{A})$ be a measurable space such that for any $x \in E$, we get that $\{x\} \in \mathcal{A}$. we can define a measure $\delta : E \times \mathcal{A} \to \{1, 0\}$ by

$$\delta_x(A) = \mathbb{1}_A(x) = \begin{cases} 1, & x \in A, \\ 0, & x \notin A. \end{cases}$$

This measure is called the *Dirac measure* or the *Dirac mass at x*. More generally, if we consider sequences $(x_n)_{n\in\mathbb{N}} \subset E$ and $(\alpha_n)_{n\in\mathbb{N}} \subset [0, \infty]$, we can define a measure $\mathscr{D}_{(x_n)}^{(\alpha_n)} : \mathcal{A} \to \bar{\mathbb{R}}_{\geq 0}$, which is defined by

$$\mathscr{D}_{(x_n)}^{(\alpha_n)}(A) := \left(\sum_{n\in\mathbb{N}} \alpha_n \delta_{x_n}\right) = \sum_{n\in\mathbb{N}} \alpha_n \delta_{x_n}(A).$$

Example A.4.0.5 (Lebesgue measure). There exists a unique measure on the measurable space $(\mathbb{R}, \mathcal{B}(\mathbb{R}))$, which is denoted by λ, such that for all open intervals $(a, b) \in \mathcal{B}(\mathbb{R})$ it is given by

$$\lambda((a, b)) = b - a.$$

Definition A.4.0.6 (Finite, σ-finite and probability measures). Let $(E, \mathcal{A})$ be a measurable space. We say that a measure μ is

(i) *finite*, if $\mu(E) < \infty$;
(ii) a *probability measure*, if $\mu(E) = 1$;
(iii) σ*-finite*, if there exists an increasing sequence (partition of the total space) $(E_n)_{n\in\mathbb{N}} \subset \mathcal{A}$, such that $E = \bigcup_{n\in\mathbb{N}} E_n$ and with $\mu(E_n) < \infty$ for all $n \in \mathbb{N}$.

Definition A.4.0.7 (Atom). Let $(E, \mathcal{A}, \mu)$ be a measure space. An element $x \in E$ is called an *atom* for μ if the set $\{x\} \in \mathcal{A}$ and

$$\mu(\{x\}) > 0.$$

Definition A.4.0.8 (Product σ-algebra). Let $(E_1, \mathcal{A}_1)$ and $(E_2, \mathcal{A}_2)$ be two measurable spaces. Then we can define the *product σ-algebra* $\mathcal{A}_1 \otimes \mathcal{A}_2$ on the product space $E_1 \times E_2$ by

$$\mathcal{A}_1 \otimes \mathcal{A}_2 := \sigma(A_1 \times A_2),$$

where $A_1 \in \mathcal{A}_1$ and $A_2 \in \mathcal{A}_2$. This is actually the σ-algebra which contains all sets of the form $(A_1 \times A_2)$.

Let $(E, \mathcal{A})$ and $(F, \mathcal{B})$ be two measurable spaces. Consider a map $f : E \to F$. Moreover, let I be an arbitrary index set and for $i \in I$, let $A_i \subset E$ and $B_i \subset F$. We can write, for $A \subset E$

$$f(A) := \{f(x) \mid x \in A\},$$

and similarly, for $B \subset F$, we can write

$$f^{-1}(B) = \{x \in E \mid f(x) \in B\}.$$

Moreover, it is easy to observe the following relations:

(i) $f\left(\bigcup_{i\in I} A_i\right) = \bigcup_{i\in I} f(A_i)$.
(ii) $f\left(\bigcap_{i\in I} A_i\right) \subseteq \bigcap_{i\in I} f(A_i)$
(where equality holds if f is injective).
(iii) $f^{-1}\left(\bigcup_{i\in I} B_i\right) = \bigcup_{i\in I} f^{-1}(B_i)$.
(iv) $f^{-1}\left(\bigcap_{i\in I} B_i\right) = \bigcap_{i\in I} f^{-1}(B_i)$.

(v) $f^{-1}(B)^{\mathsf{C}} = f^{-1}(B)^{\mathsf{C}}$.
(vi) If $\mathcal{C} \subset \mathcal{P}(F)$, then $f^{-1}(\mathcal{C}) := \{f^{-1}(C) \mid C \in \mathcal{C}\}$.

Proposition A.4.0.2. *Let E and F be two measurable spaces, where $\mathcal{B}$ is a σ-algebra on F. Then we get that*

$$\mathcal{A} := f^{-1}(\mathcal{B}) = \{f^{-1}(B) \mid B \in \mathcal{B}\}$$

is a σ-algebra on E.

Proof. First, it is obvious that $f^{-1}(F) = E$, which implies that if $F \in \mathcal{B}$, then $E \in \mathcal{A}$. Moreover, it holds that

$$f^{-1}(B)^{\mathsf{C}} = f^{-1}(B^{\mathsf{C}}) \in \mathcal{A}$$

for all $B \in \mathcal{B}$ since arbitrary unions of elements in $\mathcal{B}$ are again in $\mathcal{B}$. □

Remark A.4.0.9. It is sometimes convenient to write $\sigma(f)$ instead of $f^{-1}(B)$.

Example A.4.0.10. Let $(E, \mathcal{A})$ be a measurable space and let $F \subset E$ be a subset of E. Moreover, let $\iota : F \hookrightarrow E$ be the canonical inclusion. Then we get

$$\iota^{-1}(\mathcal{A}) = \{\iota^{-1}(A) \mid A \in \mathcal{A}\} = \{F \cap A \mid A \in \mathcal{A}\}.$$

Example A.4.0.11. Let $(E, \mathcal{A})$ be a measurable space and let $F \subset E$ be a subset of E. Moreover, let $\pi_E : E \times F \to E$ be the canonical projection. Then we get that

$$\pi_E^{-1}(\mathcal{A}) = \{\pi_E^{-1}(A) \mid A \in \mathcal{A}\} = \{A \times F \mid A \in \mathcal{A}\}.$$

Definition A.4.0.12 (Image σ-algebra). Let $(E, \mathcal{A})$ and $(F, \mathcal{B})$ be measurable spaces and let $f : E \to F$ be a map. The *image σ-algebra* of $\mathcal{A}$ by f is defined by

$$\mathcal{I} := \{I \in \mathcal{P}(F) \mid f^{-1}(I) \in \mathcal{A}\}.$$

Proposition A.4.0.3. *Let (X, d) be a metric space and let $Y \subset X$. Then the Borel σ-algebra of Y is given by*

$$\mathcal{B}(Y) = \{A \cap Y \mid A \in \mathcal{B}(X)\}.$$

Moreover, if $Y \in \mathcal{B}(X)$ then $\mathcal{B}(Y) \subset \mathcal{B}(X)$ and $\mathcal{B}(Y) = \{A \in \mathcal{B}(X) \mid A \subset Y\}$.

Proof. Let $\iota : Y \hookrightarrow X$ be the inclusion of Y into X. Then we get that

$$\mathcal{O}_Y := \{O \cap Y \mid O \in \mathcal{O}_X\} = \iota^{-1}(\mathcal{O}_X).$$

Moreover, we get that

$$\begin{aligned}\mathcal{B}(Y) &= \sigma(\mathcal{O}_Y) = \sigma(\iota^{-1}(\mathcal{O}_X)) = \iota^{-1}(\sigma(\mathcal{O}_X)) = \iota^{-1}(\mathcal{B}(X))\\ &= \{A \cap Y \mid A \subset \mathcal{B}(X)\},\end{aligned}$$

which proves the first part. The second part is easily obtained from the fact that σ-algebras are stable under finite intersections. $\square$

We have the following examples of Borel σ-algebras:

Example A.4.0.13. Let $X = \mathbb{R}_{\geq 0}$. Then we get that

$$\mathcal{B}(\mathbb{R}_{\geq 0}) = \{A \subseteq \mathbb{R}_{\geq 0} \mid A \in \mathcal{B}(\mathbb{R})\}.$$

Example A.4.0.14. Let $X = \mathbb{R}^\times := \mathbb{R} \setminus \{0\}$. Then we get that

$$\mathcal{B}(\mathbb{R}^\times) = \{A \in \mathcal{B}(\mathbb{R}) \mid 0 \notin A\}.$$

Example A.4.0.15 (Borel sets on $\bar{\mathbb{R}}$). Define $\bar{\mathbb{R}} := \mathbb{R} \cup \{-\infty, \infty\}$ and consider the map

$$\begin{aligned} f : \mathbb{R} &\to (-1, 1),\\ x &\mapsto \frac{x}{\sqrt{x^2+1}}.\end{aligned}$$

We can now consider an extension $\tilde{f}$ of f, which is defined on $\bar{\mathbb{R}}$ such that $\tilde{f}|_{\mathbb{R}} = f$ with $\tilde{f}(-\infty) = -1$ and $\tilde{f}(\infty) = 1$. Moreover, we can consider $\bar{\mathbb{R}}$ as a metric space by considering the distance d given for all $x, y \in \bar{\mathbb{R}}$ as $\mathsf{d}(x, y) := \|\tilde{f}(x) - \tilde{f}(y)\|$. We write therefore $(\bar{\mathbb{R}}, \mathsf{d})$ as a metric space. Thus, we can define the Borel σ-algebra of $\bar{\mathbb{R}}$ by the Borel sets, which are described by the metric topology of $\bar{\mathbb{R}}$. This concept is important as we will work many times with the space $\bar{\mathbb{R}}$.

Remark A.4.0.16. It is useful to note that $\bar{\mathbb{R}}$ describes a totally ordered set, since $\leq$ appears naturally as in $\mathbb{R}$. Moreover, the identity map

$$\begin{aligned}\mathrm{id}:(\mathbb{R},\mathsf{d}|_{\mathbb{R}})&\to(\mathbb{R},\|\cdot\|),\\ x&\mapsto x\end{aligned}$$

is a homeomorphism. Another useful observation is that $(\bar{\mathbb{R}},\mathsf{d})$ is a compact space and homeomorphic to the interval $[-1,1]$ and eventually $\mathbb{R}$ is an open subset of $\bar{\mathbb{R}}$.

Exercise A.4.0.17. Show that

$$\mathcal{B}(\bar{\mathbb{R}})=\sigma(\{[a,\infty]\}_{a\in\mathbb{Q}})=\sigma(\{(a,\infty]\}_{a\in\mathbb{Q}}).$$

A.5. Measurable Maps

Measure theory and the notion of integration require special structures on functions, which need to satisfy different properties, such as being measurable or bounded. The notion of a measurable map is important for the study of integration with respect to a measure and the fact that we can only consider integration with respect to a measure if the integrating function satisfies measurability. It is now important to use the σ-algebras of the underlying spaces similarly to the topological notion of continuity where the topologies of the underlying spaces are used. Let us therefore define a measurable map.

Definition A.5.0.1 (Measurable map). Let $(E,\mathcal{A})$ and $(F,\mathcal{B})$ be two measurable spaces and let $f:E\to F$ be a map. We say that f is *measurable*, if for all $B\in\mathcal{B}$ we get $f^{-1}(B)\in\mathcal{A}$.

Proposition A.5.0.1. *Let $(X,\mathcal{A}),(Y,\mathcal{B})$ and $(Z,\mathcal{G})$ be measurable spaces and consider the composition*

$$(X,\mathcal{A})\xrightarrow{f}(X,\mathcal{B})\xrightarrow{g}(Z,\mathcal{G}).$$

If f and g are both measurable, then $g\circ f$ is also measurable.

Exercise A.5.0.2. Prove Proposition A.5.0.1.

Proposition A.5.0.2. *Let $(E,\mathcal{A})$ and $(F,\mathcal{B})$ be two measurable spaces and let $f:E\to F$ be a map. Moreover, assume that there* *exists $\mathcal{C} \subset \mathcal{P}(F)$ such that $\sigma(\mathcal{C}) = \mathcal{B}$. Then we get that f is measurable if and only if for all $C \in \mathcal{C}$ we have $f^{-1}(C) \in \mathcal{A}$.*

Proof. Let us first define a σ-algebra $\mathcal{G}$ by

$$\mathcal{G} := \{B \in \mathcal{B} \mid f^{-1}(B) \in \mathcal{A}\} \supset \mathcal{C}.$$

Now, since $\mathcal{G}$ is a σ-algebra, we get that $\sigma(\mathcal{C}) \subset \mathcal{G}$ and thus $\mathcal{G} = \mathcal{B}$, which proves the claim. □

Remark A.5.0.3. Let[2] $(F, \mathcal{B}) = (\mathbb{R}, \mathcal{B}(\mathbb{R}))$. To show that f is measurable, it is enough to show either that $f^{-1}((a, b)) \in \mathcal{A}$ or $f^{-1}((-\infty, a)) \in \mathcal{A}$, for $a, b \in \mathbb{R}$ with $a < b$.

Example A.5.0.4 (Continuous maps are measurable). Assume that E and F are two metric spaces (or topological spaces), endowed with their Borel σ-algebra respectively. Then f is measurable if for every open set O of F we have $f^{-1}(O) \in \mathcal{B}(E)$. In particular, we can say that *continuous maps are measurable maps.*

Example A.5.0.5. Let $A \subset E$ be a subset of E. Then the map $\mathbb{1}_A$ is measurable if and only if $A \in \mathcal{A}$.

Remark A.5.0.6. The notion of measurability of a map $f : E \to F$, between two measurable spaces $(E, \mathcal{A})$ and $(F, \mathcal{B})$ means that $f^{-1}(\mathcal{B}) \in \mathcal{A}$. The smallest σ-algebra on E which makes f measurable is given by $f^{-1}(\mathcal{B})$ and we denote it by $\sigma(f)$. Moreover, we want to emphasize that we can write $\{f \in B\}$ for $f^{-1}(B) = \{x \in E \mid f(x) \in B\}$. Hence, we can write $\{f \geq b\}$ instead of $f^{-1}([b, \infty))$ or $\{f = b\}$ instead of $f^{-1}(\{b\})$. If f is constant and if for some $C \in F$, we have $f(x) = C$ for all $x \in E$, then f is always measurable since $f^{-1}(\mathcal{B}) = \{\emptyset, E\}$.

Lemma A.5.0.3. *Let $(E, \mathcal{A}), (F_1, \mathcal{B}_1)$ and $(F_2, \mathcal{B}_2)$ be measurable spaces and let $f_1 : E \to F_1$ and $f_2 : E \to F_2$ be two measurable maps. Then the map*

$$f : (E, \mathcal{A}) \to (F_1 \times F_2, \mathcal{B}_1 \otimes \mathcal{B}_2),$$
$$x \mapsto (f_1(x), f_2(x))$$

is measurable.

[2]We can also take $(\bar{\mathbb{R}}, \mathcal{B}(\bar{\mathbb{R}}))$.

Proof. Let us define $\mathcal{C} := \{B_1 \times B_2 \mid B_1 \in \mathcal{B}_1, B_2 \in \mathcal{B}_2\}$. Then we get, by definition of the product σ-algebra, that $\sigma(\mathcal{C}) = \mathcal{B}_1 \otimes \mathcal{B}_2$. Now, for $B_1 \times B_2 \in \mathcal{C}$ we get that $f^{-1}(B_1 \times B_2) = \underbrace{\underbrace{f^{-1}(B_1)}_{\in \mathcal{A}} \cap \underbrace{f^{-1}(B_2)}_{\in \mathcal{A}}}_{\in \mathcal{A}} \in \mathcal{A}$.

Therefore, it follows that f is measurable. □

Remark A.5.0.7. Consider $f_1 = \pi_1 \circ f$ and $f_2 = \pi_2 \circ f$ with

$$\begin{aligned} \pi_i : F_1 \times F_2 &\to F_i, \\ (y_1, y_2) &\mapsto y_i \end{aligned}$$

for $i \in \{1, 2\}$ with π_1 and π_2 being measurable. Then f_1 and f_2 are measurable.

Corollary A.5.0.4. *Let $(E, \mathcal{A})$ be a measurable space and let $f, g : E \to \mathbb{R}$ be two measurable maps, where $\mathbb{R}$ is endowed with its Borel σ-algebra $\mathcal{B}(\mathbb{R})$. Then we get that:*

(i) $f + g$,
(ii) $f \cdot g$,
(iii) $f^+ = \max\{f, 0\}$,
(iv) $f^- = \max\{-f, 0\}$,
(v) $|f|$,

are measurable, where $f = f^+ - f^-$ and $|f| = f^+ + f^-$.

Proof. We will only show (i) and leave the other points as an exercise. Note that the map $f+g$ is a composition of the map $h : x \mapsto (f(x), g(x))$ and $r : (a, b) \mapsto a + b$. The map h is clearly measurable, since f and g are measurable and the map r is clearly continuous and thus measurable. As we have seen, the composition of two measurable maps is again measurable, which shows that $f+g$ is measurable. The proof of the other points is similar. □

Remark A.5.0.8. Let us consider the field of complex numbers $\mathbb{C}$ and make the identification $\mathbb{C} \cong \mathbb{R}^2$. Then we can naturally make sense of the measurability of the map $f : E \to \mathbb{C}$, where $\mathbb{C}$ is endowed with its Borel σ-algebra $\mathcal{B}(\mathbb{C})$, by saying that f is measurable if and only if $\mathrm{Re}(f)$ and $\mathrm{Im}(f)$ are measurable.

A.6. The Theorems of Lusin and Egorov

There are two important theorems which make statements about convergence types of measurable maps. They are important to understand the behavior of a sequence of measurable maps and the relation to the concept of uniform convergence.

Theorem A.6.0.1 (Egorov). *Let $(E, \mathcal{A}, \mu)$ be a measure space. Let $f_k : E \to \bar{\mathbb{R}}$ be measurable for all $k \in \mathbb{N}$ and $f : E \to \bar{\mathbb{R}}$ be measurable and μ-a.e. finite. Moreover $f_k(x) \xrightarrow{k\to\infty} f(x)$ μ-a.e. for $x \in E$. Then for all $\delta > 0$ there exists $F \subset E$, with F compact and $\mu(E \setminus F) < \delta$ and*

$$\sup_{x\in F} |f_k(x) - f(x)| \xrightarrow{k\to\infty} 0,$$

i.e., $(f_k)_{k\in\mathbb{N}}$ converges uniformly to f in F.

Proof. Let $\delta > 0$. For $i, j \in \mathbb{N}$ set

$$C_{i,j} := \bigcup_{k=j}^{\infty} \{x \in E \mid |f_k(x) - f(x)| > 1/2\}.$$

$C_{i,j}$ is μ-measurable, because f and f_k are μ-measurable and $C_{i,(j+1)} \subset C_{i,j}$, $\forall i, j$. We also know that $f_k(x) \xrightarrow{k\to\infty} f(x)$ for μ-a.e. $x \in E$ and since $\mu(E) < \infty$ it follows that for all $i \in \mathbb{N}$

$$\lim_{j\to\infty} \mu(C_{i,j}) = \mu\left(\bigcup_{j=1}^{\infty} C_{i,j}\right) = 0.$$

So for every i there exists some number $N(i) \in \mathbb{N}$ with

$$\mu(C_{i,N(i)}) < \delta \cdot 2^{-i-1}.$$

Now set $A = E \setminus \bigcup_{i=1}^{\infty} C_{i,N(i)}$. Then we get that

$$\mu(E \setminus A) \leq \sum_{i=1}^{\infty} \mu(C_{i,N(i)}) < \delta/2,$$

and for all $i \in \mathbb{N}$ and $k \geq N(i)$

$$\sup_{x \in A} |f_k(x) - f(x)| \leq 2^{-i}.$$

Choose a $F \subset A$, where F is compact with $\mu(A \setminus F) < \delta/2$. Hence we have

$$\mu(E \setminus F) \leq \mu(E \setminus A) + \mu(A \setminus F) < \delta. \qquad \square$$

Theorem A.6.0.2 (Lusin). *Let $(E, \mathcal{A}, \mu)$ be a measure space. Let $f : E \to \bar{\mathbb{R}}$ be measurable and μ-a.e. finite. Then for all $\delta > 0$ there exists $F \subset E$, F compact with $\mu(E \setminus F) < \delta$ and $f|_F : F \to \mathbb{R}$ is continuous.*

Proof. We split the proof onto two parts.

(i) First, we are going to show the theorem for step functions of the form

$$g = \sum_{i=1}^{I} b_i \mathbb{1}_{B_i},$$

where we set $E = \bigsqcup_{i=1}^{I} B_i$ with $B_i \cap B_j = \emptyset$ for $i \neq j$. For $\delta > 0$ choose $F_i \subset B_i$ compact with

$$\mu(B_i \setminus F_i) < \delta \cdot 2^{-i}, \ \ 1 \leq i \leq I.$$

Since the sets B_i are disjoint, it follows that the sets F_i are also disjoint, because of the fact that they are also compact it follows that $\mathsf{d}(F_i, F_j) > 0$ for $i \neq j$. Therefore, we notice that g is locally constant, i.e., continuous on $F := \bigcup_{i=1}^{I} F_i \subset E$. Moreover, $F \subset E$ and

$$\mu(E \setminus F) = \mu\left(\bigcup_{i=1}^{I} (B_i \setminus F_i)\right) \leq \sum_{i=1}^{I} \mu(B_i \setminus F_i) < \delta.$$

(ii) Let $f_k : E \to \mathbb{R}$ be a step function with

$$f(x) = \lim_{k\to\infty} f_k(x), \quad x \in E,$$

where

$$f_k = \sum_{j=1}^{k} \frac{1}{j}\mathbb{1}_{A_j} = \sum_{i=1}^{I_k} b_{ik}\mathbb{1}_{B_{ik}}, \quad k \in \mathbb{N},$$

with $B_{ik} \cap B_{jk} = \emptyset$ for $i \neq j$ and $\bigsqcup_{i=1}^{I_k} B_{ik} = E$ and with

$$b_{ik} = \sum_{B_{ik}\subset A_j} \frac{1}{j}, \quad 1 \leq i \leq I_k, \quad k \in \mathbb{N}.$$

For $\delta > 0$, $g = f_k$ choose compact sets $F_k \in E$ as in part (i) with

$$\mu(E \setminus F_k) < \delta \cdot 2^{-k-1}, \quad f_k|_{F_k} : F_k \to \mathbb{R} \quad \text{continuous}, \quad k \in \mathbb{N}.$$

Choose also $F_0 \subset E$ compact with

$$\mu(E \setminus F_0) < \delta/2, \ \sup_{x\in F_0} |f_k(x) - f(x)| \xrightarrow{k\to\infty} 0.$$

Finally, let $F = \bigcap_{k=0}^{\infty} F_k \subset E$. Note that F is compact with

$$\mu(E \setminus F) \leq \mu\left(\bigcup_{k=0}^{\infty}(E \setminus F_k)\right) \leq \sum_{k=0}^{\infty} \mu(E \setminus F_k) < \delta$$

and because of the fact that $F \subset F_0$, it follows that

$$\sup_{x\in F} |f_k(x) - f(x)| \xrightarrow{k\to\infty} 0.$$

The continuity of $f_k\,|_F$, for $k \in \mathbb{N}$, gives us now the continuity of $f|_F : F \to \mathbb{R}$. □

A.7. The Limit Superior and Limit Inferior

The notion of a limit plays a very big role in measure and integration theory. The way how limits interact with integrals and how they behave under certain situations (for example changing the order of taking limits and integrating) lead to the famous limit theorems of Lebesgue integration. We need to recall the notion of the limsup and the one for the liminf in order to get a better intuition of how sequences of measurable functions behave. Let therefore $(a_n)_{n\in\mathbb{N}}$ be a sequence in $\bar{\mathbb{R}}$ and define the limit superior and the limit inferior of (a_n) as

$$\limsup_{n\to\infty} a_n := \lim_{n\to\infty} \swarrow \left(\sup_{k\geq n} a_k\right) = \inf_n \sup_{k\geq n} a_k,$$

$$\liminf_{n\to\infty} a_n := \lim_{n\to\infty} \nearrow \left(\inf_{k\geq n} a_k\right) = \sup_n \inf_{k\geq n} a_k.$$

Remark A.7.0.1. It is crucial to note that the above limits always exist in $\bar{\mathbb{R}}$.

Proposition A.7.0.1. *Let $(E, \mathcal{A})$ be a measurable space and let $(f_n)_{n\in\mathbb{N}}$ be a sequence of measurable maps such that $f_n : E \to \bar{\mathbb{R}}$. Then we get that:*

(i) $\sup_n f_n$,
(ii) $\inf_n f_n$,
(iii) $\limsup_n f_n$,
(iv) $\liminf_n f_n$,

are all measurable. In particular, if $f_n \xrightarrow{n\to\infty} f$ then f is also measurable. In general we can say that $\{x \in E \mid \lim_{n\to\infty} f_n(x) \text{ exists}\}$ is measurable.

Proof. Let us first define $f(x) := \inf_n f_n(x)$. Now we can see that it is enough to show that for all $a \in \mathbb{R}$ we have $f^{-1}([-\infty, a)) \in \mathcal{A}$. Indeed, we can observe that

$$f^{-1}([-\infty, a)) = \left\{x \in E \,\middle|\, \inf_n f_n(x) < a\right\}$$
$$= \bigcup_n \{x \in E \mid f_n(x) < a\} \in \mathcal{A},$$

and therefore we can say that $f^{-1}([-\infty, a)) = \bigcup_n \{f_n < a\}$. Moreover, we have that

$$\left\{x \in E \,\middle|\, \lim_{n\to\infty} f_n(x) \text{ exists}\right\} = \left\{x \in E \,\middle|\, \liminf_n f_n(x) = \limsup_n f_n(x)\right\}$$
$$= G^{-1}(\Delta) \in \mathcal{A},$$

where G is the map given by $G : x \mapsto (\liminf_n f_n(x), \limsup_n f_n(x))$ and Δ is the diagonal of $\bar{\mathbb{R}}^2$, which is closed and hence measurable. □

Example A.7.0.2. If a map $f : \mathbb{R} \to \mathbb{R}$ is differentiable, its derivative f' will be measurable and we can hence write it as a limit of measurable functions as

$$f'(x) = \lim_{n\to\infty} n \left(f\left(x + \frac{1}{n}\right) - f(x)\right).$$

Remark A.7.0.3. Let $(E, \mathcal{A})$ be a measurable space and let $(f_n)_{n\in\mathbb{N}}$ be a sequence of measurable functions $f_n : E \to X$ to some space X. If $\bar{\mathbb{R}}$ is described through the metric space (X, d), we get that $f_n \xrightarrow{n\to\infty} f$ implies that f is measurable. Moreover, $f_n \xrightarrow{n\to\infty} f$ if and only if for all $x \in E$ we get that $\lim_{n\to\infty} f_n(x) = f(x)$ if and only if for all $x \in E$ we get that $\lim_{n\to\infty} \mathsf{d}(f_n(x), f(x)) = 0$. If we consider a closed subset $F \subset X$, we get that

$$f^{-1}(F) = \{x \in E \mid \mathsf{d}(f(x), F) = 0\} = \left\{x \in E \,\middle|\, \lim_{n\to\infty} \mathsf{d}(f_n(x), F) = 0\right\}$$
$$= \left\{x \in E \,\middle|\, \forall p \geq 1, \exists N \in \mathbb{N} \text{ such that } \mathsf{d}(f_n(x), F) \leq \frac{1}{p}\ \forall n \geq N\right\}$$
$$= \bigcap_{p\geq 1} \bigcup_{N\in\mathbb{N}} \bigcap_{n\geq N} \left\{x \in E \,\middle|\, \mathsf{d}(f_n(x), F) \geq \frac{1}{p}\right\} \in \mathcal{A}.$$

If we consider a complete metric space (E, d), then one can show that $\{x \in E \mid f_n(x) \text{ converges}\} \in \mathcal{A}$. We leave this as an exercise.

Definition A.7.0.4 (Push-forward measure). Let $(E, \mathcal{A})$ and $(F, \mathcal{B})$ be two measurable spaces and let μ be a positive measure on $(E, \mathcal{A})$. Moreover, let $f : E \to F$ be a measurable map. Then the *push-forward* of the measure μ by f, denoted by $f_*\mu$ is defined for all $B \in \mathcal{B}$ as

$$f_*\mu(B) := \mu(f^{-1}(B)).$$

A.8. Simple Functions

After we have developed the notion of a measurable map, we need to discuss a class of very powerful and, as the name points out, *simple functions.* The advantage of these type of functions is exactly the fact that they are simple to handle and moreover one can basically proof many things for measurable functions by proving it for simple functions and deduce the general case out of that. We will later see the advantage of them being *dense* in different spaces. Let us start with the definition of a simple function.

Definition A.8.0.1 (Simple function). Let $(E, \mathcal{A})$ be a measurable space. A map $f : E \to \mathbb{R}$ is called *simple*, if it is measurable and if it takes a finite number of values. Recall again that $\mathbb{R}$ is considered as a measurable space endowed with its Borel σ-algebra $\mathcal{B}(\mathbb{R})$.

Remark A.8.0.2. By definition, one can write any simple function f as

$$f = \sum_{i \in I} \alpha_i \mathbb{1}_{A_i},$$

where I is a finite index set, α_i are real numbers and the sets $(A_i)_{i \in I}$ form an $\mathcal{A}$-measurable partition of E, i.e., $E = \bigcup_{i \in I} A_i$, $A_i \cap A_j = \emptyset$ if $i \neq j$ and $A_i \in \mathcal{A}$ for all $i \in I$.

Proof of Remark A.8.0.2. Note first that for all $x \in E$, we get that $f(x) \in \{\alpha_1, \ldots, \alpha_k\}$, where the α_i's are distinct if and only if $f^{-1}(\{\alpha_i\}) = \{x \in E \mid f(x) = \alpha_i\} = A_i \in \mathcal{A}$, since f is measurable. Hence, we get that $A_i \cap A_j = \emptyset$ for $i \neq j$ and $\bigcup_{i=1}^{k} A_i = E$.

This representation is unique if the α_i's are distinct. We can write the canonical form therefore as

$$f = \sum_{\alpha \in f(E)} \alpha \mathbb{1}_{\{f=\alpha\}}.$$

□

Remark A.8.0.3. We can observe the fact that the simple functions form a *commutative algebra*. Indeed, let $f = \sum_{i\in I} \alpha_i \mathbb{1}_{A_i}$ and $g = \sum_{j\in J} \beta_j \mathbb{1}_{B_j}$ be two simple functions (in canonical form) and let $\lambda \in \mathbb{R}$. Then we can easily obtain that

$$\lambda f + g = \sum_{\substack{i\in I \\ j\in J}} (\lambda \alpha_i + \beta_j) \mathbb{1}_{A_i \cap B_j}$$

and since $(A_i \cap B_j)_{(i,j)\in I\times J}$ forms an $\mathcal{A}$-partition of E, we get that

$$f \cdot g = \sum_{\substack{i\in I \\ j\in J}} \alpha_i \beta_j \mathbb{1}_{A_i \cap B_j}$$

is a simple function and moreover,

$$\max(f, g) = \sum_{\substack{i\in I \\ j\in J}} \max(\alpha_i, \beta_j) \mathbb{1}_{A_i \cap B_j}.$$

Theorem A.8.0.1. *Let $(E, \mathcal{A})$ be a measurable space and let $f : E \to \mathbb{R}$ be a measurable map. Then there exists a sequence of simple functions $(f_n)_{n\in\mathbb{N}}$ such that for all $x \in E$ we get*

$$f_n(x) \xrightarrow{n\to\infty} f(x).$$

Moreover, if

(i) *$f \geq 0$, we can choose an* increasing *sequence $f_n \geq 1$ (i.e., we have $0 \leq f_n \leq f_{n+1}$).*
(ii) *f is bounded, we can choose a sequence f_n such that the convergence is uniformly, i.e.,*

$$\sup_{x\in E} |f_n(x) - f(x)| \xrightarrow{n\to\infty} 0.$$

Proof. Let us first assume that $f \geq 0$. For $n \in \mathbb{N}$, we define

$$E_{n,\infty} := \{f \geq n\}, \quad E_{n,k} := \left\{\frac{k}{2^n} \leq f < \frac{k+1}{2^n}\right\},$$
$$k \in \{0, 1, \ldots, n2^n - 1\}.$$

Thus, we have $E_{n,k}, E_{n,\infty} \in \mathcal{A}$. Now define

$$f_n := \sum_{k=0}^{n2^n-1} \frac{k}{2^n} \cdot \mathbb{1}_{E_{n,k}} + n\mathbb{1}_{E_{n,\infty}},$$

and obtain that f_n is simple by construction. For $x \in E_{n,k}$ we get

$$f_{n+1}(x) = \begin{cases} f_n(x), & \frac{2k}{2^{n+1}} \leq f(x) < \frac{2k+1}{2^{n+1}}, \\ f_n(x) + \frac{1}{2^{n+1}}, & \frac{2k+1}{2^{n+1}} \leq f(x) < \frac{2(k+1)}{2^{n+1}}. \end{cases}$$

and if $x \in E_{n,\infty}$ we get that

$$f_{n+1}(x) = \begin{cases} n+1, & f(x) \geq n+1, \\ \frac{n2^{n+1}+l}{2^{n+1}}, & \frac{n2^{n+1}+l}{2^{n+1}} \leq f(x) < \frac{n2^{n+1}+l+1}{2^{n+1}}. \end{cases}$$

It follows that $0 \leq f_n(x) \leq f_{n+1}(x)$ for all $x \in E$. If furthermore $x \in \{f < n\}$, then

$$0 \leq f(x) - f_n(x) \leq 2^{-n} = \frac{1}{2^n} = \frac{k+1}{2^n} - \frac{k}{2^n} \xrightarrow{n\to\infty} 0$$

or equivalently

$$f_n(x) \xrightarrow{n\to\infty} f(x)$$

on $\{x \in E \mid f(x) < \infty\} = \bigcup_{k\geq 1}\{f < k\}$, and if $x \in \{f = \infty\} = \bigcap_{n\in\mathbb{N}}\{f \geq n\}$, then

$$f_n(x) = n \xrightarrow{n\to\infty} \infty.$$

If we have a function $f : E \to \mathbb{R}_{\geq 0}$ and we assume that there exists some $M \in (0, \infty)$ such that $0 \leq f_n(x) \leq M$ for all $x \in E$, then for

$n > M$ with $\{f \geq n\} = \emptyset$ it follows that for all $x \in E$ we get

$$0 \leq f(x) - f_n(x) \leq 2^{-n},$$

which implies that

$$\sup_{x \in E} |f(x) - f_n(x)| \leq 2^{-n} \xrightarrow{n \to \infty} 0.$$

Let us emphasize the real case. If we have a function $f : E \to \overline{\mathbb{R}}$ we have the decompositions as

$$f = f^+ - f^-, \quad |f| = f^1 + f^-, \quad \text{with } f^+, f^- \geq 0.$$

If we now take f_n^+ and f_n^- as constructed above we can obtain $f_n^+ \uparrow f^+$ and $f_n^- \uparrow f^-$ for $n \to \infty$. Moreover, we notice that for $x \in E$ the sequences $(f_n^+(x))_{n \in \mathbb{N}}$ and $(f_n^-(x))_{n \in \mathbb{N}}$ cannot be simultaneously nonzero and therefore

$$f_n = f_n^+ - f_n^- \xrightarrow{n \to \infty} f = f^+ - f^-. \qquad \square$$

A.9. Monotone Classes

A very important notion is that of a monotone class. We will see that there are many things which can be deduced by using the *monotone class lemma*.

Definition A.9.0.1 (Monotone class). Let E be some topological space and let $\mathcal{M} \subset \mathcal{P}(E)$. $\mathcal{M}$ is called a *monotone class* if the following conditions hold:

(i) $E \in \mathcal{M}$.
(ii) Let $A \in \mathcal{M}$ and $B \in \mathcal{M}$. If $A \subset B \Rightarrow B \setminus A \in \mathcal{M}$.
(iii) Let $(A_n)_{n \in \mathbb{N}} \in \mathcal{M}$. If $A_n \subset A_{n+1} \Rightarrow \bigcup_{n \in \mathbb{N}} A_n \in \mathcal{M}$.

Remark A.9.0.2. A σ-algebra is a monotone class.

Exercise A.9.0.3. Prove Remark A.9.0.2.

Remark A.9.0.4. As for σ-algebras, we notice that an arbitrary intersection of monotone classes is again a monotone class. Thus, if $\mathcal{C} \subset \mathcal{P}(E)$, we can define the monotone class generated by $\mathcal{C}$ as

$$\mathcal{M}(\mathcal{C}) = \bigcap_{\substack{\mathcal{C} \subset \mathcal{M} \\ \mathcal{M} \text{ monotone class}}} \mathcal{M}.$$

This is also by construction the smallest monotone class containing $\mathcal{C}$.

Theorem A.9.0.1 (Monotone classes lemma). *Let E be a topological space. If $\mathcal{C} \subset \mathcal{P}(E)$ is stable under finite intersection, i.e., for $A \in \mathcal{C}$ and $B \in \mathcal{C}$ we get that $A \cap B \in \mathcal{C}$, then*

$$\sigma(\mathcal{C}) = \mathcal{M}(\mathcal{C}).$$

Proof. It is obvious that $\mathcal{M}(\mathcal{C}) \subset \sigma(\mathcal{C})$ since a σ-algebra is also a monotone class. Next we want to show that $\mathcal{M}(\mathcal{C})$ is a σ-algebra to conclude that $\sigma(\mathcal{C}) \subset \mathcal{M}(\mathcal{C})$ and hence then $\mathcal{M}(\mathcal{C})$ contains $\mathcal{C}$, i.e., $\sigma(\mathcal{C})$. It is not difficult to see that a monotone class, which is stable under finite intersections, is a σ-algebra. Let us therefore show that $\mathcal{M}(\mathcal{C})$ is stable under finite intersections. First, we fix $A \in \mathcal{C}$ and define

$$\mathcal{M}_A := \{B \in \mathcal{C} \mid A \cap B \in \mathcal{M}(\mathcal{C})\}.$$

Then we get that $\mathcal{C} \in \mathcal{M}_A$ since $\mathcal{C}$ is stable under finite intersections and obviously $E \in \mathcal{M}_A$. We can also note that If $B, B' \in \mathcal{M}_A$ and $B \subset B'$, with

$$A \cap (B' \setminus B) = \underbrace{\underbrace{(A \cap B')}_{\in \mathcal{M}(\mathcal{C})} \setminus \underbrace{(A \cap B)}_{\in \mathcal{M}(\mathcal{C})}}_{\in \mathcal{M}(\mathcal{C})}$$

then $B' \setminus B \in \mathcal{M}_A$. Moreover, if $(B_n)_{n \in \mathbb{N}} \in \mathcal{M}_A$ and $A \cap \bigcup_n \underbrace{(A \cap B_n)}_{\in \mathcal{M}(\mathcal{C})}$ we get the implication

$$\bigcup_{n \geq 1} (A \cap B_n) \in \mathcal{M}(\mathcal{C}) \Rightarrow \bigcup_{n \geq 1} B_n \in \mathcal{M}_A.$$

since $(A \cap B_n)$ is increasing. Finally, we can conclude the above facts. That means if $\mathcal{M}_A$ is a monotone class containing $\mathcal{C}$, then $\mathcal{M}_A =$

$\mathcal{M}(\mathcal{C})$, which shows that for all $A \in \mathcal{C}$ and $B \in \mathcal{M}(\mathcal{C})$ we get $A \cap B \in \mathcal{M}(\mathcal{C})$. We can now apply the same idea another time. Fix $B \in \mathcal{M}(\mathcal{C})$ and define

$$\mathcal{M}_B := \{A \in \mathcal{M}(\mathcal{C}) \mid A \cap B \in \mathcal{M}(\mathcal{C})\}.$$

Now, from above, we get that $\mathcal{M}_B$ is a monotone class, i.e., $\mathcal{M}_B \subset \mathcal{M}(\mathcal{C})$ and thus for all $A, B \in \mathcal{M}(\mathcal{C})$ we get $A \cap B \in \mathcal{M}(\mathcal{C})$. Hence it follows that $\mathcal{M}(\mathcal{C})$ is stable under finite intersections and is therefore a σ-algebra. □

Corollary A.9.0.2. *Let $(E, \mathcal{A})$ be a measurable space and let μ, ν be two measures on $(E, \mathcal{A})$. Moreover, assume that there exists a family of subsets $\mathcal{C}$, which is stable under finite intersections, such that $\sigma(\mathcal{C}) = \mathcal{A}$ and $\mu(A) = \nu(A)$ for all $A \in \mathcal{C}$. Then the following conditions hold:*

(i) *If $\mu(E) = \nu(E) < \infty$, then we get $\mu = \nu$.*
(ii) *If there exists an increasing family $(E_n)_{n \in \mathbb{N}}$ with $E_n \in \mathcal{C}$ such that*

$$E = \bigcup_{n \in \mathbb{N}} E_n$$

and $\mu(E_n) = \nu(E_n) < \infty$, then it follows that $\mu = \nu$.

Proof. Let us first define the set $\mathcal{G} := \{A \in \mathcal{A} \mid \mu(A) = \nu(A)\}$. By assumption we get that $\mathcal{C} \subset \mathcal{G}$. Moreover, we note that $\mathcal{G}$ is a monotone class. Note at first that $E \in \mathcal{G}$ by assumption since $\mu(E) = \nu(E)$. Now let $A, B \in \mathcal{G}$ such that $A \subset B$ and since

$$\mu(B \setminus A) = \mu(B) - \mu(A) = \nu(B) - \nu(A) = \nu(B \setminus A),$$

we get that $(B \setminus A) \in \mathcal{G}$. Now let $(A_n)_{n \in \mathbb{N}}$ be an increasing sequence in $\mathcal{G}$. Then the fact

$$\mu\left(\bigcup_{n \in \mathbb{N}} A_n\right) = \lim_{n \to \infty} \mu(A_n) = \lim_{n \to \infty} \nu(A_n) = \nu\left(\bigcup_{n \in \mathbb{N}} A_n\right)$$

implies that $\bigcup_{n \in \mathbb{N}} A_n \in \mathcal{G}$. Moreover, since $\mathcal{G}$ is a monotone class containing $\mathcal{C}$, we get that $\mathcal{G}$ contains $\mathcal{M}(\mathcal{C})$. On the other hand, we know that $\mathcal{C}$ is stable under finite intersections and therefore it follows that $\mathcal{M}(\mathcal{C}) = \sigma(\mathcal{C})$ and that $\mathcal{G} = \sigma(\mathcal{C}) = \mathcal{A}$. Now define for all $n \in \mathbb{N}$ and $A \in \mathcal{A}$ the two sequences

$$\mu_n(A) := \mu(A \cap E_n),$$
$$\nu_n(A) := \nu(A \cap E_n).$$

Now since $\mu(E) = \nu(E)$, we get the same for the sequence elements and obtain therefore that $\mu_n = \nu_n$. Moreover, for $A \in \mathcal{A}$, we have

$$\mu(A) = \lim_{n\to\infty} \mu(A \cap E_n) = \lim_{n\to\infty} \uparrow \nu(A \cap E_n) = \nu(A).$$

Hence we get that $\mu = \nu$. □

Remark A.9.0.5. There are several applications of Corollary A.9.0.2. Let us emphasize a first one, by giving a small introduction to the Lebesgue measure. Assume that λ is a measure on the measurable space $(\mathbb{R}, \mathcal{B}(\mathbb{R}))$ such that $\lambda((a,b)) = b - a$ for $a < b$ and let $\mathcal{C}$ be the class of intervals $E_n = (-n, n)$ for $n \geq 1$. With Corollary A.9.0.2, it follows that λ is unique. We will call λ the *Lebesgue measure*. A second application is that a finite measure on $(\mathbb{R}, \mathcal{B}(\mathbb{R}))$ is uniquely characterized, for $a \in \mathbb{R}$, by the values

$$\mu((-\infty, a]) = \mu((-\infty, a)).$$

Appendix B

Basics of Integration Theory

B.1. Integration for Positive (Non-negative) Functions

In this appendix, we will introduce the integral from a different point of view, namely in terms of a measure, which describes a generalization of the already familiar Riemann integral which has several drawbacks. Let us look at one of the most standard examples where the Riemann integral approach fails. Consider the function

$$\mathbb{1}_{\mathbb{Q}}(x) := \begin{cases} 1, & x \in \mathbb{Q}, \\ 0, & x \notin \mathbb{Q}. \end{cases}$$

Using the usual techniques of Riemann integration, there is no way where we can talk about Riemann integrability of $\mathbb{1}_{\mathbb{Q}}$ and hence to define the integral

$$\int_0^1 \mathbb{1}_{\mathbb{Q}}(x)\mathrm{d}x.$$

However, with the notion of an integral with respect to a positive measure, we can also deal with such integrals as we will see. Let $(E, \mathcal{A})$ be a measurable space and let $f(x) = \sum_{i=1}^n \alpha_i \mathbb{1}_{A_i}(x)$ be a simple function, where $A_i \in \mathcal{A}$ and $\alpha_i \in \mathbb{R}$ for all $1 \leq i \leq n$ and $n \in \mathbb{N}$. Moreover, let us assume without loss of generality that

$$\alpha_1 < \alpha_2 < \cdots < \alpha_n.$$

Then $A_i = f^{-1}(\{\alpha_i\}) \in \mathcal{A}$. Let us also denote by μ a positive measure on $(E, \mathcal{A})$. Then we can define the integral for simple functions as follows:

Definition B.1.0.1 (Integral with respect to a measure (Lebesgue integral)). Assume that f takes values in $\mathbb{R}_{\geq 0}$ with $0 \leq \alpha_1 < \cdots < \alpha_n$. Then the Integral of f with respect to μ is defined by

$$\int f \mathrm{d}\mu := \sum_{i=1}^{n} \alpha_i \mu(A_i), \tag{B.1.0.1}$$

where we use the convention $0 \cdot \infty = 0$, in case that $\alpha_i = 0$ and $\mu(A_i) = \infty$. Moreover, if $f(x) = \mathbb{1}_A(x) + 0 \cdot \mathbb{1}_{A^{\mathsf{C}}}(x)$, for $A \in \mathcal{A}$, then

$$\int f \mathrm{d}\mu = \mu(A)(+\ 0 \cdot \mu(A^{\mathsf{C}})).$$

Remark B.1.0.2. It is easy to obtain that by Definition B.1.0.1, if $\int f \mathrm{d}\mu \in [0, \infty]$, then

$$f = 0 \Rightarrow \int f \mathrm{d}\mu = 0.$$

Remark B.1.0.3. The integral (B.1.0.1) is well-defined. Indeed, let $f(x) = \sum_{i=1}^{n} \alpha_i \mathbb{1}_{A_i}(x)$ be the canonical form of the simple function f, i.e., for all i we have that the α_i's are distinct, and $A_i = f^{-1}(\{\alpha_i\})$. Then, by definition, it follows that

$$\int f \mathrm{d}\mu = \sum_{i=1}^{n} \alpha_i \mu(A_i).$$

On the other hand, we can also write f as

$$f(x) = \sum_{j=1}^{m} \beta_j \mathbb{1}_{B_j}(x),$$

where $(\beta_j)_{1 \leq j \leq m}$ forms an $\mathcal{A}$-partition, with $\beta_j \geq 0$ and where the β_j's are not necessarily distinct. We want to show that the integral is still the same. Note that for each $i \in \{1, \ldots, n\}$, we get that A_i is the disjoint union of the sets B_j for which $\alpha_i = \beta_j$. Then the additivity property of the measure shows that

$$\mu(A_i) = \sum_{\{j \mid \beta_j = \alpha_j\}} \mu(B_j),$$

and therefore the integral does not depend on the representation of f.

Proposition B.1.0.1. *Let f and g be two simple, positive and measurable functions on E. Then we have the following properties:*

(i) *For all $a, b \geq 0$, we get*

$$\int (af + bg)\mathrm{d}\mu = a \int f\mathrm{d}\mu + b \int g\mathrm{d}\mu.$$

(ii) *If $f \leq g$, then*

$$\int f\mathrm{d}\mu \leq \int g\mathrm{d}\mu.$$

Proof. For (i), let us first consider simple functions

$$f = \sum_{i=1}^{n} \alpha_i \mathbb{1}_{A_i}, \qquad g = \sum_{k=1}^{m} \alpha'_k \mathbb{1}_{A'_k}.$$

Moreover, we can note that

$$A_i = \bigcup_{k=1}^{m} (A_i \cap A'_k), \qquad A'_k = \bigcup_{i=1}^{n} (A'_k \cap A_i)$$

and therefore we can write

$$f = \sum_{j=1}^{p} \beta_j \mathbb{1}_{B_j}, \qquad g = \sum_{j=1}^{p} \gamma_j \mathbb{1}_{B_j},$$

where $(B_j)_{1 \leq j \leq p}$ is an $\mathcal{A}$-partition of E obtained from a reordering of $(A_i \cap A'_k)$. Thus, we get

$$\int f\mathrm{d}\mu = \sum_{j=1}^{p} \beta_j \mu(B_j), \qquad \int g\mathrm{d}\mu = \sum_{j=1}^{p} \gamma_j \mu(B_j).$$

Hence, it follows that

$$\int (af + bg)\mathrm{d}\mu = \sum_{j=1}^{p} (a\beta_j + b\gamma_j)\mu(B_j) = a \sum_{j=1}^{p} \beta_j \mu(B_j)$$

$$+ b \sum_{j=1}^{p} \gamma_j \mu(B_j) = a \int f\mathrm{d}\mu + b \int g\mathrm{d}\mu.$$

For (ii), note that we can write g as $g = f + g - f$, where $f \geq 0$ and $g - f \geq 0$. Then we get that

$$\int g\mathrm{d}\mu = \underbrace{\int (g - f)\mathrm{d}\mu}_{\geq 0} + \int f\mathrm{d}\mu \geq \int f\mathrm{d}\mu.$$

□

Definition B.1.0.4 (Integral for measurable maps). Let $(E, \mathcal{A}, \mu)$ be a measure space. Moreover, let us denote by $\mathcal{E}^+$ the set of all nonnegative simple functions and let $f : E \to [0, \infty)$ be a measurable map. Then we can define the integral for f to be given as

$$\int f \mathrm{d}\mu := \sup_{\substack{h \leq f \\ h \in \mathcal{E}^+}} \int h \mathrm{d}\mu.$$

Remark B.1.0.5. Sometimes we have different notation for the same object. Depending on the context, we write

$$\int f \mathrm{d}\mu = \int f(x) \mathrm{d}\mu(x) = \int f(x) \mu(\mathrm{d}x).$$

Proposition B.1.0.2. *Let $(E, \mathcal{A}, \mu)$ be a measure space and let $f, g : E \to [0, \infty)$ be measurable maps. Then we have the following properties:*

(i) *If $f \leq g$, then*

$$\int f \mathrm{d}\mu \leq \int g \mathrm{d}\mu.$$

(ii) *If $\mu(\{x \in E \mid f(x) > 0\}) = 0$, then*

$$\int f \mathrm{d}\mu = 0.$$

Exercise B.1.0.6. Prove[1] Proposition B.1.0.2.

Theorem B.1.0.3 (Monotone convergence theorem). *Let $(f_n)_{n \in \mathbb{N}}$ be an increasing sequence of positive and measurable functions (with values in $[0, \infty)$), and let $f := \lim_{n \to \infty} \uparrow f_n$. Then we get that*

$$\int f \mathrm{d}\mu = \lim_{n \to \infty} \int f_n \mathrm{d}\mu.$$

[1]Note that it is enough to prove it for simple functions.

Proof. We know that since $f_n \leq f$, we have $\int f_n \mathrm{d}\mu \leq \int f \mathrm{d}\mu$ and hence

$$\lim_{n\to\infty} \int f_n \mathrm{d}\mu \leq \int f \mathrm{d}\mu.$$

We need to show that

$$\int f \mathrm{d}\mu \leq \lim_{n\to\infty} \int f_n \mathrm{d}\mu.$$

Define $h := \sum_{i=1}^{m} \alpha_i \mathbb{1}_{A_i}$, such that $h \leq f$, and let $a \in [0,1)$. Moreover, define

$$E_n := \{x \in E \mid ah(x) \leq f_n(x)\}.$$

We can immediately observe that E_n is measurable[2] and since $f_n \uparrow f$ as $n \to \infty$ and $a < 1$, we can deduce that

$$E = \bigcup_{n\geq 1} E_n, \qquad E_n \subset E_{n+1}.$$

Furthermore, we can show that $f_n \geq a\mathbb{1}_{E_n} h$. To see this, wee need to emphasize the following two cases:

(i) If $x \in E_n$, then $a\mathbb{1}_{E_n}(x)h(x) = ah(x)$ and by definition of E_n we get that $f_n(x) \geq ah(x)$.

(ii) If $x \notin E_n$, then $a\mathbb{1}_{E_n}(x)h(x) = 0$ and thus $f_n(x) \geq 0$ holds since f_n is positive.

Hence, it follows that

$$\int f_n \mathrm{d}\mu \geq \int a\mathbb{1}_{E_n} h \mathrm{d}\mu = a \int \mathbb{1}_{E_n} h \mathrm{d}\mu = a \sum_{i=1}^{m} \alpha_i \mu(A_i \cap E_n).$$

Since $E_n \uparrow E$ and $A_i \cap E_n \uparrow A_i$ as $n \to \infty$, we have $\mu(A_i \cap E_n) \uparrow \mu(A_i)$ as $n \to \infty$. Thus, we get that

$$\lim_{n\to\infty} \int f_n \mathrm{d}\mu \geq a \sum_{i=1}^{m} \alpha_i \mu(A_i) = a \int h \mathrm{d}\mu.$$

---

[2]We leave this as an exercise.

Note that the left-hand side does not depend on a. Now we let $a \to 1$ and obtain then

$$\lim_{n\to\infty} \int f_n \mathrm{d}\mu \geq \int h \mathrm{d}\mu.$$

This is now true for every $h \in \mathcal{E}^+$ with $h \leq f$ and the left-hand side does not depend on h. Therefore, we have

$$\lim_{n\to\infty} \uparrow \int f_n \mathrm{d}\mu \geq \sup_{\substack{h\in\mathcal{E}^+ \\ h\leq f}} \int h \mathrm{d}\mu = \int f \mathrm{d}\mu.$$

Recall that for any positive measurable map f (i.e., with values in $[0, \infty)$) there exists an increasing sequence $(f_n)_{n\in\mathbb{N}}$ of simple positive functions such that $f = \lim_{n\to\infty} f_n$. □

Proposition B.1.0.4. *Let $(E, \mathcal{A}, \mu)$ be a measure space. Then we have the following properties*:

(i) *If f and g are two positive and measurable functions on E and $a, b \in \mathbb{R}_{\geq 0}$, then*

$$\int (af + bg) \mathrm{d}\mu = a \int f \mathrm{d}\mu + b \int g \mathrm{d}\mu.$$

(ii) *If $(f_n)_{n\in\mathbb{N}}$ is a sequence of measurable and positive functions on E, then*

$$\int \sum_n f_n \mathrm{d}\mu = \sum_n \int f_n \mathrm{d}\mu.$$

Proof. For (i), take two positive sequences $(f_n)_{n\in\mathbb{N}}$ and $(g_n)_{n\in\mathbb{N}} \geq 0$ of simple functions such that $f_n \uparrow f$ and $g_n \uparrow g$ as $n \to \infty$. Then we get that

$$\begin{aligned}\int (af + bg) \mathrm{d}\mu &= \lim_{n\to\infty} \int (af_n + bg_n) \mathrm{d}\mu \\ &= \lim_{n\to\infty} \left(a \int f_n \mathrm{d}\mu + b \int g_n \mathrm{d}\mu \right) = a \int f \mathrm{d}\mu + b \int g \mathrm{d}\mu,\end{aligned}$$

where the first and last equality is due to Theorem B.1.0.3. Now (ii) is an immediate consequence of Theorem B.1.0.3. Indeed, define

$$g_N := \sum_{n=1}^{N} f_n,$$

such that $g_N \geq 0$ and $\lim_{N\to\infty} \uparrow g_N = \sum_{n=1}^{\infty} f_n$. If we now apply Theorem B.1.0.3 to $(g_N)_{N\in\mathbb{N}}$, we get

$$\begin{aligned}\int \sum_{n=1}^{\infty} f_n \mathrm{d}\mu &= \int \lim_{N\to\infty} g_N \mathrm{d}\mu = \lim_{N\to\infty} \int g_N \mathrm{d}\mu \\ &= \lim_{N\to\infty} \int \sum_{n=1}^{N} f_n \mathrm{d}\mu \overset{(i)}{=} \lim_{N\to\infty} \sum_{n=1}^{N} \int f_n \mathrm{d}\mu = \sum_{n=1}^{\infty} \int f_n \mathrm{d}\mu,\end{aligned}$$

where the second equality is due to Theorem B.1.0.3. This ends the proof. □

Example B.1.0.7. Consider the Dirac measure δ_x for $x \in E$ and let $f : E \to \mathbb{R}_{\geq 0}$ be a measurable map. Then we get that

$$\int f \mathrm{d}\delta_x = f(x).$$

Example B.1.0.8. Consider the natural numbers $\mathbb{N}$ as a measure space together with the σ-algebra $\mathcal{P}(\mathbb{N})$ and the *counting measure* on $\mathbb{N}$, which is the measure μ satisfying that for all $A \subset \mathbb{N}$ we get $\mu(A) = |A|$. Then for a measurable map $f : \mathbb{N} \to \mathbb{R}_{\geq 0}$ we get

$$\int f \mathrm{d}\mu = \sum_{n\in\mathbb{N}} f(n).$$

Moreover, we also get that for every positive sequence $(a_{n,k})_{n\in\mathbb{N},k\in\mathbb{N}} \geq 0$ we have

$$\sum_{k\in\mathbb{N}} \sum_{n\in\mathbb{N}} a_{n,k} = \sum_{n\in\mathbb{N}} \sum_{k\in\mathbb{N}} a_{n,k}.$$

Corollary B.1.0.5. *Let $(E, \mathcal{A})$ be a measurable space and let f be a positive measurable map on E. Moreover, let us define for $A \in \mathcal{A}$ a map ν on E by*

$$\nu(A) := \int_E \mathbb{1}_A(x) f(x) \mathrm{d}\mu = \int_A f(x) \mathrm{d}\mu.$$

Then ν is a measure on $(E, \mathcal{A})$, which is called the measure with density f with respect to μ and we write

$$\nu = f \circ \mu.$$

Remark B.1.0.9. It is clear that if $\mu(A) = 0$ then $\nu(A) = 0$ for all $A \in \mathcal{A}$.

Proof of Corollary B.1.0.5. First of all, note that $\nu(\emptyset) = 0$ follows from Proposition B.1.0.2. Now let $(A_n)_{n \in \mathbb{N}} \in \mathcal{A}$ be a sequence of measurable sets with $A_n \cap A_m = \emptyset$ for all $n \neq m$. Then we get that

$$\begin{aligned}
\nu\left(\bigcup_{n \geq 1} A_n\right) &= \int \mathbb{1}_{\bigcup_{n \geq 1} A_n}(x) f(x) \mathrm{d}\mu \\
&= \int \sum_{n \geq 1} \mathbb{1}_{A_n}(x) f(x) \mathrm{d}\mu \\
&= \sum_{n \geq 1} \int \mathbb{1}_{A_n}(x) f(x) \mathrm{d}\mu \\
&= \sum_{n \geq 1} \nu(A_n),
\end{aligned}$$

which actually shows that ν is a measure on E. □

Remark B.1.0.10. We usually say that a property is true μ-almost everywhere and we write μ-a.e. (or a.e. if it is clear for which measure), if this property holds on a set $A \in \mathcal{A}$ with $\mu(A^{\mathsf{C}}) = 0$. For example, if f and g are both measurable maps on E, then $f = g$ a.e. means that

$$\mu(\{x \in E \mid f(x) \neq g(x)\}) = 0.$$

Proposition B.1.0.6. *Let $(E, \mathcal{A}, \mu)$ be a measure space and let $f, g : E \to \mathbb{R}$ be measurable and positive functions. Then the following properties hold:*

(i) $\mu(\{x \in E \mid f(x) > a\}) \leq \frac{1}{a} \int f \mathrm{d}\mu$, *for all* $a > 0$.
(ii) *If* $\int f \mathrm{d}\mu < \infty$, *then* $f < \infty$ *a.e.*
(iii) $\int f \mathrm{d}\mu = 0$, *if and only if* $f = 0$ *a.e.*
(iv) *If* $f = g$ *a.e., then* $\int f \mathrm{d}\mu = \int g \mathrm{d}\mu$.

Proof. (i) Define the set $A_a := \{x \in E \mid f(x) > a\}$. Then we can observe that $f(x) \geq a \mathbb{1}_{A_a}(x)$ for all $x \in A_a$. Hence, it follows that

$$\int f(x) \mathrm{d}\mu \geq \int a \mathbb{1}_{A_a}(x) \mathrm{d}\mu$$

and

$$\mu(A_a) \leq \frac{1}{a} \int f \mathrm{d}\mu.$$

(ii) For $n \geq 1$, define the sets $A_n := \{x \in E \mid f(x) \geq n\}$ and $A_\infty := \{x \in E \mid f(x) = \infty\} = \bigcap_{n \in \mathbb{N}} A_n$. Then we get that

$$\mu(A_\infty) = \lim_{n \to \infty} \mu(A_n),$$

which implies that

$$\underbrace{\mu(A_n) \leq \frac{1}{n} \int f(x) \mathrm{d}\mu}_{\xrightarrow{n \to \infty} 0} < \infty.$$

(iii) We have already seen that if $f = 0$ a.e., then $\int f \mathrm{d}\mu = 0$. Therefore, it is enough to show the other direction. Let $n > 1$. Then we get that

$$\mu(B_n) \leq n \underbrace{\int f \mathrm{d}\mu}_{0} = 0.$$

Thus, we get that

$$\mu(\{x \in E \mid f(x) > 0\}) = \mu\left(\bigcup_{n \geq 0} B_n\right) \leq \sum_{n \geq 1} \mu(B_n) = 0.$$

(iv) Let us first introduce a special notation at this point. We will write $f \vee g$ for $\sup(f, g)$ and $f \wedge g$ for $\inf(f, g)$. Now assume that $f = g$ a.e., which implies that $f \vee g = f \wedge g$ and hence we get that

$$\int (f \vee g)\mathrm{d}\mu = \int (f \wedge g)\mathrm{d}\mu + \int (f \vee g - f \wedge g)\mathrm{d}\mu = \int (f \wedge g)\mathrm{d}\mu.$$

Because of the fact that

$$f \wedge g \leq f \leq f \vee g \Rightarrow \int f\mathrm{d}\mu = \int (f \vee g)\mathrm{d}\mu = \int (f \wedge g)\mathrm{d}\mu,$$

$$f \wedge g \leq g \leq f \vee g \Rightarrow \int g\mathrm{d}\mu = \int (f \vee g)\mathrm{d}\mu = \int (f \wedge g)\mathrm{d}\mu,$$

we finally get that

$$\int f\mathrm{d}\mu = \int g\mathrm{d}\mu.$$

□

Lemma B.1.0.7 (Fatou). *Let $(f_n)_{n\in\mathbb{N}}$ be a sequence of real valued, measurable and positive functions on a measure space $(E, \mathcal{A}, \mu)$. Then we get that*

$$\int \liminf_{n\to\infty} f_n \mathrm{d}\mu \leq \liminf_{n\to\infty} \int f_n \mathrm{d}\mu.$$

Proof. Recall that we actually have

$$\liminf_{n\to\infty} f_n = \lim_{n\to\infty} \nearrow \left(\inf_{k\geq n} f_n\right), \quad \left(= \sup_n \inf_{k\geq n} f_k\right).$$

Using Theorem B.1.0.3, we get that

$$\int \liminf_{n\to\infty} f_n \mathrm{d}\mu = \lim_{n\to\infty} \int \inf_{k\geq n} f_k \mathrm{d}\mu.$$

Now for all $p \geq k$, we have

$$\inf_{n\geq k} f_n \leq f_p,$$

$$\Rightarrow \int \inf_{n\geq k} f_n \mathrm{d}\mu \leq \int f_p \mathrm{d}\mu,$$

$$\Rightarrow \int \inf_{n\geq k} f_n \mathrm{d}\mu \leq \inf_{p\geq k} \int f_p \mathrm{d}\mu,$$

$$\Rightarrow \lim_{k\to\infty} \int \inf_{n\geq k} f_n \mathrm{d}\mu \leq \lim_{k\to\infty} \uparrow \inf_{p\geq k} \int f_p \mathrm{d}\mu = \liminf_{n\to\infty} \int f_n \mathrm{d}\mu.$$

This proves the lemma. □

B.2. Integrable Functions

Definition B.2.0.1 (Integrable). Let $(E, \mathcal{A}, \mu)$ be a measure space and let $f : E \to \mathbb{R}$ be a measurable map. We say that f is *integrable* with respect to μ if

$$\int |f| \mathrm{d}\mu < \infty.$$

Moreover, if f is integrable, we define its integral by

$$\int f \mathrm{d}\mu := \int f^+ \mathrm{d}\mu - \int f^- \mathrm{d}\mu.$$

Remark B.2.0.2. Note that we always have

$$\int f^{\pm} \mathrm{d}\mu \leq \int |f| \mathrm{d}\mu.$$

For instance, if we consider the function $f(x) := \frac{\sin(x)}{x}$, we get

$$\int_{\mathbb{R}_{\geq 0}} \frac{\sin(x)}{x} \mathrm{d}x = \frac{\pi}{2}, \quad \text{but} \quad \int_{\mathbb{R}_{\geq 0}} \left| \frac{\sin(x)}{x} \right| \mathrm{d}x = \infty.$$

Remark B.2.0.3. We will denote by $\mathcal{L}^1(E, \mathcal{A}, \mu)$ the space of integrable (and measurable) functions. Moreover, we denote by $\mathcal{L}^1_+(E, \mathcal{A}, \mu)$ the same space, but containing only positive functions.

Proposition B.2.0.1. *Let $(E, \mathcal{A}, \mu)$ be a measure space. Then the following properties hold:*

(i) *If $f \in \mathcal{L}^1(E, \mathcal{A}, \mu)$, then $\left|\int f \mathrm{d}\mu\right| \leq \int |f| \mathrm{d}\mu$.*
(ii) *$\mathcal{L}^1(E, \mathcal{A}, \mu)$ is a vector space and the map $f \mapsto \int |f| \mathrm{d}\mu$ is a linear form.*
(iii) *If $f, g \in \mathcal{L}^1(E, \mathcal{A}, \mu)$ and $f \leq g$, then $\int f \mathrm{d}\mu \leq \int g \mathrm{d}\mu$.*
(iv) *If $f, g \in \mathcal{L}^1(E, \mathcal{A}, \mu)$ and $f = g$ a.e., then $\int f \mathrm{d}\mu = \int g \mathrm{d}\mu$.*

Proof. (i) Let $f \in \mathcal{L}^1(E, \mathcal{A}, \mu)$. Then we get that

$$\begin{aligned}\left|\int f \mathrm{d}\mu\right| &= \left|\int f^+ \mathrm{d}\mu - \int f^- \mathrm{d}\mu\right| \\ &\leq \left|\int f^+ \mathrm{d}\mu\right| + \left|\int f^- \mathrm{d}\mu\right| \\ &= \int f^+ \mathrm{d}\mu + \int f^- \mathrm{d}\mu \\ &= \int (f^+ + f^-)\, \mathrm{d}\mu \\ &= \int |f| \mathrm{d}\mu.\end{aligned}$$

(ii) Indeed, $\mathcal{L}^1(E, \mathcal{A}, \mu)$ is a linear space and for $f \in \mathcal{L}^1(E, \mathcal{A}, \mu)$ we get that the map

$$f \mapsto \int f \mathrm{d}\mu \tag{B.2.0.1}$$

is a linear form. First we want to show that $\int af \mathrm{d}\mu = a \int f \mathrm{d}\mu$ for some $a \in \mathbb{R}$. Let therefore $a \in \mathbb{R}$ and consider the case where $a \geq 0$ and the one where $a < 0$ as follows:

$$\underline{a \geq 0}: \quad \begin{aligned}\int (af) \mathrm{d}\mu &= \int (af)^+ \mathrm{d}\mu - \int (af)^- \mathrm{d}\mu \\ &= a \int f^+ \mathrm{d}\mu - a \int f^- \mathrm{d}\mu \\ &= a \int f \mathrm{d}\mu.\end{aligned}$$

$$\underline{a < 0}: \quad \begin{aligned}\int (af) \mathrm{d}\mu &= \int (af)^+ \mathrm{d}\mu - \int (af)^- \mathrm{d}\mu \\ &= (-a) \int f^- \mathrm{d}\mu - \int f^+ \mathrm{d}\mu \\ &= a \int f \mathrm{d}\mu.\end{aligned}$$

If f and g are in $\mathcal{L}^1(E, \mathcal{A}, \mu)$, the inequality $|f + g| \leq |f| + |g|$ implies that $(f + g) \in \mathcal{L}^1(E, \mathcal{A}, \mu)$. One has to check the linearity of the map (B.2.0.1). It is easy to obtain the following implications:

$$
\begin{aligned}
&(f+g)^+ - (f+g)^- = (f+g) = f^+ - f^- + g^+ - g^-, \\
&\Rightarrow (f+g)^+ + f^- + g^- = (f+g)^- + f^+ + g^+, \\
&\Rightarrow \int (f+g)^+ \mathrm{d}\mu + \int f^- \mathrm{d}\mu + \int g^- \mathrm{d}\mu = \int (f+g)^- \mathrm{d}\mu \\
&\quad + \int f^+ \mathrm{d}\mu + \int g^+ \mathrm{d}\mu, \\
&\Rightarrow \int (f+g)^+ \mathrm{d}\mu - \int (f+g)^- \mathrm{d}\mu = \int f^+ \mathrm{d}\mu - \int f^- \mathrm{d}\mu \\
&\quad + \int g^+ \mathrm{d}\mu - \int g^- \mathrm{d}\mu, \\
&\Rightarrow \int (f+g) \mathrm{d}\mu = \int f \mathrm{d}\mu + \int g \mathrm{d}\mu.
\end{aligned}
$$

(iii) Let $f, g \in \mathcal{L}^1(E, \mathcal{A}, \mu)$ with $f \leq g$. Note that we can write $g = f + (g - f)$ and by assumption $g - f \geq 0$, which also implies that $\int (g - f)\mathrm{d}\mu \geq 0$. Therefore, we have

$$\int g \mathrm{d}\mu = \int f \mathrm{d}\mu + \int (g - f)\mathrm{d}\mu \geq \int f \mathrm{d}\mu,$$

which proves the claim.

(iv) Let $f, g \in \mathcal{L}^1(E, \mathcal{A}, \mu)$ with $f = g$ a.e., which also implies that $f^+ = g^+$ a.e. and $f^- = g^-$ a.e. Thus, it follows that

$$\int f^+ \mathrm{d}\mu = \int g^+ \mathrm{d}\mu \quad \text{and} \quad \int f^- \mathrm{d}\mu = \int g^- \mathrm{d}\mu.$$

Therefore, we get that $\int f \mathrm{d}\mu = \int g \mathrm{d}\mu$.

Exercise B.2.0.4. Show that if $f, g \in \mathcal{L}^1(E, \mathcal{A}, \mu)$ and $f \leq g$ a.e., then

$$\int f \mathrm{d}\mu \leq \int g \mathrm{d}\mu.$$

B.2.1. Extension to the complex case

Let $(E, \mathcal{A}, \mu)$ be a measure space and let $f : E \to \mathbb{C}(\cong \mathbb{R}^2)$ be a measurable map, which basically means that $\mathrm{Re}(f)$ and $\mathrm{Im}(f)$ are both measurable maps. We say that f is integrable if $\mathrm{Re}(f)$ and $\mathrm{Im}(f)$ are both integrable, or equivalently

$$\int |f| d\mu < \infty,$$

and we write $f \in \mathcal{L}^1_{\mathbb{C}}(E, \mathcal{A}, \mu)$. This is simply because of the fact that

$$\int f \mathrm{d}\mu = \int (\mathrm{Re}(f) + \mathrm{Im}(f)) \mathrm{d}\mu = \int \mathrm{Re}(f) \mathrm{d}\mu + \int \mathrm{Im}(f) \mathrm{d}\mu.$$

Moreover, the properties (i), (ii) and (iv) of Proposition B.2.0.1 also hold for the complex case.

B.3. Lebesgue's Dominated Convergence Theorem

An important question of integration theory is whether the interchanging of limits and integrals is actually possible and under which condition on the sequence of maps on some measure space. We have already seen the monotone convergence theorem (Theorem B.1.0.3) and Fatou's lemma (Lemma B.1.0.7), giving some simple conditions for such an interchange. Another way of achieving the same with different conditions is due to Lebesgue, who gave a more general condition, which we will discuss in this section.

Theorem B.3.0.1 (Lebesgue's dominated convergence theorem). *Let $(E, \mathcal{A}, \mu)$ be a measure space and let $(f_n)_{n \in \mathbb{N}}$ be a sequence of functions in $\mathcal{L}^1(E, \mathcal{A}, \mu)$ (respectively, $\mathcal{L}^1_{\mathbb{C}}(E, \mathcal{A}, \mu)$). Moreover, assume that the following hold:*

(i) *There exists a measurable map f on E with values in $\mathbb{R}$ (respectively, $\mathbb{C}$) such that for all $x \in E$ we have*

$$\lim_{n \to \infty} f_n(x) = f(x).$$

(ii) *There exists a positive and measurable map* $g : E \to \mathbb{R}_{\geq 0}$ *such that*

$$\int g \mathrm{d}\mu < \infty$$

and such that $|f_n| \leq g$ *a.e. for all* $n \in \mathbb{N}$.

Then $f \in \mathcal{L}^1(E, \mathcal{A}, \mu)$ *(respectively,* $\mathcal{L}^1_{\mathbb{C}}(E, \mathcal{A}, \mu)$*) and we have*

$$\lim_{n\to\infty} \int |f_n - f| \mathrm{d}\mu = 0$$

and hence

$$\lim_{n\to\infty} \int f_n \mathrm{d}\mu = \int f \mathrm{d}\mu.$$

Proof. Let us first assume some stronger assumptions:

(1) $\lim_{n\to\infty} f_n(x) = f(x)$ for all $x \in E$.
(2) There exists a positive and measurable map $g : E \to \mathbb{R}_{\geq 0}$ such that

$$\int g \mathrm{d}\mu < \infty$$

and $|f_n(x)| \geq g(x)$ for all $x \in E$.

Consider the general case where (1) and (2) hold. Now define the set

$$A := \left\{ x \in E \;\middle|\; f_n(x) \xrightarrow{n\to\infty} f(x) \quad \text{and} \quad \forall n, |f_n(x)| \leq g(x) \right\}.$$

Then $\mu(A^{\mathsf{C}}) = 0$. Let us now apply the first part of the proof to

$$\tilde{f}_n(x) := \mathbb{1}_A(x) f_n(x),$$
$$\tilde{f}(x) := \mathbb{1}_A(x) f(x).$$

Then we have that $f = \tilde{f}$ a.e. and $f_n = \tilde{f}_n$ a.e. Therefore, we get the following equalities:

$$\int f_n \mathrm{d}\mu = \int \tilde{f}_n \mathrm{d}\mu,$$

$$\int f \mathrm{d}\mu = \int \tilde{f} \mathrm{d}\mu,$$

$$\lim_{n\to\infty} \int f_n \mathrm{d}\mu = \int f \mathrm{d}\mu,$$

$$\lim_{n\to\infty} \int |f - f_n| \mathrm{d}\mu = 0.$$

Since now $|f| \leq g$ and $\int g \mathrm{d}\mu < \infty$, we get that $\int |f| \mathrm{d}\mu < \infty$. Hence, we have

$$|f - f_n| \leq 2g \quad \text{and} \quad |f - f_n| \xrightarrow{n\to\infty} 0.$$

Applying Lemma B.1.0.7, we can observe that

$$\liminf_{n\to\infty} \int 2g - |f - f_n| \mathrm{d}\mu \geq \int \liminf_{n\to\infty} 2g - |f - f_n| \mathrm{d}\mu,$$

and therefore we get that

$$\liminf_{n\to\infty} \int 2g - |f - f_n| \mathrm{d}\mu \geq \int 2g \mathrm{d}\mu,$$

which implies that

$$\int 2g \mathrm{d}\mu - \limsup_{n\to\infty} \int |f - f_n| \mathrm{d}\mu \geq \int 2g \mathrm{d}\mu.$$

It is now easy to observe that $\limsup_{n\to\infty} \int |f - f_n| \mathrm{d}\mu \leq 0$, which basically implies that $\lim_{n\to\infty} \int |f - f_n| \mathrm{d}\mu = 0$ and thus we can finally deduce that

$$\left| \int f \mathrm{d}\mu - \int f_n \mathrm{d}\mu \right| = \left| \int (f - f_n) \mathrm{d}\mu \right| \leq \int |f - f_n| \mathrm{d}\mu \xrightarrow{n\to\infty} 0$$

which simply means that

$$\lim_{n\to\infty} \int f_n \mathrm{d}\mu = \int f \mathrm{d}\mu. \qquad \square$$

B.4. Parameter Integrals

In this section, we want to consider a special case of an integral. Namely, we want to look at integrals of the form

$$\int f(u,x)\mathrm{d}\mu,$$

for some $u \in E$, which actually gives rise to a map

$$F : E \to \mathbb{R},$$

$$u \mapsto F(u) = \int f(u,x)d\mu.$$

Example B.4.0.1 (Gamma function). The *Gamma function* is defined as the map

$$\Gamma : \mathbb{C} \to \mathbb{R},$$

$$s \mapsto \Gamma(s) = \int_0^\infty t^{s-1}\mathrm{e}^{-t}\mathrm{d}t,$$

The question is, whether this integral converges for all $s \in \mathbb{C}$. This is certainly not the case and only possible if $\mathrm{Re}(s) > 0$. An important functional equation is given as

$$\Gamma(n+1) = n!$$

where $n \in \mathbb{N}$. This function is of great importance in analytic number theory and mathematical physics. It can be also used to express the volume of the unit ball in $\mathbb{R}^d$.

Theorem B.4.0.1. *Let $(E, \mathcal{A}, \mu)$ be a measure space. Let (U, d) be a metric space and let $f : U \times E \to \mathbb{R}$ (or $\mathbb{C}$) with $u_0 \in U$. Moreover, assume that the following properties hold:*

(i) *The map $x \mapsto f(u,x)$ is measurable for all $x \in U$.*
(ii) *The map $u \mapsto f(u,x)$ is continuous at u_0 a.e.*
(iii) *There exists a measurable function $g \in \mathcal{L}^1(E, \mathcal{A}, \mu)$ such that for all $u \in U$*

$$|f(u,x)| \leq g(x) \quad a.e.$$

Then the map $F(u) = \int_E f(u,x)\mathrm{d}\mu$ is well-defined and continuous at u_0.

Proof. From (iii) it follows that the map $x \mapsto f(u,x)$ is integrable for every $u \in U$, and so $F(u)$ is well-defined. Take a sequence $(u_n)_{n\in\mathbb{N}} \in U$ such that $u_n \xrightarrow{n\to\infty} u_0$, which basically means that $\mathsf{d}(u_n, u_0) \xrightarrow{n\to\infty} 0$. Then by continuity from (ii), we get that $f(u_n, x) \xrightarrow{n\to\infty} f(u_0, x)$ a.e. and from (iii) we can apply Theorem B.3.0.1 to obtain that

$$\lim_{n\to\infty} F(u_n) = \lim_{n\to\infty} \int f(u_n, x)\mathrm{d}\mu = \int f(u_0, x)\mathrm{d}\mu = F(u_0). \qquad \square$$

Remark B.4.0.2. Observe that $F(u)$ is continuous if F is continuous at every point $u \in U$.

Example B.4.0.3 (Fourier analysis).

(1) Let μ be a measure on $(\mathbb{R}, \mathcal{B}(\mathbb{R}))$ such that $\mu(\{x\}) = 0$ for all $x \in \mathbb{R}$. Moreover, let $\phi \in \mathcal{L}^1(\mathbb{R}, \mathcal{B}(\mathbb{R}), \mu)$. Then the map

$$F(u) = \int_{\mathbb{R}} \mathbb{1}_{(-\infty,u]}(x)\phi(x)\mathrm{d}\mu = \int_{(-\infty,u]} \phi(x)\mathrm{d}\mu$$

is continuous. Here we have $f(u,x) = \mathbb{1}_{(-\infty,u]}(x)\phi(x)$. The map $u \mapsto f(u,x)$ is continuous at u_0 for all $x \in \mathbb{R} \setminus \{u_0\}$ for some $u_0 \in \mathbb{R}$. However, we have $\mu(\{u_0\}) = 0$, which implies that the map $u \mapsto f(u,x)$ is a.e. continuous at u_0 with $|f(u,x)| \leq \phi(x)|$ and $|\phi|$ is integrable by assumption.

(2) Consider now the Lebesgue measure λ and $\phi \in \mathcal{L}^1(\mathbb{R}, \mathcal{B}(\mathbb{R}), \lambda)$. Moreover, define

$$\hat{\phi}(u) := \int_{\mathbb{R}} \mathrm{e}^{iux}\phi(x)\mathrm{d}\lambda,$$

which is called the *Fourier transform* of ϕ. The map $u \mapsto \mathrm{e}^{iux}\phi(x)$ is actually continuous for all $x \in \mathbb{R}$, which implies that $\hat{\phi}(u)$ is continuous at any $u \in \mathbb{R}$.

(3) Let again $\phi \in \mathcal{L}^1(\mathbb{R}, \mathcal{B}(\mathbb{R}), \lambda)$ and $h : \mathbb{R} \to \mathbb{R}$ be a continuous and bounded map. The *convolution* of h and ϕ is defined by

$$(h * \phi)(u) := \int_{\mathbb{R}} h(u - x)\phi(x) \mathrm{d}\lambda(x).$$

The map $(h * \phi)$ is continuous at any $u \in \mathbb{R}$. Moreover, the map $u \mapsto h(u-x)\phi(x)$ is continuous for all $x \in \mathbb{R}$. Since h is bounded, there exists a constant $k > 0$, such that $|h(x)| \leq k$ for all $x \in \mathbb{R}$.

B.5. Differentiation of Parameter Integrals

Theorem B.5.0.1 (Differentiation of integrals). *Let $I \in \mathbb{R}$ be a real interval and $(E, \mathcal{A}, \mu)$ a measure space. Let $f : I \times E \to \mathbb{R}$ (or $\mathbb{C}$) and let $u_0 \in I$. Moreover, assume that the following properties hold:*

(i) *The map $x \mapsto f(u, x)$ is in $\mathcal{L}^1(E, \mathcal{A}, \mu)$ for all $u \in I$.*
(ii) *The map $u \mapsto f(u, x)$ is a.e. differentiable at u_0 with derivation denoted by $\partial_u f(u_0, x)$.*
(iii) *There exists a map $g \in \mathcal{L}^1(E, \mathcal{A}, \mu)$ such that for all $u \in I$*

$$|f(u, x) - f(u_0, x)| \leq g(x)|u - u_0|.$$

Then the map $F(u) := \int_E f(u, x)\mathrm{d}\mu$ is differentiable at u_0 and its derivative is given by

$$F'(u) = \int_E \partial_u f(u_0, x)\mathrm{d}\mu.$$

Remark B.5.0.1. It is often useful to replace (ii) and (iii) with the following points respectively:

- The map $u \mapsto f(u, x)$ is a.e. differentiable at any point in I,
- There exists a map $g \in \mathcal{L}^1(E, \mathcal{A}, \mu)$ such that $|\partial_u f(u, x)| \leq g(x)$ a.e. for all $u \in I$.

Moreover, if f is differentiable on a interval $[a, b]$, there exists a $\theta_{a,b}$, by the mean value theorem, such that

$$f'(\theta_{a,b}) = \frac{f(b) - f(a)}{b - a}.$$

If f' is also bounded on $[a, b]$, i.e., there is a constant M such that $|f'(x)| \leq M$, then

$$\left|\frac{f(b) - f(a)}{b - a}\right| \leq M.$$

Proof of Theorem B.5.0.1. Let $(u_n)_{n\geq 1}$ be a sequence in I such that $u_n \xrightarrow{n\to\infty} u_0$ and assume that $u_n \neq u_0$ for all $n \geq 1$. Now define the sequence

$$\phi_n(x) := \frac{f(u_n, x) - f(u_0, x)}{u_n - u_0}.$$

Then we can obtain that

$$\lim_{n\to\infty} \phi_n(x) = \partial_u f(u_0, x) \quad \text{a.e.}$$

Now (iii) allows us to use Theorem B.3.0.1. Hence, we get that

$$\lim_{n\to\infty} \frac{F(u_n) - F(u_0)}{u_n - u_0} = \lim_{n\to\infty} \int_E \phi_n(x) \mathrm{d}\mu = \int_E \partial_u f(u_0, x) \mathrm{d}\mu. \qquad \square$$

Example B.5.0.2 (Fourier analysis).

(i) Let $\phi \in \mathcal{L}^1(\mathbb{R}, \mathcal{B}(\mathbb{R}), \lambda)$ be an integrable map such that

$$\int_{\mathbb{R}} |x\phi(x)| \mathrm{d}\lambda < \infty.$$

Then its Fourier transform is given by

$$\hat{\phi}(u) = \int_{\mathbb{R}} \underbrace{\mathrm{e}^{iux} \phi(x)}_{f(u,x)} \mathrm{d}\lambda,$$

which is differentiable and the derivative of it is then

$$\hat{\phi}'(u) = i \int_{\mathbb{R}} \mathrm{e}^{iux} x\phi(x) \mathrm{d}\lambda.$$

Therefore, we can write

$$|\partial_u f(u, x)| = |i\mathrm{e}^{iux} x\phi(x)| = |x\phi(x)|.$$

(ii) Let $\phi \in \mathcal{L}^1(\mathbb{R}, \mathcal{B}(\mathbb{R}), \lambda)$ and $h : \mathbb{R} \to \mathbb{R}$ be a bounded C^1-map with bounded derivative h'. Then the convolution $(h * \phi)$ is differentiable and its derivative is given by $(h * \phi)' = h' * \phi'$. Recall that the convolution $(h * \phi)$ is given by

$$(h * \phi)(u) = \int_{\mathbb{R}} \underbrace{h(u - x)\phi(x)}_{f(u,x)} \, \mathrm{d}\lambda.$$

Moreover, $|h(u - x)\phi(x)| \leq M|\phi(x)|$, where M is such that $|h(x)| \leq M$ and $|h'(x)| \leq M$ for all $x \in \mathbb{R}$. Then we get that

$$\partial_u f(u, x) = h'(u - x)\phi(x) \quad \text{and} \quad |\partial_u f(u, x)| \leq M|\phi(x)|.$$

Exercise B.5.0.3. Let μ be a measure on $(\mathbb{R}, \mathcal{B}(\mathbb{R}))$ such that for all $x \in \mathbb{R}$, $\mu(\{x\}) = 0$. Moreover, let $\phi \in \mathcal{L}^1(\mathbb{R}, \mathcal{B}(\mathbb{R}), \mu)$ such that

$$\int_{\mathbb{R}} |x\phi(x)| \mathrm{d}\mu < \infty.$$

Furthermore, for $u \in \mathbb{R}$, define

$$F(u) := \int_{\mathbb{R}} (u - x)^+ \phi(x) \mathrm{d}\mu.$$

Show that F is differentiable and that

$$F'(u) = \int_{(-\infty, u]} \phi(x) \mathrm{d}\mu.$$

Bibliography

[1] M. R. Adams, V. Guillemin, *Measure Theory and Probability*, The Wadsworth and Brooks/Cole Mathematics Series, Birkhäuser Basel, 1996.
[2] P. Barbe, M. Ledoux, *Probabilité*, EDP Sciences-Collection: Enseignement SUP-Maths, 2007.
[3] P. Billingsley, *Probability and Measure*, Wiley Series in Probability and Mathematical Statistics, third edn. Wiley-Interscience Publication, 1995.
[4] L. Breiman, *Probability (Classics in Applied Mathematics)*, Society for Industrial and Applied Mathematics (SIAM), 1968.
[5] M. Briane, G. Pagés, *Théorie de l'Intégration*, fourth edn. Vuibert, 2006.
[6] K. L. Chung, *A Course in Probability Theory*, Probability and Mathematical Statistics, A Series of Monographs and Textbooks, second edn. Elsevier Academic Press, 1974.
[7] D. L. Cohn, *Measure Theory*, Birkhäuser Advanced Texts Basler Lehrbücher, second edn. Birkhäuser Basel, 2013.
[8] J. L. Doob, *Measure Theory*, Graduate Texts in Mathematics, Vol. 143, Springer-Verlag New York, 1994.
[9] R. M. Dudley, *Real Analysis and Probability*, second edn. Cambridge University Press, 2002.
[10] R. Durrett, *Probability, Theory and Examples*, Cambridge Series in Statistical and Probabilistic Mathematics, Vol. 49, second edn. Cambridge University Press, 2019.
[11] M. Einsiedler, T. Ward, *Functional Analysis, Spectral Theory and Applications*, Graduate Texts in Mathematics, Vol. 127, Springer International Publishing, 2017.
[12] W. Feller, *An Introduction to Probability Theory and Its Applications*, Vol. 1, Wiley Series in Probability and Statistics, 1991.

[13] J.-F. Le Gall, *Intégration, Probabilités et Processus Aléatoires*, FIMFA, Département Mathématiques et Applications, Ecole normale supérieur de Paris, 2006.
[14] G. Grimmett, D. Stirzaker, *Probability and Random Processes*, third edn. Oxford University Press, 2001.
[15] J. Neveu, *Bases Mathématiques du Calcul des Probabilité*, The Mathematical Gazette, Vol. 50(371), Cambridge University Press, 1964.
[16] J. Pitman, *Probability*, Springer Texts in Statistics, Springer-Verlag New York, 1993.
[17] D. Revuz, *Measure et Intégration*, Hermann, 1997.
[18] D. Revuz, *Probabilité*, Hermann, 1997.
[19] W. Rudin, *Real and Complex Analysis*, third edn. McGraw Hill Education Ltd., 1986.
[20] D. W. Stroock, *A Concise Introduction to the Theory of Integration*, second edn. Birkhäuser Basel, 1994.
[21] D. W. Stroock, *Probability Theory: An Analytic View*, second edn. Cambridge University Press, 2011.

Index

Printed in the United States
by Baker & Taylor Publisher Services